Variational Theories for Liquid Crystals

APPLIED MATHEMATICS AND MATHEMATICAL COMPUTATION

Editors

R.J. Knops and K.W. Morton

This series presents texts and monographs at graduate and research level covering a wide variety of topics of current research interest in modern and traditional applied mathematics, in numerical analysis and computation.

1 Introduction to the Thermodynamics of Solids *J.L. Ericksen* (1991)

2 Order Stars *A. Iserles and S.P. Nørsett* (1991)

3 Material Inhomogeneities in Elasticity *G. Maugin* (1993)

4 Bivectors and Waves in Mechanics and Optics *Ph. Boulanger and M. Hayes* (1993)

5 Mathematical Modelling of Inelastic Deformation *J.F. Besseling and E. van der Geissen* (1993)

6 Vortex Structures in a Stratified Fluid; Order from chaos *Sergey I. Voropayev and Yakov D. Afanasyev* (1994)

7 Numerical Hamiltonian Problems *J.M. Sanz-Serna and M.P. Calvo* (1994)

(Full details concerning this series, and more information on titles in preparation are available from the publisher.)

Variational Theories for Liquid Crystals

Epifanio G. Virga

Faculty of Engineering, University of Pisa, Italy

CRC Press
Taylor & Francis Group
Boca Raton London New York

CRC Press is an imprint of the
Taylor & Francis Group, an **informa** business
A CHAPMAN & HALL BOOK

First published 1994 by Chapman & Hall

Published 2019 by CRC Press
6000 Broken Sound Parkway NW, Suite 300
Boca Raton, FL 33487-2742

First issued in paperback 2019

ISBN 13: 978-0-367-44906-3 (pbk)
ISBN 13: 978-0-412-39880-3 (hbk)

First edition 1994

Softcover reprint of the hardcover 1st edition 1994

Typeset in 10/12 Times by Thomson Press (India), New Delhi.

DOI 10.1007/978-1-4899-2867-2

A catalogue record for this book is available from the British Library

Library of Congress Catalog Card Number: 94-70268

Dedicated to the Memory of my Father,
Pasquale Virga (1915–1991)

Contents

Colour plates appear between pp 178–179

Preface

The programme of this book was projected in a survey written by J.L. Ericksen for *SIAM News* (see issue No. 4, November 1987).

After a sentence inoculating the reader with the germs of hope:

> Clearly, [liquid crystals] are [materials] of growing importance from a technological point of view, and their development entails a healthy interaction between science and technology,

Ericksen remarks that applied mathematicians should heed:

> To a large extent, the research has been done by non-mathematicians. Frank Leslie, a mathematician who is well known to workers in the field for his important contributions, is a notable exception. However, from a mathematical viewpoint, development has been slow, although the situation is beginning to change.

This faint vein of optimism was justified by the outcomes of the programme presented by J.L. Ericksen himself and D. Kinderlehrer as part of the activities during the academic year 1984–85 at the Institute for Mathematics and its Applications at Minneapolis. As Ericksen says,

> expert analysts were [then] introduced to and became fascinated with some of the [unsolved] issues. They have already proved deep and beautiful results with interesting physical implications. In particular, we are now seeing the rapid development of a mathematically sound, nonlinear static theory of defects, something that does not exist for any other kind of material.

In the autumn of 1985 I found myself on the edges of this activity, and I felt the ripples emanating from it. I was so fascinated by it that my major scientific interest has been in the mathematical theories of liquid crystals ever since.

This book stems from an advanced course on mathematical physics that I taught for three years at the University of Pavia. It concerns equilibrium of liquid crystals, strictly from a mathematical point of view. It is written in the style of modern *continuum mechanics*, and it aims to profit from the rigorous methods of this branch of mathematics so as to gain a deeper insight into matters which belong traditionally within the realms of both physics and chemistry.

Most of the mathematical methods employed throughout the book belong to the *calculus of variations*. Because flows of liquid crystals are dissipative, I do not treat them.

Essentially two variational theories of liquid crystals are explained in this book: one, due to Zocher, Oseen and Frank, is classical, the other, put forward by Ericksen, is rather new as to the mathematical formulation, though it has been anticipated in the physical literature of the past two decades. The new theory provides a better explanation of defects in liquid crystals, especially of those concentrated on lines and surfaces, which are beyond the scope of the classical theory.

This book reviews the major achievements of the classical theory as well as the latest discoveries of the new theory (see Chapters 3 and 6). Chapter 4 concerns bifurcations in liquid crystals; they are relevant to various devices in which exchanges of stability are produced by switching electric or magnetic fields on and off. Chapter 5 is devoted entirely to free-boundary problems, which concern droplets of liquid crystals whose shape is to be determined. Two introductory chapters on physical and mathematical preliminaries are meant to make the book readily accessible to both mathematicians and physicists (see Chapters 1 and 2). Going through Chapter 2 also offers the reader an opportunity to become familiar with the notation employed throughout the book.

The bibliography contains two families of entries: *References* and *Further reading*. The former family consists of the papers and books cited here: they are my sources. The latter family consists of papers and books that influenced me, perhaps indirectly, while I was writing the various sections of this book.

E.G. Virga
Pisa

Acknowledgements

I first owe my gratitude to Messrs. J.L. Ericksen, W. Noll and C. Truesdell, who have sustained my effort with their suggestions and their enthusiasm: without their persistent help I would often have been at a loss.

Many friends have helped me in writing this book, either spontaneously or at my request. Messrs. P. Biscari, J.L. Ericksen, J.-B. Fournier, M. Hayes, D. Kinderlehrer, F.M. Leslie, I. Stewart, C. Truesdell, P. Vannucci, P. Villaggio and Mrs. A. Napoli have read parts of the manuscript or the whole of it, catching mistakes and suggesting improvements. Messrs. R. Cohen, O.D. Lavrentovich and M. Luskin have granted permission to reproduce colourful pictures of their own, which ornament the book. Messrs. R. Rosso and A. Vicini, who were students of mine in Pavia, allowed me to make use of their notes. Mr. P. Biscari produced the figures that represent graphs of functions, while Mr. L. Pagni produced all the others. Mr. I. Demi has taken half a dozen of the photographs reproduced here. Mrs. G. Cai has carefully typed the manuscript, being kind enough to excuse my frequent changes of mind.

I thank all of them, and I hope that their work will not have been in vain.

1

Physical background

This book mainly concerns the variational theories of liquid crystals, phrased in the language of modern continuum mechanics.

In this chapter I outline the phenomenological premisses upon which the following chapters are built. Here the development is merely descriptive; it aims to make the reader familiar with the most relevant features of this fascinating state of matter.

A fairly complete account of both chemical and physical properties of liquid crystals can be found in the books by Gray (1962), De Gennes (1974) and Chandrasekhar (1977), which have already become classic. I also refer the reader to the major review papers of Brown and Shaw (1957), Stephen and Straley (1974) and Ericksen (1976), as well as to the nice introductory booklet by Collings (1990). Two other papers, addressed to a broad learned public, deserve notice for their clarity, namely the article of Fergason (1964) and the collective work by the Orsay Group (1971).

1.1 Liquid crystals

Though the name **liquid crystal** is likely to be familiar to most readers, it might sound self-contradictory to those who first hear it. Most people intuitively associate the word crystal with something solid.

How can a crystal be liquid?

G. Friedel raised this issue in 1926 (*cf.* p. 1052 of Brown and Shaw (1957)), proposing to adopt the name **mesomorphs** or **mesoforms** instead of liquid crystals since, as we shall see shortly below, these substances are indeed to be regarded as intermediate states of matter. Nevertheless, these terms found little acceptance, like others such as **paracrystals**. The term liquid crystal has been widely used since O. Lehmann coined it in the late 19th century, maybe just because it sounds self-contradictory. Be that as it may; we shall stand on the side of tradition and call the substances treated in this book liquid crystals.

Liquid crystals were discovered in 1888 by F. Reinitzer and O. Lehmann. Reinitzer's seminal paper appeared in Volume **9** (1888) of *Monatshefte für Chemie* (Reinitzer (1888)), and a translation into English has recently been published in *Liquid Crystals* (Reinitzer (1989)) to celebrate the centenary of the first paper. The attribution of the discovery has long been disputed. I do not venture any opinion about this slippery matter, which I leave to the historians of science. Several papers on the history of liquid crystals could help the interested reader to understand the wealth of the early literature, namely Kelker (1973), Kelker and Knoll (1989) and Sackmann (1989).

Reinitzer studied different derivatives of cholesterol which are solid at room temperature. In heating them up he observed peculiar behaviour: it appeared that these substances have, as it were, *two* melting points. Here is, in Reinitzer's words, the description of the experiment he performed with cholesteryl benzoate:

> With respect to the melting point, a significant deviation from Schulze's data was noted. He found it to be 150–151°. However, despite continued careful purification I was able to find only 145.5° (corr. 146.6°). However, it struck me that the substance, in this case, melted not into a clear transparent but always into a cloudy, only translucent liquid, which I initially considered to be a sign of impurities, although both microscopic and crystallographic examinations of the compound revealed no sign of non-uniformity. Upon closer examination, it was then also noted that when it was heated to higher temperatures, the clouding suddenly vanished. This happens at 178.5°C (corr. 180.6). At the same time I found that the substance heated thus high displayed, upon cooling, colour phenomena quite similar to those already described for the acetate. This remarkable phenomenon of the presence of two melting points, if one may express it thus, and the appearance of the colour phenomenon were primarily what made me think that here and in the case of acetate physical isomerism must be present, for which reason I requested Professor Lehmann in Aachen to make a closer investigation of these circumstances.
>
> (*cf.* pp. 15–16 of Reinitzer (1989)).

The cloudy substance observed by Reinitzer was what in modern terms is called a **cholesteric liquid crystal**, while the higher melting point, that at which it ceased to be cloudy, is nowadays called the *clearing point*. We shall see below which microscopic changes occur in such a substance when it exhibits a cloudy appearance and what happens when it reaches its clearing point.

Before proceding further we need to make our terminology precise.

Liquid crystals are **mesophases**, that is intermediate states of matter, which flow like nearly incompressible viscous fluids, and yet retain several features, especially optical, characteristic of crystals. Most liquid crystals are organic substances, which we call **mesogenic**, in that they give rise to mesophases under appropriate circumstances. More generally, the quality of being mesogenic does not pertain only to pure substances, but also to mixtures, suspensions and other types of aggregates.

There are essentially two distinct ways to induce the liquid crystalline phase in a mesogenic substance, namely by changing its temperature or its concentration in a solvent. Correspondingly, we say that there are two types of **mesomorphisms**: we call the former **thermotropic**, and the latter **lyotropic**.

Thousands of liquid crystals, both thermotropic and lyotropic, have now been discovered or synthesized. Besides cholesteryl benzoate, other thermotropic liquid crystals are *e.g.* *p*-azoxyanisole (PAA) and *p*-methoxybenzylidene-*p*-*n*-butylaniline (MBBA), which exhibit mesophases in the temperature ranges 118°C to 135°C, and 21°C to 41°C, respectively. Many lyotropic liquid crystals are indeed biological systems, such as aqueous solutions of both **tobacco mosaic** and **cucumber virus** at a concentration of a few per cent. Other substances which can be liquid crystals are **polymers**. Polymeric molecules are usually very long and complex; they are **macromolecules** which can be broken into elementary units, called **monomers**, regularly repeated along a chain. Polymeric liquid crystals are both thermotropic and lyotropic; among these latter there are for example many derivatives of **cellulose**.

In 1922 G. Friedel proposed to classify liquid crystals into three wide categories, which he called **nematic**, **cholesteric** and **smectic**. This classification has met with wide support and will also be employed here.

So far we have just roughly outlined some qualitative features of liquid crystals, without offering any distinctive criterion which can lead to an understanding of why one substance is mesogenic, while another is not. Such a criterion must indeed be based on the microscopic properties of these substances, as indicated by the fact that the molecules of all liquid crystals are found to be strongly elongated in one direction. Also the differences between the above three classes are better understood if they are related to the different shapes of the molecules and their arrangement in space. We give below a qualitative description of these microscopic features, separately for the three classes of mesophases.

1.1.1 Nematics

The adjective **nematic** comes from the Greek *νῆμα*, which means 'thread'. The same etymology is shared by the noun **nematode**, which according to the *Oxford English Dictionary* designates a 'slender unsegmented worm'.

The molecules of nematic liquid crystals closely resemble rods whose typical dimensions are 5 Å by 20 Å. These slender molecules enjoy a further symmetry: if, so to say, *head* and *tail* were exchanged, they would still seem alike. For any straight molecule this property entails a mirror symmetry with respect to a plane orthogonal to its axis.

In this chapter our development will often be concerned with molecules. The reader must be warned that we envisage a rather naive microscopic model of liquid crystals. The properties that here we attribute to the molecules should indeed be regarded as pertaining to *ideal* molecules; nevertheless, they apply to the *average* behaviour of the real molecules: various asymmetries seem to cancel out because they occur somewhat randomly in the different molecules. It is hard to make these ideas more precise, as is discussed by Frank (1980).

The nematic phase manifests itself when the interaction between neighbouring molecules tends to make them parallel to one another, so as to induce a partial *orientational order* at the microscopic scale. In a pure substance this ordering effect competes against the disordering effect of temperature, whereas in a solution it also competes against the random interpenetration of the different species.

The appearance of the nematic phase in both thermotropic and lyotropic liquid crystals results from the delicate equilibrium between two contrary tendencies, the one in favour of orientational order and the other against it. So long as a scant degree of orientation persists among the molecules, the substance does not cease to be a liquid crystal. When the orientational order is completely lost, no direction is any longer privileged because the molecules are randomly oriented; hence the substance becomes **isotropic**. For thermotropic nematics this happens at a specific temperature T_{NI}, which marks the onset of the isotropic phase; it is the temperature that we referred to above as the clearing point. For lyotropic liquid crystals the isotropic phase takes over with no change in the temperature; the role of T_{NI} is played by a critical value of the concentration.

The first melting point observed by Reinitzer is usually denoted by T^*. It is properly *the* melting point of the substance, the temperature at which the liquid phase succeeds the solid one.

While in a nematic liquid crystal even a faint order must reign among the molecular long axes, no order whatsoever is demanded of the centres of mass of the molecules. Nematic liquid crystals are completely free to flow; the positions in space of their molecules are free from any constraint. A *positional order* prevails only below T^*, when the molecules are bound to prescribed points in space.

Both PAA and MBBA, the liquid crystals mentioned above, are nematic at room temperature.

The molecules of genuine nematic liquid crystals cannot possess any permanent dipole moment, since it would destroy their mirror symmetry. Thus, nematic liquid crystals are not ferroelectric materials. If polar molecules with opposite dipole moments are equally mixed together, the substance which they constitute may be a nematic liquid crystal, but still it cannot be ferroelectric.

If there are no external fields to induce specific molecular orientations, the orienting interactions would incline all molecules in a nematic phase to be aligned along an average direction. It is as if the molecules, though shaken by the thermal disturbances, were to comply with an equilibrium criterion which makes them tend to be closely packed. They behave like a bunch of slender rods which must be packed into the least volume. We may say that in the absence of any external field the molecules of a nematic liquid crystal acquire a **natural orientation** in which their axes are nearly parallel. Clearly, every straight line in space has the same probability of designating the average direction of a natural orientation.

1.1.2 Cholesterics

Cholesteric liquid crystals are not much different from nematic ones, so much so that Collings (1990) (*cf.* p. 14) proposes to call them **chiral nematics**. The prefix *chiro-*, which comes from the Greek $\chi\varepsilon\acute{\iota}\rho$, refers to *hand*. We now see how hands come about in our discussion.

Cholesteric molecules resemble helical springs, which may be either right-handed or left-handed. We also say that they may have opposite **chiralities**, one the *enantiomorph* of the other. Clearly, in a pure substance all molecules must have one and the same chirality. They exhibit a privileged direction, which is the axis of the helices. Most of the above intuitive ideas about the microscopic behaviour of nematics apply also to cholesterics: the role played there by the axis of the rods is to be taken here by the axis of the helices. There are, however, two new relevant features which make the difference between nematics and cholesterics.

As above, only the orientation of the axis of the molecules matters in their interactions; that is, if a molecule is turned upside-down its interactions with the neighbouring molecules remain unaffected. In this sense, the molecules of cholesterics are also invariant under exchange of head and tail, but this property cannot be phrased in terms of mirror symmetries, since reflections reverse chirality.

This is the first novelty borne by cholesterics. The second is their natural orientation. In the absence of any external field distorting the molecular axes, they still arrange themselves as if they were to reduce the free volume between adjacent molecules. Whoever has attempted to pack closely a handful of screws is likely to have seen that their axes do not prefer to be parallel to one another.

This analogy makes it easier to understand the natural orientations of cholesterics: the molecular axes tend to be aligned along a preferred direction which fails to be the same at all points in space, but varies, reflecting the chirality of the molecules. Such a direction lies everywhere parallel to a given plane, say $\mathscr{P}$; it does not vary on each plane parallel to $\mathscr{P}$, but progressively rotates as one of these planes is shifted onto another. The axis orthogonal to $\mathscr{P}$ is called the *cholesteric axis* of the natural orientation. Clearly, since $\mathscr{P}$ is arbitrary, any straight line in space is likely to be the cholesteric axis of a natural orientation.

To be more specific, suppose that the z-axis of a system of right-handed Cartesian co-ordinates, which we denote by (x, y, z), is the cholesteric axis of a natural orientation, so that the plane (x, y) here takes the place of $\mathscr{P}$.[1] If in such a frame we denote by (n_x, n_y, n_z)

[1] The mathematical language of this book will be set up in the following chapter. Here we adhere to an old-fashioned notation, which is presumably familiar to most readers.

the components of a unit vector which lies along the preferred direction of the molecules, a natural orientation of cholesterics can be described by the following formulae:

$$n_x = \cos(\tau z + \varphi_0),\ n_y = \sin(\tau z + \varphi_0),\ n_z = 0, \tag{1.1}$$

where τ is a scalar, which may be either positive or negative, and φ_0 is an angle in the interval $[0, 2\pi[$.

While φ_0 depends only on the frame of reference employed to write (1.1), and so has no intrinsic significance, τ, which we call *twist*, is a material modulus, and depends on the temperature in a way which is characteristic of each cholesteric liquid crystal.

The tip of the vector in (1.1) describes a helix which winds around the z-axis; the distance between any two adjacent spirals is

$$P = \frac{2\pi}{|\tau|}, \tag{1.2}$$

which is called the *natural pitch*. A typical value of P is 6×10^3 Å. When τ is positive the above helix is right-handed; when τ is negative it is left-handed. It may happen that τ vanishes at some critical temperature; the liquid crystal then becomes nematic and the field defined by (1.1) accordingly reduces to a constant.

Though there is no evidence that cholesterol itself is a liquid crystal, many of its compounds certainly are, as we have already learned from the Reinitzer quotation. Of course, they are all cholesterics.

Mixing together substances with opposite chiralities may give rise to nematic liquid crystals. This is indeed the clearest indication that being nematic or cholesteric is a quality which does not pertain to the single molecule, but is rather a collective property which accounts for averages taken over many molecules.

1.1.3 Smectics

Smectic liquid crystals are essentially different from both nematics and cholesterics. They owe their name to σμῆγμα, which is the Greek for *soap*, since a soap bubble provides the most common example of a smectic liquid crystal. This book will not cover smectics, because I think that a sound mathematical theory able to describe them is still lacking. Thus, the account of their qualitative features will be shorter than above.

One fact must, however, be noted. Smectic liquid crystals exhibit a close interplay between positional and orientational ordering of molecules. The intermolecular forces are strong enough to favour preferred positions of the molecules, besides preferred orientations of their axis. Such a feature gives these substances some stiffness, and yet they are able to flow.

Often the positional order results in two-dimensional stratified structures, which slide upon each other, and rank nearly all molecules of the substance. In a natural orientation the structures are usually planes within which the orientational order of the molecules is somehow reminiscent of that pertaining to nematics and cholesterics. Hence smectics are sometime referred to as **lamellar mesophases**.

There are many such mesophases, which differ from one another in the arrangement of molecules within each layer. For example, in smectics A the molecules of each layer are randomly distributed, but their long axes tend to be aligned orthogonally to the layer. In smectics B, on the other hand, molecules tend to occupy specific positions like tesserae in a mosaic. Smectics C are like smectics A, but their molecules are tilted at a definite angle to the normal of each layer. There are also other smectic mesophases, which we do not describe here; they all share a lamellar structure which assigns the molecules different special arrangements in space.

The same liquid crystal may exhibit different smectic mesophases at different temperatures. If it can also experience another mesophase besides the smectic ones, it must occur at higher temperatures because, as the above discussion indicates, smectics are the liquid crystals closest to solids. For example, 4-n-pentylbenzenethio-4′-n-decyloxybenzoate, often abbreviated as $\overline{10}S5$, is a smectic C liquid crystal between 60°C and 63°C. It is smectic *A* between 63°C and 80°C, where it becomes nematic: it remains so until the temperature reaches 86°C, where the transition to the isotropic phase occurs. Cholesteryl myristate becomes smectic A at 71°C and it remains so until the temperature reaches 79°C; it is cholesteric until 85°C and then it becomes an isotropic liquid. Both these substances are solid before they become smectic.

I do not know of any liquid crystal that exhibits all three main mesophases.[2] When smectics are heated sufficiently, they became either nematic or cholesteric before reaching the isotropic phase.

---

[2] Indeed chemists do not know of any thermotropic transition of a cholesteric to a nematic. It seems that a chiral substance is always cholesteric, even if it happens to prefer some nematic-like order.

1.2 Early theories

The intuitive ideas outlined in the preceding section suggest a simple picture of the basic phenomena which give rise to liquid crystals. They will be given a sound mathematical form in Chapter 3 within the continuum theory proposed by F. Frank in 1958. In this section we briefly comment on two earlier theories which in the 1930s were competing. The interest in such a controversy is not merely historical, since one of these theories is to be regarded as the natural precursor of Frank's. These theories are the **swarm theory** and the **distortion theory**.

The swarm theory was initially proposed by E. Bose in 1909 (*cf.* the detailed list of references given by Brown and Shaw (1957)). It applies only to nematics and cholesterics and, in one word, regards them as **polycrystals**.

According to this theory the molecules of liquid crystals group themselves in colonies, called swarms, within which they lie approximately in the same direction. Nevertheless, in the absence of any external action, the orientation of the molecules in one swarm is completely independent of that in another.

The swarms are dispersed in an isotropic melt, where the molecules do not exhibit any preferred orientation. The composition of a single swarm is not constant in time: molecules easily migrate from one swarm to the next or go astray in the midst. Also the shape of each swarm changes in time and it is subject to external forces. Ornstein and Kast (1993) estimated that the average number of molecules which constitute a swarm is approximately 1×10^5 and that the order of magnitude for the diameter of the swarms is 1×10^{-6} cm.

In general, the interactions between swarms are much weaker than those among molecules in a single swarm, and only external forces such as electric or magnetic fields are able to orient the swarms.

The swarm theory can easily explain the cloudy appearance of liquid crystals: when the swarms are completely disordered and their orientations randomly distributed, the light that traverses the substance is scattered in all directions, like sunlight on a misty day. To describe also the interaction of a liquid crystal with a solid boundary, such as that of a glass container, the swarms close to the boundary are taken to be oriented up to a depth of a few hundreds of millimetres by the electric double layers and the capillary forces exerted at the surface of contact between the melt and the solid boundary (see Ornstein and Kast (1933)).

In Ornstein and Kast's words the main features of this theory are summarized as follows:

> Liquid crystals are aggregates of sub-microscopic homogeneous ranges, the so-called swarms, in which the molecules all lie parallel. The structure is to be visualized as of a polycrystalline nature; the swarms have an individual existence of considerable duration. The swarms lie perfectly irregularly; their axes are distributed at random. Any further arrangement of the swarms into larger homogeneous ranges is caused by external forces ... [The] resistence to the turning of the swarms arises from the way they knock against one another and are deformed, the magnitude of which depends on the elastic properties of the swarms. The swarm formation comes about, in the first place, because of the unsymmetrical shape of the molecules.
> (*cf.* p. 944 of Ornstein and Kast (1933)).

The swarm theory appeared to be so successful that Brown and Shaw in their review paper published in 1957 still regarded it as the more widely accepted, though they warn us that

> even today there is no one theoretical interpretation of the mesomorphic structure that completely explains all the experimental data. (*cf.* p. 1053 of Brown and Shaw (1957)).

Another theory was proposed by H. Zocher in 1927.[3] It was known as the distortion theory. In 1933 Zocher applied this theory to nematic liquid crystals subject to a magnetic field; more generally, he aimed to give a mathematical treatment of the distortion produced on liquid crystals by external fields.

Zocher's theory was based on three major assumptions, which he stated as follows:

> (1) The whole substance tends to take up a position such that the axial direction at every point is the same. (2) Any force acting so as to disturb this state where the directions are uniform causes a distortion in which the direction changes continuously until a restoring force of an elastic nature holds the applied

[3] An earlier reference to a continuum theory can be found in Oseen (1921).

force in equilibrium. (3) At surfaces of solid bodies (*e.g.* glass or metal) the positions at first assumed are almost unchangeable. (*cf.* p. 945 of Zocher (1933)).

Thus, in Zocher's view the orientation of the molecules changes continuously in space and it reacts in an elastic fashion to the external distorting actions; when these latter are absent the natural orientation prevails everywhere. Furthermore, a solid boundary is able to prescribe the orientation of the molecules in a way which is not easily affected by the external fields acting on the bulk of the substance.

Though Zocher was rather successful in explaining the interactions of nematics with magnetic fields, his theory did not prevail over the swarm theory. It is, however, relevant to our development in that it anticipates Frank's continuum theory, which is nowadays well established (*cf.* Chapter 3 below).

Frank first understood that the conflict between the swarm theory and the distortion theory is illusory, for they have indeed different objects: while the former essentially aims at a statistical theory of liquid crystals, the latter lays down the premisses for a continuum theory. A synthesis of these points of view requires a deeper understanding of both the difference and the interplay between a statistical and a continuum description of liquid crystals. In the following section, leaving aside the historical development of these ideas, we come to represent in modern mathematical terms the degree of order prevailing in liquid crystals.

1.3 Order parameters

A sound continuum theory of liquid crystals must be based on a clear definition of the quantity that describes on a macroscopic scale the co-operative habit of the molecules outlined in Section 1.1. For nematics and cholesterics such a habit results in a degree of orientational order, while for smectics it also entails a positional order. Though intuitive, such a notion of order is to be made precise: it clearly pertains to the microscopic description of liquid crystals, but it must also affect the macroscopic features characteristic of these substances.

A clear distinction of two scales of length, one microscopic and the other macroscopic, would prevent any ambiguous interpretation of the premisses upon which we build the continuum theories studied in this book.

Our development below applies only to nematics and cholesterics, and so it concerns only orientational order.

Liquid crystals constitute material bodies, which take up shapes in space. In modern continuum mechanics the concepts of *body* and *shape* have been given precise mathematical definitions, which the reader can find in Sections I.2, I.3 and II.1 of Truesdell (1991). A continuous body can essentially be identified with the collection of the material elements, or **material points**, which constitute it. The classical theory of such bodies regards each of their shapes merely as the set of the points in space occupied by their material points.

Bodies consisting of liquid crystals are not described within such a theory, because they are **bodies with microstructure**, for which further parameters of a microscopic origin have mechanical significance also on a macroscopic scale. These latter are often called **order parameters**, for they attempt to reflect various degrees of order, possibly relative to different scales of length. Capriz (1989) has recently proposed one mathematical scheme which includes several special theories of bodies with microstructure. The reader is referred to his tract for a quick and effective introduction to the continuum theory of what are regarded as **new materials**, among which liquid crystals are presumably the oldest.

Here we do not pursue any systematic treatment of liquid crystals as special materials endowed with microstructure; rather, we define the order parameters suitable for them through elementary considerations of statistical mechanics.

1.3.1 Statistical distribution

Let $\mathscr{B}$ be the region in space occupied by a liquid crystal, that is, its shape. Each point of $\mathscr{B}$ is indeed the place of a material point; we think of it as a large collection of elongated molecules whose orientation is described below along the lines followed by Ericksen (1991) (similar ideas were in part implied in Section 2.1 of De Gennes (1974)).

Let a point p of $\mathscr{B}$ be given and let N be the number of molecules that we imagine contained in it. We assume that their centres of mass are distributed completely at random, while their orientations, which we describe by a vector of the unit sphere $\mathbb{S}^2$, obey a distribution law whose probability density is denoted by f. We take f as an integrable function of $\mathbb{S}^2$ into $\mathbb{R}^+$, the positive real line. Thus, if S is any subset of $\mathbb{S}^2$, the probability of finding one molecule oriented within S is given by

$$\mathrm{p}[S] := \int_S f\,da, \tag{1.3}$$

where a denotes the area measure on $\mathbb{S}^2$.[4] Correspondingly, the number of molecules which are *expected* to be oriented within S is given by

$$\mathrm{n}[S] := N\mathrm{p}[S].$$

Clearly, the function f is such that

$$\mathrm{p}[\mathbb{S}^2] = \int_{\mathbb{S}^2} f\,da = 1, \tag{1.4}$$

which amounts to saying that the probability of finding one molecule of p however oriented in $\mathbb{S}^2$ is 1.

We further assume that f is an even function, that is

$$f(\mathbf{l}) = f(-\mathbf{l}) \quad \text{for all } \mathbf{l} \in \mathbb{S}^2. \tag{1.5}$$

Such an assumption expresses in a statistical language the peculiar fact, often illuminated in the intuitive description of Section 1.1, that in both nematics and cholesterics one cannot distinguish between the *head* and *tail* of a molecule.

Equation (1.5) has two interesting consequences, which we now examine. First, for any subset S of $\mathbb{S}^2$ let its **opposite** S^- be defined thus

$$S^- := \{\mathbf{l} \in \mathbb{S}^2 \mid -\mathbf{l} \in S\},$$

so that the union of S and S^- is radially symmetric with respect to the centre of $\mathbb{S}^2$. Clearly, (1.5) implies that

$$\mathrm{p}[S^-] = \mathrm{p}[S] \quad \text{for all } S \in \mathbb{S}^2. \tag{1.6}$$

Second, let $\mathbf{m}$ be the vector defined as follows:

$$\mathbf{m} := \int_{\mathbb{S}^2} \mathbf{l} f(\mathbf{l})\,da, \tag{1.7}$$

---

4 Here our development becomes somewhat more mathematical. The reader who wishes to become acquainted with the mathematical language generally employed in this book may consult the first three sections of the following chapter.

so that it represents the first moments of the probability density f. By (1.5), we easily show that

$$\mathbf{m} = \mathbf{0}. \tag{1.8}$$

Thus, the first *gross* information about the probability distribution defined by (1.3) comes from its second moments, which are represented by the second-order tensor

$$\mathbf{M} := \int_{\mathbb{S}^2} \mathbf{l} \otimes \mathbf{l} f(\mathbf{l}) da. \tag{1.9}$$

We now derive the main properties of $\mathbf{M}$. Taking the trace of both sides of (1.9), since $\mathbf{l}$ ranges over the unit sphere, we get

$$\operatorname{tr} \mathbf{M} = \int_{\mathbb{S}^2} \operatorname{tr}(\mathbf{l} \otimes \mathbf{l}) f(\mathbf{l}) da = \int_{\mathbb{S}^2} f(\mathbf{l}) da = 1. \tag{1.10}$$

Likewise, since the tensor $\mathbf{l} \otimes \mathbf{l}$ is symmetric for every vector $\mathbf{l}$, we have that

$$\mathbf{M}^T = \int_{\mathbb{S}^2} (\mathbf{l} \otimes \mathbf{l})^T f(\mathbf{l}) da = \mathbf{M}.$$

Thus, $\mathbf{M}$ is a symmetric tensor.

1.3.2 Averages

Let $\mathbf{e}$ be a unit vector. $\mathbf{M}$ is a linear mapping which transforms $\mathbf{e}$ into $\mathbf{Me}$. The inner product of $\mathbf{e}$ and $\mathbf{Me}$, which we write as $\mathbf{e} \cdot \mathbf{Me}$, can be given an interesting statistical interpretation. It easily follows from (1.9) that

$$\mathbf{e} \cdot \mathbf{Me} = \int_{\mathbb{S}^2} (\mathbf{l} \cdot \mathbf{e})^2 f(\mathbf{l}) da,$$

and so it is plain that $\mathbf{e} \cdot \mathbf{Me}$ is the average of $(\mathbf{l} \cdot \mathbf{e})^2$, for which we employ the following notation:

$$\mathbf{e} \cdot \mathbf{Me} =: \langle (\mathbf{l} \cdot \mathbf{e})^2 \rangle.$$

Clearly, if we call ϑ the angle between the orientation of one molecule and the direction defined by $\mathbf{e}$, we also have

$$\mathbf{e} \cdot \mathbf{Me} = \langle \cos^2 \vartheta \rangle.$$

This discussion immediately leads us to the following inequalities:

$$0 \leqslant \mathbf{e} \cdot \mathbf{M}\mathbf{e} \leqslant 1,$$

where the lower and the upper bounds are approached when f is such that the molecules are nearly all orthogonal to $\mathbf{e}$ or parallel to it.

1.3.3 Isotropic distribution

Suppose that f is constant. By (1.4), this amounts to saying that

$$f = f_0 \equiv \frac{1}{4\pi}.$$

The function f_0 describes the microscopic arrangement in which the molecules are equally distributed in all directions, and so they lack any preferred orientation. In a word, this distribution is *isotropic*.

Let $\mathbf{M}_0$ be the second-moments tensor associated with f_0. We now show that $\mathbf{M}_0$ must be proportional to the identity. Let $\mathbf{R}$ be any rotation about the centre of $\mathbb{S}^2$, which clearly maps $\mathbb{S}^2$ into itself. Moreover, changing the variable $\mathbf{l}$ into $\mathbf{R}\mathbf{l}$ in the definition (1.9) does not change $\mathbf{M}_0$ since f_0 is constant, and so

$$\mathbf{M}_0 = \int_{\mathbb{S}^2} \mathbf{R}\mathbf{l} \otimes \mathbf{R}\mathbf{l} f_0 \, da,$$

whence, since $\mathbf{R}\mathbf{l} \otimes \mathbf{R}\mathbf{l} = \mathbf{R}(\mathbf{l} \otimes \mathbf{l})\mathbf{R}^T$, we arrive at

$$\mathbf{M}_0 = \mathbf{R}\mathbf{M}_0\mathbf{R}^T. \tag{1.11}$$

Every rotation $\mathbf{R}$ satisfies the equation:

$$\mathbf{R}\mathbf{R}^T = \mathbf{I},$$

$\mathbf{I}$ being the identity tensor. Since $\mathbf{R}$ can be chosen as any rotation, we see that $\mathbf{M}_0$ must commute with all of them:

$$\mathbf{M}_0\mathbf{R} = \mathbf{R}\mathbf{M}_0 \quad \text{for all rotations } \mathbf{R}. \tag{1.12}$$

As is shown in the following chapter (*cf.* both Lemma 2.17 and Theorem 2.18) a tensor obeys (1.12) and does not vanish whenever it is proportional to the identity. Since by (1.10) the trace of $\mathbf{M}_0$ is 1, we thus reach the conclusion that

$$\mathbf{M}_0 = \tfrac{1}{3}\mathbf{I}. \tag{1.13}$$

It follows from (1.13) that

$$\langle \cos^2 \vartheta \rangle = \tfrac{1}{3}$$

when the distribution of molecules is isotropic.

1.3.4 Order tensor

Following De Gennes (1974) (*cf.*, in particular, pp. 30–32), we call the second-order tensor defined by

$$\mathbf{Q} := \mathbf{M} - \mathbf{M}_0 \tag{1.14}$$

the **order tensor**. It measures how the second-moments tensor associated with a given probability density deviates from its isotropic value. $\mathbf{Q}$ vanishes when $f = f_0$, and this clearly means that there is no orientational order at the points of $\mathscr{B}$ where it happens. In general, $\mathbf{Q}$ should be interpreted as a measure of the local degree of orientational order in liquid crystals.

Such an interpretation, however, requires caution, since $\mathbf{Q}$ may also vanish when $f \neq f_0$. We illustrate this by an example. Consider a probability density f which possesses an axis of symmetry, that is, assume that there is an element $\mathbf{e}$ of $\mathbb{S}^2$ such that

$$f(\mathbf{R}\mathbf{l}) = f(\mathbf{l}) \quad \text{for all rotations } \mathbf{R} \text{ about } \mathbf{e}. \tag{1.15}$$

If $\mathbf{R}$ is as in (1.15), changing the variable $\mathbf{l}$ into $\mathbf{R}\mathbf{l}$ in (1.9) leads us to

$$\mathbf{M} = \int_{\mathbb{S}^2} \mathbf{R}\mathbf{l} \otimes \mathbf{R}\mathbf{l} f(\mathbf{R}\mathbf{l})\, da = \int_{\mathbb{S}^2} \mathbf{R}(\mathbf{l} \otimes \mathbf{l})\mathbf{R}^T f(\mathbf{l})\, da = \mathbf{R}\mathbf{M}\mathbf{R}^T. \tag{1.16}$$

Thus, $\mathbf{M}$ commutes with all rotations about $\mathbf{e}$, and so by a formula which will be proved in Section 2.5 (*cf.*, in particular, Theorem 2.20 and Lemma 2.19), it must be such that

$$\mathbf{M} = \alpha \mathbf{e} \otimes \mathbf{e} + \beta(\mathbf{I} - \mathbf{e} \otimes \mathbf{e}), \tag{1.17}$$

where α and β are both scalars. It is easily seen that the tensor in (1.17) obeys (1.10) whenever

$$\beta = \tfrac{1}{2}(1 - \alpha), \tag{1.18}$$

and so (1.17) can also be given the form

$$\mathbf{M} = \tfrac{1}{2}(3\alpha - 1)\mathbf{e} \otimes \mathbf{e} + \tfrac{1}{2}(1 - \alpha)\mathbf{I}. \tag{1.19}$$

If in (1.19) we set $\alpha = \tfrac{1}{3}$, $\mathbf{M}$ reduces to $\mathbf{M}_0$. Now, computing $\mathbf{e} \cdot \mathbf{M}\mathbf{e}$

from (1.19), we arrive at the following formula for α:

$$\alpha = \langle \cos^2 \vartheta \rangle,$$

and so $\alpha = \frac{1}{3}$ also when f is such that the molecules are all nearly distributed on a cone around **e** whose semi-angle is $\vartheta = \arccos 3^{-1/2}$. Thus, we cannot infer from $\mathbf{M} = \mathbf{M}_0$ that $f = f_0$.

To discriminate between the above conical distribution and the isotropic one, we should compute higher moments of the probability density. Here, however, we shall not do so; rather, we take the attitude of regarding as equivalent any two probability densities which share the same second-moments tensor, and so we say that *all distribution laws which obey*

$$\mathbf{M} = \mathbf{M}_0$$

are isotropic. This is indeed a crude approximation if viewed from the standpoint of a statistical theory, but it bears the significant advantage of leading to a continuum theory in which the degree of microscopic order is merely described by a second-order tensor, namely $\mathbf{Q}$ in (1.14).

Let $\mathbf{Q}$ be given in the space of the symmetric tensors. It follows from (1.10) and (1.13) that

$$\operatorname{tr} \mathbf{Q} = 0. \tag{1.20}$$

Moreover, by the Spectral Theorem (*cf.* Section 2.2), it can be represented as follows:

$$\mathbf{Q} = \lambda_1 \mathbf{e}_1 \otimes \mathbf{e}_1 + \lambda_2 \mathbf{e}_2 \otimes \mathbf{e}_2 + \lambda_3 \mathbf{e}_3 \otimes \mathbf{e}_3, \tag{1.21}$$

where λ_1, λ_2 and λ_3 are its eigenvalues and $\mathbf{e}_1$, $\mathbf{e}_2$, $\mathbf{e}_3$ are the corresponding eigenvectors. By (1.20), λ_1, λ_2 and λ_3 satisfy

$$\lambda_3 = -(\lambda_1 + \lambda_2), \tag{1.22}$$

and so the three eigenvalues of $\mathbf{Q}$ are equal only if they all vanish. When this is not the case only two possibilities can arise: either $\lambda_1 = \lambda_2$ or $\lambda_1 \neq \lambda_2$.

When $\lambda_1 = \lambda_2$, by (1.22), (1.21) reads as follows:

$$\mathbf{Q} = \lambda_1(\mathbf{e}_1 \otimes \mathbf{e}_1 + \mathbf{e}_2 \otimes \mathbf{e}_2 - 2\mathbf{e}_3 \otimes \mathbf{e}_3),$$

whence, since the identity tensor can also be written as

$$\mathbf{I} = \mathbf{e}_1 \otimes \mathbf{e}_1 + \mathbf{e}_2 \otimes \mathbf{e}_2 + \mathbf{e}_3 \otimes \mathbf{e}_3,$$

we arrive at

$$\mathbf{Q} = s(\mathbf{n} \otimes \mathbf{n} - \tfrac{1}{3}\mathbf{I}), \tag{1.23}$$

where we have set

$$\mathbf{n} = \mathbf{e}_3 \quad \text{and} \quad s = -3\lambda_1.$$

Likewise, when $\lambda_1 \neq \lambda_2$ we easily give (1.21) the following form

$$\mathbf{Q} = -(s_1 \mathbf{n}_1 \otimes \mathbf{n}_1 + s_2 \mathbf{n}_2 \otimes \mathbf{n}_2) + \tfrac{1}{3}(s_1 + s_2)\mathbf{I}, \tag{1.24}$$

where we have employed the notation

$$\mathbf{n}_1 = \mathbf{e}_1,\ \mathbf{n}_2 = \mathbf{e}_2,\ s_1 = -2\lambda_1 - \lambda_2,\ s_2 = -\lambda_1 - 2\lambda_2.$$

Clearly, when s_1 and s_2 are equal (1.23) and (1.24) coincide.

We say that the liquid crystal is **uniaxial** when $\mathbf{Q}$ is as in (1.23), and that it is **biaxial** when $\mathbf{Q}$ is as in (1.24). Strictly speaking, these definitions apply only to the single point of $\mathscr{B}$ that we are exploring, and so a liquid crystal could be uniaxial at one point and biaxial at another.

A biaxial liquid crystal becomes uniaxial when s_1 and s_2 are equal but different from zero, or when either s_1 or s_2 vanishes. Furthermore, both uniaxial and biaxial liquid crystals become isotropic whenever $\mathbf{Q}$ vanishes.

1.3.5 Fresnel's ellipsoid

Uniaxial and **biaxial** are terms borrowed from the language of optics. To understand their meaning fully, we must digress slightly.

To follow the entire development below the reader should be already familiar with the most elementary phenomena connected with the propagation of light both in a vacuum and in material media. (A clear and simple introduction to this matter, also in view of its application to liquid crystals, can be found in Chapter 4 of Collings (1990).) Thus, basic definitions such as those of **index of refraction** and **polarization of light** will not be recalled here. We shall consider only monochromatic, plane light waves.

In an optically isotropic material, linearly polarized light can propagate in all directions without suffering any change in its polarization. This is not the case for optically anisotropic materials. Nonetheless, there is always at least one direction of propagation such that every polarization orthogonal to it would travel undistorted, if excited. The best way to illustrate such a feature is to draw the classical Fresnel's ellipsoid. Here we follow in the main the neat presentation of it given in Section 6.10 of Rossi (1957).

Three mutually orthogonal axes and three scalars greater than 1 are associated with every homogeneous optical medium. We identify the axes with three vectors of $\mathbb{S}^2$, $\mathbf{e}_1$, $\mathbf{e}_2$, and $\mathbf{e}_3$; we denote the scalars by n_1, n_2 and n_3, and we define the second-order tensor

$$\mathbf{F} := n_1\mathbf{e}_1 \otimes \mathbf{e}_1 + n_2\mathbf{e}_2 \otimes \mathbf{e}_2 + n_3\mathbf{e}_3 \otimes \mathbf{e}_3. \tag{1.25}$$

Thus n_1, n_2 and n_3 are the eigenvalues of $\mathbf{F}$, which correspond to the eigenvectors $\mathbf{e}_1$, $\mathbf{e}_2$ and $\mathbf{e}_3$, respectively; they are called **principal indices of refraction** of the material. **Fresnel's ellipsoid** is the surface in the three-dimensional Euclidean space $\mathscr{E}$ represented by

$$\mathscr{F}_o := \{x \in \mathscr{E} \mid (x - o)\cdot\mathbf{F}(x - o) = 1\}, \tag{1.26}$$

where o is a given point of $\mathscr{E}$.

If linearly polarized light propagates in the direction of $\mathbf{e}_3$, as the wave travels the polarization fails to change only in two cases, namely when the direction of polarization is parallel to $\mathbf{e}_1$ and when it is parallel to $\mathbf{e}_2$; the speeds of propagation are then c/n_1 and c/n_2, respectively, and so they are different if n_1 and n_2 are different.[5] When the polarization of light is in the plane orthogonal to $\mathbf{e}_3$, but lies in neither of the directions $\mathbf{e}_1$ and $\mathbf{e}_2$, the optical vector that represents it can be seen as the superposition of two vectors, one oscillating in the direction of $\mathbf{e}_1$ and the other in that of $\mathbf{e}_2$. Since when $n_1 \neq n_2$ these two polarizations propagate at different speeds, the optical vector rotates around $\mathbf{e}_3$ as the wave travels through the medium, and so in general it gives rise to an **elliptic** polarization. Analogous considerations apply also to waves propagating along $\mathbf{e}_1$ or $\mathbf{e}_2$. The reader should recall this discussion when in the next section we come to explain the birefringence of liquid crystals.

To understand how Fresnel's ellipsoid illustrates the changes that occur in the linear polarization of light when it propagates in a direction other than $\mathbf{e}_1$, $\mathbf{e}_2$ and $\mathbf{e}_3$, we consider first a simpler case. We assume that n_1 and n_2 are equal, but different from n_3. In such a case the medium is called **optically uniaxial**, and $\mathbf{e}_3$ is its **optic axis**.

Let $\mathbf{e}$ be a vector of $\mathbb{S}^2$. We denote by $\mathscr{P}_\mathbf{e}$ the plane through o

---

[5] As usual, c denotes here the speed of light in a vacuum.

orthogonal to **e**. If $\mathbf{e} \neq \mathbf{e}_3$ we give the name **principal section** of Fresnel's ellipsoid to the plane through o on which both **e** and $\mathbf{e}_3$ lie.

The intersection between $\mathscr{F}_o$ and $\mathscr{P}_\mathbf{e}$ is generally an ellipse with one axis orthogonal to the principal section; it is a circle only when $\mathbf{e} = \mathbf{e}_3$. The length of the semi-axis orthogonal to the principal section is just n, the common value of n_1 and n_2, while the length of the other semi-axis, which we denote by n', is intermediate between n and n_3. If the polarization of light is orthogonal to the principal section, it propagates undistorted at the speed c/n, whereas it propagates at the speed c/n' if it lies on the principal section (*cf.* Figure 1.1).

It should be noted, however, that in the latter case the optical vector is no longer orthogonal to the direction of propagation. Why this is so can be explained by the electromagnetic theory of light. In a uniaxial medium the electric field and the electric displacement are not parallel if the former field is neither parallel nor orthogonal to the optic axis (*cf.* Section 2.5 of Chapter 2, for example). It is shown in Section 8.13 of Rossi (1957) that for a plane electromagnetic wave propagating in a uniaxial medium along a direction neither parallel nor orthogonal to the optic axis, the electric displacement oscillates in a direction orthogonal to the direction of propagation, whereas

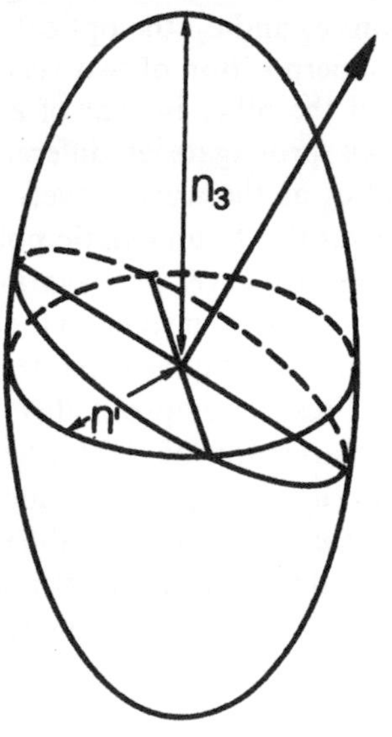

Figure 1.1

the electric field, which conventionally designates the optical vector, does not.

Since for a uniaxial medium Fresnel's ellipsoid is symmetric about the optic axis, the wave polarized in the direction orthogonal to the principal section travels at the same speed for all directions of propagation, and so it is called the **ordinary wave**. The wave polarized on the principal section is called instead the **extraordinary wave**. A wave whose polarization is neither parallel nor orthogonal to the principal section is elliptically polarized in a plane which is no longer orthogonal to the direction of propagation.

For an optically uniaxial medium the principal index of refraction relative to the optic axis is called **extraordinary**, while that relative to any direction orthogonal to the optic axis is called **ordinary**. They are often denoted by n_e and n_o, respectively.

So far, as is customary in optics, we have regarded the wavelength of light as prescribed. When, for the same anisotropic medium, we study the propagation of light with different wavelengths, the above construction based on Fresnel's ellipsoid still applies: only the principal indices of refraction change, while the principal axes, being characteristic of the material, do not.

For example, quartz is an optically uniaxial crystal: for the wavelength of the yellow line of sodium (5893 Å) its extraordinary and ordinary indices of refraction are $n_e = 1.553$ and $n_o = 1.544$ (*cf.* Table 4 of Appendix 4 to Rossi (1957)).

A medium is **optically biaxial** if n_1, n_2 and n_3 are all different from one another. We do not go into further details concerning the above geometric construction, but are content to outline one qualitative feature of it. When all three semi-axes of Fresnel's ellipsoid have different length, there are precisely two planes whose intersections with the ellipsoid are circles; they both pass through the axis of intermediate length. If light propogates orthogonally to one of these planes, its linear polarization remains unaffected. These special directions of propagation are called **principal optic axes**, and in general they are not at right angles. These materials are called biaxial because there are precisely two such axes.

1.3.6 Optic axis

Here we assume that when the order tensor **Q** *is represented as in* (1.23) *the liquid crystal is optically uniaxial and* **n** *is its optic axis* *whenever* $s \neq 0$. *Similarly, we assume that when* **Q** *is as in* (1.24) *the liquid crystal is optically biaxial.*

Though these assertions could be better justified within a more refined molecular model, giving the order tensor defined above an optical interpretation is to be regarded as a genuine assumption in the continuum description of liquid crystals which is adopted throughout this book.

Q is a macroscopic field which varies in space. Here we shall only consider liquid crystals which are uniaxial at all points, and so **Q** will always be taken as in (1.23). Nevertheless, it may vanish somewhere in the region occupied by the liquid crystal, and this, as we learned above, would imply that the material has there become isotropic. Completing our optical interpretation of **Q**, *we assume that the medium also becomes optically isotropic whenever s vanishes.*

In a liquid crystal, unlike any common uniaxial crystal, the optic axis varies in space, but it need not be defined in the whole region occupied by the body: indeed **n** *is not defined wherever s vanishes.*

We shall come back to this issue shortly below; here we take s as having a constant non-zero value. If the order tensor **Q** is prescribed, formula (1.23) does not define *one* vector **n** of $\mathbb{S}^2$, but rather the tensor $\mathbf{n} \otimes \mathbf{n}$, namely the axis of **n**. This should be the proper meaning of sentences such as '**n** is physically undistinguished from its opposite'. The truth is that **n**, considered as a member of the unit sphere, does not adequately represent the optic axis of a liquid crystal or, expressed differently, that even when s is constant throughout the material, $\mathbb{S}^2$ is not the appropriate manifold to describe the local degree of order.

A glance at (1.23) suffices to show that for any s, either constant or not, the manifold fit for liquid crystals is that of all traceless symmetric second-order tensors with two equal eigenvalues. A systematic treatment of such a manifold is still lacking, and will not be attempted in this book. We rather employ s and **n** as basic fields to describe liquid crystals: the former will be taken in an interval of the real line, while the latter will lie on $\mathbb{S}^2$. We thus adhere to a well established tradition, but we must be aware that it might sometimes be misleading.

If we think of **n** as the unit vector in (1.23), we can interpret it as designating an **average orientation** of the molecules, though by no means can they be regarded as lying all parallel to it. Thus, we often call **n** the **orientation** or **director**, instead of the **optic axis**. All these terms are synonyms in this book.

The features of the natural molecular orientations described in Section 1.1 will hereafter be ascribed to **n**, and interpreted also in optical terms (*cf.* the following section).

We have already seen that even when s is constant in space the optic axis may still vary. Nevertheless, it is to be noted that both the ordinary and the extraordinary index of refraction depend only on the temperature and the wavelength of light. For example, Stephen and Straley (1974) report on p. 700 the following data for PAA near 125°C:

$$n_o = 1.565, \quad n_e = 1.829,$$

and for MBBA near 25°C,

$$n_o = 1.54, \quad n_e = 1.75;$$

in both cases the wavelength is that of the yellow line of sodium.

1.3.7 Degree of orientation

The scalar s in (1.23) is the **degree of orientation** of uniaxial liquid crystals. We have learned above that when it vanishes there is no orientational order among the molecules and the medium also becomes optically isotropic.

When **Q** is as in (1.23), **M** has the following form by (1.14) and (1.13):

$$\mathbf{M} = s\mathbf{n}\otimes\mathbf{n} + \tfrac{1}{3}(1-s)\mathbf{I}. \tag{1.27}$$

Hence we arrive at

$$\langle(\mathbf{l}\cdot\mathbf{n})^2\rangle = \mathbf{n}\cdot\mathbf{M}\mathbf{n} = \tfrac{1}{3}(2s+1),$$

and so

$$s = \tfrac{1}{2}(3\langle(\mathbf{l}\cdot\mathbf{n})^2\rangle - 1), \tag{1.28}$$

which can also be written as

$$s = \tfrac{3}{2}\langle\cos^2\vartheta - \tfrac{1}{3}\rangle,$$

where ϑ denotes the angle between **n** and the long axis of a single molecule.

Since $0 \leqslant (\mathbf{l}\cdot\mathbf{n})^2 \leqslant 1$, we see from (1.28) that s is subject to

$$-\tfrac{1}{2} \leqslant s \leqslant 1. \qquad (1.29)$$

By (1.22) the lower bound of s represents the state of microscopic order in which all molecules in a material element are orthogonal to $\mathbf{n}$, but otherwise disordered. Likewise, the upper bound of s corresponds to the state of perfect microscopic order in which all molecules lie parallel to $\mathbf{n}$. Both these extreme cases are ideal and unlikely to be actually attained.

By (1.27), the eigenvalues of $\mathbf{M}$ are $\frac{1}{3}(1-s)$, $\frac{1}{3}(1-s)$ and $\frac{1}{3}(1+2s)$, and so $\mathbf{M}$ is **positive definite**, as was to be expected, unless s takes one of the above limiting values. Likewise, the eigenvalues of $\mathbf{Q}$ are $-\frac{1}{3}s$, $-\frac{1}{3}s$, and $\frac{2}{3}s$, and so by (1.29) the magnitude of $\mathbf{Q}$ is subject to[6]

$$|\mathbf{Q}|^2 = \tfrac{2}{3}s^2 \leqslant \tfrac{2}{3}.$$

1.3.8 Defects

Let the orientation $\mathbf{n}$ be given in the region $\mathscr{B}$ occupied by a liquid crystal. If the mapping into $\mathbb{S}^2$ which assigns it is not everywhere smooth, we say that a **defect** of $\mathbf{n}$ occurs wherever it is discontinuous. In principle, a defect may be concentrated in isolated points, lines, or surfaces. The classical continuum theory of liquid crystals takes s as constant throughout $\mathscr{B}$. We shall see in Chapter 6 that only point defects are well explained within such a theory, whereas both line and surface defects are not.

The need for a unified treatment of defects has prompted Ericksen (1991) to amend the classical theory. In the new theory that he proposes, the changes in space of the degree of orientation can **relax** the energy of defects, which are indeed to be identified with the regions in $\mathscr{B}$ where s vanishes: there the liquid crystal becomes isotropic and the optic axis is not defined. Thus, the new theory interprets defects as localized transitions of the liquid crystal to its isotropic phase, although no change in the temperature is involved.

The germ of this idea essentially goes back to Fan (1971), whose work was built upon the works of De Gennes (1969) and Fan and Stephen (1970). Pursuing this line of thought, Ericksen (1991) has established a new continuum theory for liquid crystals, which will

---

[6] The **magnitude** of a tensor is defined below in Section 2.2.1 of Chapter 2.

be illustrated in Chapter 6 together with the types of defects that have been explained so far.

In the new theory s and $\mathbf{n}$ are fields defined thus:

$$s:\mathscr{B}\to[-\tfrac{1}{2},1],\quad \mathbf{n}:\mathscr{B}\backslash\mathscr{S}(s)\to\mathbb{S}^2,$$

where

$$\mathscr{S}(s):=\{p\in\mathscr{B}\,|\,s(p)=0\}$$

is the **singular set**, where defects may occur.

Such a description of liquid crystals is to some extent reminiscent of that proposed by Leslie (1968), who employed a vector of variable length to represent both the average direction of molecules and their degree of orientation. What makes Ericksen's description different from Leslie's is that the degree of orientation may well be negative in the former, and so unfit to be represented as the length of a vector.

Ericksen (1976) reported the experimental values of s observed away from defects by Brown, Doane and Neff in 1970; these values range in the interval

$$0.3\leqslant s\leqslant 0.7.$$

1.3.9 Smectics

The statistical interpretation of the order parameters upon which we have built this section applies to both nematic and cholesteric liquid crystals. It has served here as a mere motivation for the foundation of a continuum theory, and so the specific form of the probability density f has remained unknown. When it comes to smectics, however, the microscopic picture is rather different in that a spatial order is also present besides the orientational one, and the probability density must accordingly depend also on the space variables. Such a dependence has always been given a special form in all the attempts made so far to bring smectics within a macroscopic theory phrased in the language of order parameters (*cf. e.g.* Pikin and Indenbom (1978)).

1.4 Optical properties

I describe in this section the most relevant features exhibited by liquid crystals when they are traversed by polarized light. There are indeed many optical properties of liquid crystals worth mentioning, but it is beyond the scope of this book to describe all of them. I prefer to single out a few issues, which may also help the reader inexperienced in crystal optics to interpret the colourful pictures which often occur in the physical literature.

1.4.1 Birefringence

We saw in the preceding section that liquid crystals are optically uniaxial materials whose optic axis may vary in space. To illustrate the birefringence of a liquid crystal we consider the case when it occupies a layer $\mathscr{B}$ of thickness d, which can be represented as follows:

$$\mathscr{B} = \{p \in \mathscr{E} \,|\, p - o = x\mathbf{e}_1 + y\mathbf{e}_2 + z\mathbf{e}_3, \quad x, y \in]0, l[, \, z \in]0, d[\}, \quad (1.30)$$

where o is a given point of the three-dimensional Euclidean space $\mathscr{E}$ and $\mathbf{e}_1, \mathbf{e}_2, \mathbf{e}_3$ is an orthonormal basis. We further assume the optic axis in $\mathscr{B}$ to be everywhere orthogonal to $\mathbf{e}_3$ and independent of the co-ordinate parallel to it, that is

$$\mathbf{n}(p)\cdot\mathbf{e}_3 = 0, \, (\nabla\mathbf{n}(p))\mathbf{e}_3 = 0 \quad \text{for all} \quad p \in \mathscr{B}. \quad (1.31)$$

We apply the term *planar* to all orientation fields that obey (1.31). For simplicity, we first consider the case when $\mathbf{n}$ is constant in $\mathscr{B}$;

$$\mathbf{n} = \mathbf{e}_1. \quad (1.32)$$

Let a plane, linearly polarized wave propagating along $\mathbf{e}_3$ be incident on the face of $\mathscr{B}$ at $z = 0$. We represent the polarization of such a wave by the vector

$$\mathbf{p} = e_x\mathbf{e}_1 + e_y\mathbf{e}_2, \quad (1.33)$$

which within the electromagnetic theory of light is to be interpreted as the electric field of the wave. Both components of $\mathbf{p}$ vary in time according to the formulae

$$e_x = a_x \cos\omega t, \; e_y = a_y \cos\omega t, \quad (1.34)$$

where ω is the *angular frequency* of the wave and a_x, a_y can be given the following form

$$a_x = a\cos\alpha, \; a_y = a\sin\alpha, \quad (1.35)$$

a being a positive scalar and α an angle in the interval $[0, \pi[$, which represent the *amplitude* of the wave and the angle between the optic axis and the direction of polarization, respectively.

We can think of the incident wave as the superposition of two plane waves with different polarizations, namely $\mathbf{p}_1 := e_x\mathbf{e}_1$ and $\mathbf{p}_2 := e_y\mathbf{e}_2$. The principal axes of Fresnel's ellipsoid are parallel to $\mathbf{e}_1$, $\mathbf{e}_2$ and $\mathbf{e}_3$; the ordinary index of refraction n_o is relative to both $\mathbf{e}_2$ and $\mathbf{e}_3$, while the extraordinary index of refraction n_e is relative to $\mathbf{e}_1$. Since the wave travels along $\mathbf{e}_3$, the principal section of Fresnel's ellipsoid is the plane defined by $\mathbf{e}_1$ and $\mathbf{e}_3$, and so the polarization $\mathbf{p}_1$ is that of the extraordinary wave, while $\mathbf{p}_2$ is that of the ordinary wave. As we learned in the preceding section, they travel at different speeds, namely c/n_e and c/n_o, respectively. Thus, at all points of $\mathscr{B}$ with the same co-ordinate z the polarizations $\mathbf{p}_1$ and $\mathbf{p}_2$ are the same as those of the incident beam computed at two differently delayed times:

$$\mathbf{p}_1 = a_x \cos\omega\left(t - \frac{n_e z}{c}\right)\mathbf{e}_1, \ \mathbf{p}_2 = a_y \cos\omega\left(t - \frac{n_o z}{c}\right)\mathbf{e}_2, \tag{1.36}$$

where a_x and a_y are again as in (1.35).

Defining the *frequency* ν of the incoming wave as

$$\nu := \frac{\omega}{2\pi} \tag{1.37}$$

and its *wavelength in a vacuum* λ_0 as

$$\lambda_0 := \frac{c}{\nu}, \tag{1.38}$$

we readily give (1.36) the following form

$$\mathbf{p}_1 = a_x \cos\varphi_x(t, z)\mathbf{e}_1, \ \mathbf{p}_2 = a_y \cos\varphi_y(t, z)\mathbf{e}_2, \tag{1.39}$$

where

$$\varphi_x(t, z) := 2\pi\left(\nu t - \frac{n_e z}{\lambda_0}\right), \ \varphi_y(t, z) := 2\pi\left(\nu t - \frac{n_o z}{\lambda_0}\right). \tag{1.40}$$

Thus, the **phase difference** between the ordinary wave and the extraordinary wave is delivered by the function defined by

$$\varphi(z) := \varphi_y(t, z) - \varphi_x(t, z) = 2\pi(n_e - n_o)\frac{z}{\lambda_0}. \tag{1.41}$$

Clearly, when $n_e > n_o$, as is the case for most liquid crystals, $\varphi(z)$ is positive for all $z \in]0, d[$.

It is to be noted that equation (1.41) applies only when α is neither

0 nor $\pi/2$, since in these cases one or the other of the orthogonal vectors in (1.39) vanishes, and so the phase difference is not defined.

Futhermore, for any $z \in]0, d[$, $\varphi(z)$ does not depend on α, and so it is the same for all polarizations of the incoming wave. By (1.41), when the wave emerges from the face of $\mathscr{B}$ at $z = d$, the phase difference is given by

$$\varphi(d) = 2\pi(n_e - n_o)\frac{d}{\lambda_0}, \tag{1.42}$$

and so in general the outgoing wave has an **elliptic polarization**. As is explained for example in Section 6.6 of Rossi (1957), the ellipse described by the optical vector is tangent to the rectangle with semi-axes a_x and a_y, disposed along $\mathbf{e}_1$ and $\mathbf{e}_2$, respectively.

We now discuss a few special cases where (1.42) describes situations qualitatively different from one another.

First, let $d(n_e - n_o) = \lambda_0/4$ so that $\varphi(d) = \pi/2$. In such a case the semi-axes of the ellipse described by the optical vector of the outgoing wave are parallel to $\mathbf{e}_1$ and $\mathbf{e}_2$. Furthermore, if $\alpha = \pi/4$ then $a_x = a_y$ and the polarization becomes circular.

Second, let $d(n_e - n_o) = \lambda_0/2$ so that $\varphi(d) = \pi$. Then also the outgoing wave is linearly polarized and its optical vector makes the angle 2α with that of the incoming wave so that they oscillate at right angles when $\alpha = \pi/4$.

Third, let $d(n_e - n_o) = \frac{3}{4}\lambda_0$ so that $\varphi(d) = \frac{3}{2}\pi$. In this case the semi-axes of the ellipse described by the optical vector are again parallel to $\mathbf{e}_1$ and $\mathbf{e}_2$, as for $\varphi(d) = \pi/2$, but now the ellipse is traced in the opposite sense.

Finally, let $d(n_e - n_o) = \lambda_0$ so that $\varphi(d) = 2\pi$. In such a case the polarization of the outgoing wave is just the same as that of the incoming one. When $d(n_e - n_o)$ keeps increasing by multiples of $\lambda_0/4$, the special cases outlined above occur again in the same order.

The foregoing discussion is based on the premiss that the phase difference φ is well defined. When this is not the case, that is when the polarization of the incoming wave is either parallel or orthogonal to the optic axis, the polarization of the outgoing wave is just the same as that of the incoming one.

In summary, there are precisely three cases when the polarization of the wave is the same on the two sides of the layer $\mathscr{B}$: namely when it is parallel or orthogonal to the optic axis, or when $d(n_e - n_o)$ is a multiple of λ_0. This conclusion has been reached here under the assumption that $\mathbf{n}$ is constant throughout $\mathscr{B}$, but it in fact holds whenever $\mathbf{n}$ obeys (1.31) and $|\nabla\mathbf{n}|$ is not too large. In general, for a given incoming wave the first or the second of the three cases above may occur at different points of the face of $\mathscr{B}$ exposed to light. This criterion will suffice below to explain the main qualitative features of the optical patterns which are generated when $\mathscr{B}$ is placed between two crossed polarizers, orthogonal to $\mathbf{e}_3$.

1.4.2 Extinction branches

Let $\mathscr{B}$ be as in (1.30). We consider here the class of orientation fields that are represented, in terms of a given function ψ, in the form

$$\mathbf{n}(p) = \cos\psi(\vartheta)\mathbf{e}_1 + \sin\psi(\vartheta)\mathbf{e}_2, \quad \text{for all } p\in\mathscr{B}, \tag{1.43}$$

where ϑ and the co-ordinates (x, y, z) of p are related through the formula

$$\tan\vartheta = \frac{y}{x}. \tag{1.44}$$

Clearly, every field in (1.43) obeys (1.31) and it is constant on the straight lines emanating from o in the plane orthogonal to $\mathbf{e}_3$. All these fields, except that with ψ constant, exhibit a defect along the segment

$$\mathscr{S} = \{p\in\mathscr{B} \mid p - o = z\mathbf{e}_3\},$$

which is often called a **disclination**, since there the **inclination** of the optic axis suffers a discontinuity.

In particular, we take ψ to be given by

$$\psi(\vartheta) = S\vartheta + \psi_0, \tag{1.45}$$

where both S and ψ_0 are scalars, the former of which, if it does not vanish, is also referred to as the **strength** of the disclination.

Let $\mathscr{C}$ be any loop which encloses o on the face of $\mathscr{B}$ at $z=0$. Along such a path ϑ is increased by 2π on a single turn, and so by (1.45) ψ is correspondingly altered by the angle

$$\psi(2\pi) - \psi(0) = 2\pi S. \tag{1.46}$$

If $\mathbf{n}$ and its opposite are taken to represent the same orientation of the optic axis at any point of $\mathscr{B}$, then it follows from (1.46) that

$$2\pi S = m\pi, \tag{1.47}$$

where m is an integer. Thus, S must be equal to half an integer. We shall see in Section 3.7 how the integer m is related to what is called **Frank's index**.

We now suppose that the layer $\mathscr{B}$ is placed between two polarizers whose axes are at right angles, and both orthogonal to $\mathbf{e}_3$. A plane wave travelling along $\mathbf{e}_3$ becomes linearly polarized when it traverses one polarizer; then its polarization changes while it propagates through $\mathscr{B}$, as we have learned above, so that it is generally **elliptic** when the outgoing wave encounters the other polarizer, which is conventionally called the **analyser**. This latter always lets one component of the optical vector go through, and so $\mathscr{B}$ is transparent at all points that alter the linear polarization of the incoming wave. This is indeed a distinctive feature of all optically anisotropic media: being transparent between crossed polarizers.

Furthermore, since $\mathbf{n}$ here varies in space, the brightness of the layer does also, and there are also points of $\mathscr{B}$ that look black. These are the points through which the linear polarization of the wave propagates unchanged, that is where $\mathbf{n}$ is either parallel or orthogonal to the axis of the polarizer. Here, of course, we have left aside the case when $d(n_e - n_o)$ is a multiple of λ_0, because the whole layer would then look black.

Thus, we derive the following qualitative criterion which applies also to situations more general than that envisaged above: *When a liquid crystal layer with a planar orientation is inserted between two crossed polarizers, it is opaque precisely where the optic axis is parallel either to the polarizer or to the analyser.*

We now examine in more detail the opaque zones produced in $\mathscr{B}$ by the disclinations described by (1.45). Here when a point obeys the above criterion, so do all other points in the same half-line from o. Thus, the opaque zones of $\mathscr{B}$ are indeed radial branches emanating from the disclination, which are called **extinction branches** because they intercept light.

Let $\delta\vartheta$ be the angle between two adjacent extinction branches. It must be related to an increment of ψ equal to $\pi/2$, that is

$$\delta\psi = |S|\delta\vartheta = \frac{\pi}{2},$$

whence it follows that

$$\delta\vartheta = \frac{\pi}{2|S|}. \tag{1.48}$$

If N is the number of extinction branches of the disclination, we readily derive from (1.48) that

$$N = \frac{2\pi}{\delta\vartheta} = 4|S|. \tag{1.49}$$

For example, when S is either $+1$ or -1 there are 4 extinction branches at right angles to each other: the black pattern is then a Maltese cross. Plate 1.I, which is reproduced below by courtesy of O.D. Lavrentovich as are all plates of this chapter except the last two, illustrates Maltese crosses in spherical droplets with a point defect in their centres. Though a detailed explanation of Plate 1.I would demand more knowledge of crystal optics, we can still understand its main features by use of formulae (1.48) and (1.49) with $|S| = 1$.

In general, the convergence to a point of different extinction branches is a sign that there is a defect at that point. The number of such branches then gives information about the structure of the defect through (1.49), which yields $|S|$.

Also the sign of S can easily be inferred. Let $\mathbf{e}_1$ and $\mathbf{e}_2$ be parallel to the axes of the polarizer and the analyser, respectively. When the liquid crystal layer is rotated by the angle β about $\mathbf{e}_3$, while both the polarizer and the analyser are kept fixed, the function ψ that describes the orientation field in (1.43) becomes

$$\psi^*(\vartheta) = \psi(\vartheta) + \beta,$$

where ϑ is defined as in (1.44). In particular, if ψ is as in (1.45), changing ϑ into

$$\vartheta^* := \vartheta + \frac{\beta}{S}, \tag{1.50}$$

we arrive at

$$\psi^*(\vartheta) = \psi(\vartheta^*). \tag{1.51}$$

It follows from (1.50) and (1.51) that the extinction branches and the layer rotate by different angles, namely β/S and β. Such a discrepancy gives us a simple criterion to detect the sign of S: it is positive when the extinction branches and the layer rotate likewise, negative otherwise.

The reader is referred to Section 3.5 of Chandrasekhar (1977) to learn more about the optical properties of liquid crystals which reveal the structure of defects in a layer with planar orientation fields.

When the optic axis within the layer is not planar, new patterns of extinction branches arise around defects. This is the case for the **hybridly aligned nematic films** studied by Lavrentovich and Nastishin (1990). They consider a layer of nematic liquid crystal whose thickness is between 1 μm and 20 μm. The sample is illuminated from below and observed from above through a polarizing microscope. The lower and upper faces of the layer are treated so that the optic axis is tangent to the former and normal to the latter. Thus, the orientation field is prescribed on the upper face, but it is somehow **degenerate** at the lower one, because it is there free to vary, though remaining planar.

Plates 1.II and 1.III show patterns around defects observed by Lavrentovich and Nastishin (1990): in both of them there are extinction branches grouped together in a narrow sector. Equation (1.48), which requires the angle between any two adjacent branches to be the same, is not valid here, and so (1.49) also fails to apply. Although an integer *m* may always be assigned to any disclination, here it is not easily related to the number of extinction branches.

Close examination of how the patterns in Plates 1.II and 1.III change when the film is rotated while the polarizers are held fixed has led Lavrentovich and Nastishin to conclude that in a sector

Figure 1.2

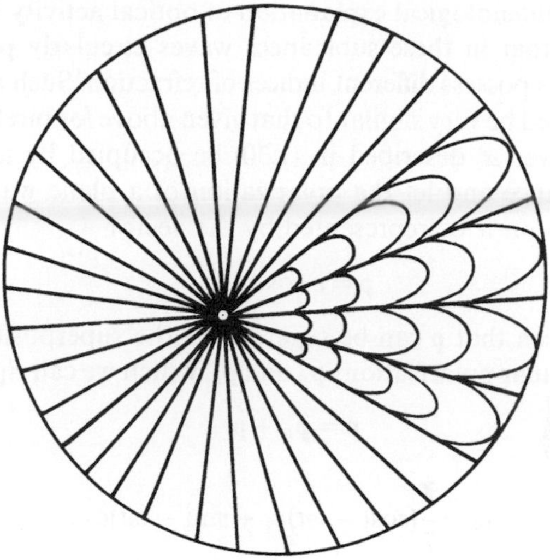

Figure 1.3

around the defects the integral lines of **n** projected onto the plane of the film look like the petals of a daisy. Figures 1.2 and 1.3 illustrate the sketches of such lines for the defects depicted in the middle of Plates 1.II and 1.III.

Plate 1.IV shows a colourful picture shot by Lavrentovich, which exhibits more defects. Such an optical pattern becomes more complex as the thickness of the film is made smaller. Plates 1.V and 1.VI illustrate two beautiful examples of these patterns, as does the plate on the frontispiece of this book, which is taken from the same series.

1.4.3 Optical activity

We mainly follow here the simple phenomenological description of optical activity given in Section 6.15 of Rossi (1957). There are substances, mostly organic, which rotate the optical vector of the monochromatic plane waves that travel through them. Such substances are called **optically active**. A common example is any aqueous solution of sugar; in such a case the angle by which the polarization changes is proportional to the concentration. There are also optically active uniaxial crystals, such as quartz. Here, however, the optical activity is not easy to observe, because it clashes with the birefringence of the medium.

The phenomenological explanation of optical activity is based on the premiss that in these substances waves circularly polarized in opposite ways possess different indices of refraction. Such an explanation will indeed be very similar to that given above for birefringence.

Let the layer $\mathscr{B}$ described in (1.30) be occupied by an optically active substance and let the polarization of a plane wave incident orthogonally on it be represented by the vector

$$\mathbf{p} = e \cos \omega t \mathbf{e}_1. \tag{1.52}$$

It is easily seen that $\mathbf{p}$ can be regarded as the superposition of two opposite circular polarizations $\mathbf{p}_R$ and $\mathbf{p}_L$, which we call **right** and **left**:

$$\mathbf{p} = \mathbf{p}_R + \mathbf{p}_L,$$

where

$$\begin{aligned} \mathbf{p}_R &= \frac{e}{2}(\cos(-\omega t)\mathbf{e}_1 + \sin(-\omega t)\mathbf{e}_2), \\ \mathbf{p}_L &= \frac{e}{2}(\cos \omega t \mathbf{e}_1 + \sin \omega t \mathbf{e}_2). \end{aligned} \tag{1.53}$$

We now assume that two different indices of refraction, which we denote by n_R and n_L, are associated with the propagation in the medium of $\mathbf{p}_R$ and $\mathbf{p}_L$. Thus, the argument that led us to (1.36) above now shows that at all points of $\mathscr{B}$ with the same co-ordinate z the right and left polarizations evolve in time according to the formulae

$$\mathbf{p}_R = \frac{e}{2}(\cos \varphi_R(t, z)\mathbf{e}_1 + \sin \varphi_R(t, z)\mathbf{e}_2),$$

$$\mathbf{p}_L = \frac{e}{2}(\cos \varphi_L(t, z)\mathbf{e}_1 + \sin \varphi_L(t, z)\mathbf{e}_2),$$

where

$$\varphi_R(t, z) := -\omega\left(t - \frac{n_R z}{c}\right), \quad \varphi_L(t, z) := \omega\left(t - \frac{n_L z}{c}\right).$$

Since $\mathbf{p}_R$ and $\mathbf{p}_L$ share the same length at all times, the resulting polarization $\mathbf{p} = \mathbf{p}_R + \mathbf{p}_L$ lies along the direction that bisects the angle between them and so makes the angle

$$\alpha := \frac{\varphi_R + \varphi_L}{2} = \frac{\omega(n_R - n_L)z}{2c}$$

with $\mathbf{e}_1$. Moreover, it is an easy exercise to see that the length of $\mathbf{p}$

is given by

$$|\mathbf{p}| = e\sqrt{\frac{1+\cos(\varphi_R-\varphi_L)}{2}}.$$

Thus, we arrive at the following formula for the polarization of the wave travelling in $\mathscr{B}$:

$$\mathbf{p} = e\sqrt{\frac{1+\cos(\varphi_R-\varphi_L)}{2}}(\cos\alpha\mathbf{e}_1+\sin\alpha\mathbf{e}_2). \tag{1.54}$$

It is to be noted that α depends only on z, and so in the layer the wave is linearly polarized at all times, and when it exits its polarization is rotated by the angle

$$\alpha_0 = \pi(n_R - n_L)\frac{d}{\lambda_0}, \tag{1.55}$$

with respect to the polarization of the incident wave (*cf.* (1.37) and (1.38) above). The angle in (1.55) is **negative** whenever the polarization rotates **clockwise** with respect to an observer towards whom the wave travels and positive otherwise.

The ratio α_0/d depends only on the substance in the layer and the wavelength of the incident beam; customarily, its opposite is called the **optical activity** of the substance, and so it is **positive** when the polarization rotates **clockwise** while it gets closer to the observer. For example, the optical activity of quartz is 21.72 degrees per mm when $\lambda_0 = 5893$ Å (*cf. e.g.* Table 5 of Appendix 4 to Rossi (1957)).

It should be noted that in general both n_R and n_L depend on λ_0 in a complicated manner, and so also does the optical activity.

When a substance is optically active, at close inspection it generally reveals a chiral structure. This is for example the case with quartz, in the lattice of which the silicon atoms are arranged on helices which wind around the oxygen atoms. Also the molecules of glucose, which give rise to optically active solutions when dispersed in water, are chiral. These substances, like any chiral one, may exist in two opposite enantiomorphs, and so there are two phases of quartz, which are customarily called α and β, as well as two phases of glucose, namely dextrose and levulose. Such enantiomorphs have opposite optical activities.

Cholesteric liquid crystals are also chiral substances, and so we expect them to exhibit optical activity. This is indeed the case, but the frequency of the incoming light plays a central role here: the optical properties of cholesterics are qualitatively different in three different ranges of the frequency of light, which depend also on the natural pitch of the substance.

Suppose that the layer $\mathscr{B}$ described by (1.30) above is occupied by a cholesteric liquid crystal in its natural orientation with the cholesteric axis parallel to $\mathbf{e}_3$, so that the optic axis is represented by

$$\mathbf{n} = \cos \tau z \mathbf{e}_1 + \sin \tau z \mathbf{e}_2$$

(*cf.* equation (1.1) above).

Let a linearly polarized plane wave be incident on $\mathscr{B}$ in the direction of $\mathbf{e}_3$; in this case the **critical wavelength** is given by

$$\lambda_c := P, \tag{1.56}$$

where P is the natural pitch of the helical structure (*cf.* (1.2) above). Equation (1.56) is **Bragg's relation** for normal incidence (*cf. e.g.* p. 188 of Sommerfeld (1953)), provided we identify $\mathbf{n}$ and $-\mathbf{n}$ as optically equivalent.

Let n_o and n_e be the ordinary and the extraordinary indices of refraction of the medium. The critical angular frequencies are defined for the ordinary and the extraordinary wave:

$$\omega_o := \frac{2\pi c}{n_o \lambda_c}, \quad \omega_e := \frac{2\omega c}{n_e \lambda_c}. \tag{1.57}$$

Let ω_m and ω_M be the minimum and the maximum of ω_o and ω_e. Whenever the angular frequency ω of the incident wave is either greater than ω_M or smaller than ω_m, the layer is optically active. For $\omega < \omega_m$ the rotation of the optical vector is opposite to the helical structure, so that the optical activity is positive whenever the cholesteric helix is right-handed and negative otherwise. On the other hand, for $\omega > \omega_M$ the optical vector and the cholesteric helix rotate in the same way, and so the optical activity is negative when the cholesteric helix is right-handed, and positive otherwise.

These qualitative properties are briefly explained in Section XVIII.B of Stephen and Straley (1974) within the electromagnetic theory of light.

The optical activity of cholesterics is generally much greater than that of any other substance: its typical magnitude is of thousands of degrees per mm. For example, Brown (1973) reports the following values of the optical activity of (−)2-methyl-butyl-4-(cyanobenzylidene-amino)-cinnamate at 70°C: 5300 and 27 650 degrees per mm, for wave-lengths of 6580 Å and 4520 Å, respectively.

Following the brief but neat presentation of Section XVIII of Stephen and Straley (1974), we outline below other optical properties of cholesterics, which are explained in more detail in both Section 4.1 of Chandrasekhar (1977) and Section 6.1 of De Gennes (1974). I refer the reader to these books for a comprehensive treatment of the optical properties of cholesterics.

1.4.4 Selective reflection

Now let the angular frequency of the incident wave ω fall between ω_m and ω_M, and suppose, just to be specific, that the cholesteric helix in the layer $\mathscr{B}$ is right-handed. As we have learned above, a linearly polarized wave can be seen as the superposition of two waves with opposite circular polarizations. The left-polarized wave is transmitted nearly unchanged by the layer, apart from a negligible component which is reflected by the interface exposed to light, whereas the right-polarized wave does not propagate in the layer and is completely reflected as a *right-polarized* wave. This latter feature is quite remarkable, as any ordinary mirror reflects a circularly polarized wave with reverse polarization.

If, on the other hand, the cholesteric helix is left-handed and ω is still between ω_m and ω_M, the layer reflects completely any left-polarized wave.

Thus, the interval $[\omega_m, \omega_M]$ defines the **reflection band**: every circularly polarized wave whose angular frequency falls within this band is completely reflected whenever its polarization matches the chirality of the cholesteric helix.

Generally, for a cholesteric liquid crystal the ordinary and extraordinary indices of refraction do not differ by much, and so by (1.57) the reflection band is very narrow. To see this better, we set

$$\bar{\omega} := \tfrac{1}{2}(\omega_o + \omega_e) \quad \text{and} \quad \delta\omega := |\omega_e - \omega_o|,$$

whence we easily arrive at

$$\frac{\delta\omega}{\bar{\omega}} = 2\frac{|n_e - n_o|}{n_e + n_o}. \tag{1.58}$$

If we take $n_e = 1.57$ and $n_o = 1.51$, which are values quite common for a cholesteric liquid crystal (*cf. e.g.* Teucher, Ko and Labes (1972)), we get $(\delta\omega/\bar{\omega}) \approx 0.04$.

Moreover, $\bar{\omega}$ generally falls within the visible spectrum, and so does the reflection band. It is then easy to explain why a cholesteric layer illuminated by white light has a pure colour. Each monochromatic component of the incident beam can be regarded as the superposition of two incoherent circularly polarized waves with opposite polarizations. Only the components whose angular frequencies are in the reflection band come back to the observer, and moreover they all have the same circular polarization, which agrees with the cholesteric helix as above. Thus, the narrower the reflected band, the sharper is the colour shown by the layer.

When the angle of incidence of the beam relative to the cholesteric axis is different from zero, both $\bar{\omega}$ and $\delta\omega$ are slightly affected (*cf. e.g.* p. 698 of Stephen and Straley (1974)) and so the characteristic colour shown by the layer changes accordingly. It is to be noted in passing that for oblique incidence there are also several reflection bands, which depend on the angle of incidence.

It seems that the hard outer wings of some beetles behave like cholesteric layers, and so the iridescence of these insects may be explained through a number of reflection bands. Such an interpretation is also confirmed by the experimental evidence that the light reflected by them is circularly polarized (*cf. e.g.* Gaubert (1924)).

We have already remarked in Section 1.1 that the natural pitch of cholesterics is very sensitive to the temperature, as the reflection band must be, since by (1.56) and (1.57) both $\bar{\omega}$ and $\delta\omega$ depend on the pitch. We learn for example on p. 1059 of Brown (1973) that there are cholesteric liquid crystals capable of showing a visible change of colour even when the temperature changes by just 1×10^{-3} °C.

Plates 1.VII and 1.VIII illustrate the coloured patterns produced on a cholesteric layer heated by brief contact with the fingers of the photographer. (We read on the back of this gadget that the monster depicted on the front side is a tyrannosaur, but it has long been extinct, so they say.)

1.4.5 Adiabatic approximation

When a monochromatic beam is incident on a cholesteric layer along the cholesteric axis with a wavelength much smaller than the natural pitch of the helical structure, the propagation of light inside the layer can easily be described through an approximation, often called **adiabatic**, which we now illustrate.

Suppose that the layer $\mathscr{B}$ is described again by (1.30) above and that the incoming wave is linearly polarized and propagates along $\mathbf{e}_3$. The optical vector at the interface $z=0$ can be regarded as the superposition of two vectors vibrating with the same frequency along the optic axis and the direction orthogonal to it. In the adiabatic approximation these two vibrations rotate with the optic axis so as to be, respectively, parallel and orthogonal to it everywhere in $\mathscr{B}$. The phase difference between these two components of the polarization is the same as if the optic axis were not twisted in $\mathscr{B}$, and so it can be computed as it was above for a nematic layer.

Thus, when the incident wave is polarized either along the optic axis at $z=0$ or at right angles to it, the phase difference vanishes everywhere in $\mathscr{B}$, and the wave emerges from the layer polarized along the optic axis at $z=d$ or at right angles to it. When, on the other hand, the optical vector of the incident wave makes an angle between 0 and $\pi/2$ with the direction of the optic axis at $z=0$, the phase difference of the two orthogonal components of the polarization in $\mathscr{B}$ is given by (1.41), and so when the wave emerges from the layer it is in general elliptically polarized.

Such an approximate description of the polarization in $\mathscr{B}$ is accurate only when the wavelength is small compared with the pitch of the cholesteric structure. This limiting case was first studied in 1911 by M.C. Mauguin (*cf. e.g.* p. 195 of Chandrasekhar (1977)), who considered a cell within which a nematic liquid crystal is uniformly twisted between two plates. If the total twist is less than a complete turn and if the thickness of the cell is much larger than the wavelength of the light propagating through it, the approximate description outlined above can be fully justified within the electromagnetic theory of light.

The reader should recall this approximation when we come to describe in Section 4.5 of Chapter 4 below the electro-optical effect on which many modern liquid-crystal displays are based. In most of these devices the total twist imposed on the nematic by the plates bounding the cell is a quarter of a complete turn and the thickness of the layer is about 10 μm. Berreman (1973) has computed numerically the solutions to the equations of the electromagnetic theory of light for these cells: the results of his analysis agree quite well with the adiabatic approximation.

1.5 References

Berreman, D.W. (1973) Optics in smoothly varying anisotropic planar structures: Application to liquid-crystal twist cells, *J. Opt. Soc. Am.*, **63**, 1374–1380.

Brown, G.H. (1973) Structure, properties, and some applications of liquid crystals, *J. Opt. Soc. Am.*, **63**, 1505–1514.

Brown, G.H. and Shaw, W.G. (1957) The mesomorphic state. Liquid crystals, *Chem. Rev.*, **57**, 1049–1157.

Capriz, G. (1989) *Continua with microstructure*, Springer, New York.

Chandrasekhar, S. (1977) *Liquid crystals*, Cambridge University Press, Cambridge, Second edn, 1992.

Collings, P.J. (1990) *Liquid crystals. Nature's delicate phase of matter.* Adam Hilger, Bristol.

De Gennes, P.G. (1969) Phenomenology of short-range-order effects in the isotropic phase of nematic materials, *Phys. Lett.*, **30A**, 454–455.

De Gennes, P.G. (1974) *The physics of liquid crystals*, Clarendon Press, Oxford, Second edn, 1993, with J. Prost.

Ericksen, J.L. (1976) Equilibrium theory of liquid crystals, in: *Advances in liquid crystals*, Vol. 2, G.H. Brown ed., Academic Press, New York.

Ericksen, J.L. (1991) Liquid crystals with variable degree of orientation, *Arch. Rational Mech. Anal.*, **113**, 97–120.

Fan, C.-P. (1971) Disclination lines in liquid crystals, *Phys. Lett.*, **34A**, 335–336.

Fan C.-P. and Stephen, M.J. (1970), Isotropic–nematic phase transition in liquid crystals, *Phys. Rev. Lett.*, **25**, 500–503.

Fergason, J.L. (1964) Liquid crystals, *Scientific American*, **211**, 76–85.

Frank, F.C. (1980) In: Liquid Crystals, S. Chandrasekhar ed., Hayden & Sons, Philadelphia, 1–6.

Gaubert, P. (1924) Sur la polarisation circulaire de la lumière réfléchie par les insectes, *C.R. Hebd. Séanc. Acad. Sci. Paris*, **179**, 1148–1150.

Gray, G.W. (1962) *Molecular properties of liquid crystals*, Academic Press, London.

Kelker, H. (1973) History of liquid crystals, *Mol. Cryst. Liq. Cryst.*, **21**, 1–48.

Kelker, H. and Knoll, P.M. (1989) Some pictures of the history of liquid crystals, *Liq. Crystals*, **5**, 19–42.

Lavrentovich, O.D. and Nastishin, Yu.A. (1990) Defects in degenerate hybrid aligned nematic liquid crystals, *Europhys. Lett.*, **12**, 135–141.

Leslie, F.M. (1968) Some constitutive equations for liquid crystals, *Arch. Rational Mech. Anal.*, **28**, 265–283.
Ornstein, L.S. and Kast, W. (1933) New arguments for the swarm theory of liquid crystals, *Trans. Faraday Soc.*, **29**, 931–944.
Orsay Group (1971) Les cristaux liquides, *La Recherche*, **2**, 433–441.
Oseen, C.W. (1921) Versuch einer kinetischen Theorie der kristallinischen Flüssigkeiten, *Gl. Svenska Vetenskapsakademiens Handlingar*, **61**, No. 16.
Pikin, S.A. and Indenbom, V.L. (1978) Thermodynamic states and symmetry of liquid crystals, *Sov. Phys. Usp.*, **21**, 487–501.
Reinitzer, F. (1888) Beiträge zur Kenntniss des Cholesterins, *Monatsh. Chem.*, **9**, 421–441.
Reinitzer, F. (1989) Contributions to the knowledge of cholesterol, *Liq. Crystals*, **5**, 7–18. Translation of Reinitzer (1988).
Rossi, B. (1957) *Optics*, Addison-Wesley, Reading, Massachusetts.
Sackmann, H. (1989) Smectic liquid crystals. A historical review, *Liq. Crystals*, **5**, 43–55.
Sommerfeld, A. (1953) *Optics*, Volume IV of *Lectures on theoretical physics*, Academic Press, New York.
Stephen, M.J. and Straley, J.P. (1974) Physics of liquid crystals, *Rev. Mod. Phys.*, **46**, 617–704.
Teucher, I., Ko, K. and Labes, M.M. (1972) Birefringence and optical rotatory dispersion of a compensated cholesteric liquid crystal, *J. Chem. Phys.*, **56**, 3308–3311.
Truesdell, C.A. (1991) *A first course in rational continuum mechanics*, Vol. 1, second edition, corrected, revised and augmented, Academic Press, New York.
Zocher, H. (1933) The effect of a magnetic field on the nematic state, *Trans. Faraday Soc.*, **29**, 945–957.

1.6 Further reading

Liquid crystals

Blinov, L.M. (1983) *Electro-optical and magneto-optical principles of liquid crystals*, J. Wiley & Sons, Chichester.
De Jeu, W.H. (1980) *Physical properties of liquid crystalline materials*, Gordon & Breach, New York.
Donald, A.M. and Windle, A.H. (1992) *Liquid crystalline polymers*, Cambridge University Press, Cambridge.

Dubois, J.C. (1982a) Liquid crystals: polymorphism, molecular structure and synthesis, in: Cristalli liquidi, *Atti della I Scuola Nazionale del GNCL, Rende*, 10–21 Settembre 1981, CLU, Torino, 5–33.

Dubois, J.C. (1982b) Structure des mesophases, in: Cristalli liquidi, *Atti della I Scuola Nazionale del GNCL, Rende*, 10–21 Settembre 1981, CLU, Torino, 35–51.

Friedel, G. (1922) Les états mésomorphes de la matière, *Ann. Phys. (Paris)*, **18**, 273–474.

Gray, G.W. and Winsor, P.A. eds. (1974) *Liquid crystals and plastic crystals*, Volume 1: *Preparation, constitution and applications*. Volume 2: *Physico-chemical properties and methods of investigation*, Ellis Horwood, Chichester.

Leadbetter, A.J. (1987) Structural classification of liquid crystals, in: *Thermotropic liquid crystals, Critical reports on applied chemistry*, Vol. 22, G.W. Gray ed., J. Wiley & Sons, Chichester, 1–27.

Priestley, E.B., Wojtowicz, P.J. and Sheng, P. eds. (1974) *Introduction to liquid crystals*, Plenum Press, New York.

Vertogen, G. and De Jeu, W.H. (1988) *Thermotropic liquid crystals, fundamentals*, Springer, Berlin.

Early theories

Lehmann, O. (1890) Einige Fälle von Allotropie, *Z. Krist.*, **18**, 464–467.

Oseen, C.W. (1931) Probleme für die Theorie der anisotropen Flüssigkeiten, *Z. Kristallographie*, **79**, 173–185.

Zocher, H. (1925) Über freiwillige Strukturbildung in Solen, Z. Anorg. Chem., **147**, 91–110.

Order parameters

Blenk, S., Ehrentraut, H. and Muschik, W. (1991a) Statistical foundation of macroscopic balances for liquid crystals in alignment tensor formulation, *Physica A*, **174**, 119–138.

Blenk, S., Ehrentraut, H. and Muschik, W. (1991b) Orientation-balances for liquid crystals and their representation by alignment tensors, *Mol. Cryst. Liq. Cryst.*, **204**, 133–141.

Blenk, S., Ehrentraut, H. and Muschik, W. (1992) Macroscopic constitutive equations for liquid crystals induced by their mesoscopic orientation distribution, *Int. J. Eng. Sci.*, **30**, 1127–1143.

Denn, M.M. and Reimer, J.A. (1991) Rheology of thermotropic nematic liquid crystalline polymers, in: *Nematics. Mathematical and physical aspects*, J.-M. Coron, J.-M. Ghidaglia and F. Hélein eds., Kluwer Academic Publishers, Dordrecht, 107–112.

Doi, M. (1981) Molecular dynamics and rheological properties of concentrated solutions of rodlike polymers in isotropic and liquid crystalline phases, *J. Polymer Sci.*, **19**, 229–243.

Ericksen, J.L. (1984) A thermodynamic view of order parameters for liquid crystals, in: *Orienting Polymers, Lecture Notes in Mathematics No. 1063*, Springer, Berlin, 27–36.

Hess, S. (1986) Transport phenomena in anisotropic fluids and liquid crystals, *J. Non-Equilib. Thermodyn.*, **11**, 175–193.

Hess, S. and Pardowitz, I. (1981) On the unified theory of non-equilibrium phenomena in the isotropic and nematic phases of a liquid crystal; spatially inhomogeneous alignment, *Z. Naturforsch.*, **36a**, 554–558.

Hess, S., Schwazl, J.F. and Baalss, D. (1990) Anisotropy of the viscosity of nematic liquid crystals and of oriented ferro-fluids via non-equilibrium molecular dynamics, *J. Phys. Condens. Matter*, **2**, SA279–SA284.

Kaiser, P. and Hess, S. (1991) Relaxation of the alignment tensor in the isotropic and nematic phases of liquid crystals in the presence of an external field, *J. Non-Equilib. Thermodyn.*, **16**, 381–398.

Kilian, A. and Hess, S. (1989) Derivation and application of an algorithm for the numerical calculation of the local orientation of nematic liquid crystals, *Z. Naturforsch.*, **44a**, 693–703.

Marrucci, G. (1985) Rheology of liquid crystalline polymers, *Pure & Appl. Chem.*, **57**, 1545–1552.

Weider, T., Stottut, U., Loose, W. and Hess, S. (1991) Order in fluids: shear-induced anisotropy in dense fluids of spherical particles and in gases of rotating molecules, *Physica A*, **174**, 1–14.

Wissbrun, K.F. (1981) Rheology of rod-like polymers in the liquid crystalline state, *J. Rheol.*, **25**, 619–662.

Optical properties

Lavrentovich, O.D. and Pergamenshchik, V.M. (1990) Periodic domain structures in thin hybrid nematic layers, *Mol. Cryst. Liq. Cryst.*, **179**, 125–132.

Lipson, S.G. and Lipson, H. (1981) *Optical physics*, second edition. Cambridge University Press, Cambridge.

2

Mathematical preliminaries

In this chapter I recall all the basic mathematical tools that will be extensively employed in this book. For the most standard of these I do not give detailed proofs: some can be found in more general books such as Halmos (1958), Bowen and Wang (1976) and Noll (1987); others are simple exercises.

This chapter can be divided into two parts. The first half is devoted to general facts and serves as an introduction to the mathematical terminology used throughout the book; it is a repertoire of results and methods to be called upon in the proofs of many theorems given below. The second half contains theorems more germane to the topic of this book and opens the way to the development of the following chapters.

The reader need not go through all this chapter before proceeding further: most of the results stated or proved here are briefly recalled in the body of the book, when they are needed. It might be sensible not to read this chapter systematically, though for the relevance of the methods illustrated here it cannot be regarded by any means as a misplaced appendix.

2.1 Points, vectors and tensors

The physical events we consider in this book take place in a three-dimensional Euclidean space. To make our development rigorous we give a formal definition of such a space.

Definition. 2.1 We say that $\mathscr{E}$ is a **three-dimensional Euclidean space** if there is an inner-product linear space $\mathscr{V}$ of dimension 3 associated with it such that:

(a) The elements of $\mathscr{V}$, called **translations**, are mappings of $\mathscr{E}$ into itself:

$$\mathbf{v} \in \mathscr{V}, \quad \mathbf{v}: \mathscr{E} \to \mathscr{E};$$

(b) The sum of two elements of $\mathscr{V}$ satisfies:

$$(\mathbf{u} + \mathbf{v})(p) = \mathbf{u}(\mathbf{v}(p)) \quad \text{for all } \mathbf{u}, \mathbf{v} \in \mathscr{V} \text{ and all } p \in \mathscr{E};$$

(c) For any two given elements p and q of $\mathscr{E}$ there is exactly one $\mathbf{v} \in \mathscr{V}$ such that

$$\mathbf{v}(p) = q.$$

2.1.1 Points and vectors

In what follows we select once and for all a three-dimensional Euclidean space $\mathscr{E}$, whose elements we call **points**, which is to be identified with the ordinary space where we live. We also call $\mathscr{V}$ the **translation space** of $\mathscr{E}$ and its members **vectors.** $\mathscr{V}$ is a linear space over $\mathbb{R}$, as are all linear spaces we shall consider in the following.

By part (c) of the above definition, a vector is a mapping of $\mathscr{E}$ onto itself which is completely determined by the value taken on a single point of $\mathscr{E}$. Figure 2.1 certainly looks familiar to most readers. It shows that the point p is changed into q by applying $\mathbf{v}$ to it, but also that p' is changed into q' by applying to it precisely the same vector $\mathbf{v}$, that is

$$q = \mathbf{v}(p), \quad q' = \mathbf{v}(p'). \tag{2.1}$$

We shall never again use formulae like (2.1) in this book. The

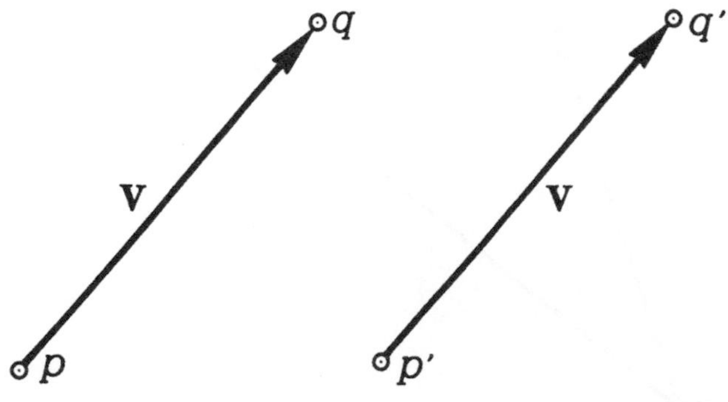

Figure 2.1

following notation has already become customary:

$$\mathbf{v}(p) =: p + \mathbf{v}. \tag{2.2}$$

It defines the addition of a vector $\mathbf{v}$ to a point p as the operation that produces a point by translating p through $\mathbf{v}$.

Likewise, condition (c) suggests how to define the difference of two points of $\mathscr{E}$: $q - p \in \mathscr{V}$ is defined to be *the* vector that translates p to q, so that, by (2.2),

$$(q - p) + p = q. \tag{2.3}$$

Accordingly, the null vector of $\mathscr{V}$ can be regarded as the difference of two equal points of $\mathscr{E}$:

$$\mathbf{0} = p - p \quad \text{for any } p \in \mathscr{E}.$$

It is to be noted that whereas both the difference of two points and the addition of a vector to a point are well defined, there is no way of defining either the addition of two points or the difference of a point from a vector.

We are now in a position to interpret better part (b) of the above definition. Let p be a point of $\mathscr{E}$. If $\mathbf{v}$ and $\mathbf{u}$ are vectors of $\mathscr{V}$ we set

$$q = p + \mathbf{v} \quad \text{and} \quad x = q + \mathbf{u}.$$

Thus,

$$x = p + \mathbf{w} \quad \text{with} \quad \mathbf{w} = \mathbf{u} + \mathbf{v}.$$

Figure 2.2 shows how the addition of two vectors of $\mathscr{V}$ agrees with

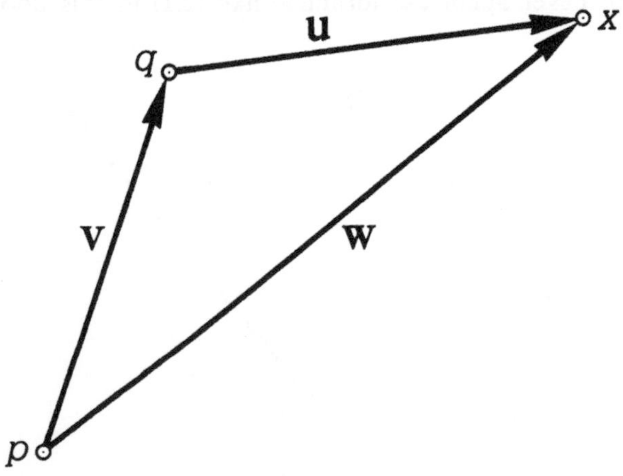

Figure 2.2

the **rule of composition of forces** that everybody has encountered when studying elementary mechanics. Drawing a similar diagram, the reader will easily convince himself that

$$\mathbf{u}+\mathbf{v}=\mathbf{v}+\mathbf{u} \quad \text{for all } \mathbf{u},\mathbf{v}\in\mathscr{V},$$

as it should be, $\mathscr{V}$ being a linear space.

The inner product of any two vectors of $\mathscr{V}$, say $\mathbf{u}$ and $\mathbf{v}$, will be denoted by

$$\mathbf{u}\cdot\mathbf{v}.$$

The **length** of a vector $\mathbf{u}$ is defined by

$$|\mathbf{u}|=(\sqrt{\mathbf{u}\cdot\mathbf{u}}). \tag{2.4}$$

The **unit sphere** of $\mathscr{V}$ is the set $\mathbb{S}^2$ of all vectors whose length is 1:

$$\mathbb{S}^2:=\{\mathbf{v}\in\mathscr{V}\,|\,|\mathbf{v}|=1\}.$$

In the Euclidean space $\mathscr{E}$ a **distance** is naturally induced by (2.4): it is the function $d:\mathscr{E}\times\mathscr{E}\rightarrow\mathbb{R}^+$ defined by

$$d(p,q):=|p-q| \quad \text{for all } p,q\in\mathscr{E}.$$

We apply the term **basis** of $\mathscr{V}$ to any list $e=(\mathbf{e}_1,\mathbf{e}_2,\mathbf{e}_3)$ of three vectors of $\mathbb{S}^2$ that are mutually orthogonal, that is such that

$$\mathbf{e}_i\cdot\mathbf{e}_j=\delta_{ij} \quad \text{for } i,j=1,2,3,$$

where δ_{ij} is the **Kronecker symbol**:

$$\delta_{ij}:=\begin{cases}1 & \text{if } i=j\\ 0 & \text{if } i\neq j.\end{cases}$$

Thus, if $e=(\mathbf{e}_1,\mathbf{e}_2,\mathbf{e}_3)$ is a basis of $\mathscr{V}$ every vector $\mathbf{u}$ of $\mathscr{V}$ can be written in the form

$$\mathbf{u}=\sum_{i=1}^{3}u_i\mathbf{e}_i,$$

where

$$u_i:=\mathbf{u}\cdot\mathbf{e}_i, \quad \text{for } i=1,2,3,$$

are the **Cartesian components** of $\mathbf{u}$ in the basis e.

Usually, a list such as e is called an *orthonormal* basis of $\mathscr{V}$; here the only basis of $\mathscr{V}$ we consider are orthonormal and we call them just *bases* for brevity.

Let $\mathscr{U}$ be a subset of $\mathscr{V}$. We denote by span $\{\mathscr{U}\}$ the subset of all vectors in $\mathscr{V}$ that are linear combinations of vectors in $\mathscr{U}$. That is, if $\mathbf{v}$ is given in span $\{\mathscr{U}\}$, there are vectors in $\mathscr{U}$, say $\mathbf{u}_1,\dots,\mathbf{u}_n$, and scalars in $\mathbb{R}$, say $\lambda_1,\dots,\lambda_n$, such that

$$\mathbf{v}=\sum_{i=1}^{n}\lambda_i\mathbf{u}_i.$$

Clearly,

$$\mathscr{V}=\operatorname{span}\{e\},$$

for any basis e of $\mathscr{V}$.

2.1.2 Tensors

In general, two types of tensors appear in this book: **second-order tensors** and **fourth-order tensors**. Occasionally, we also encounter **third-order tensors**: they are briefly considered at the end of this section.

Second-order tensors are linear mappings of $\mathscr{V}$ into itself; they constitute the set

$$L(\mathscr{V}):=\{\mathbf{L}:\mathscr{V}\to\mathscr{V}\,|\,\mathbf{L}\text{ is linear}\}. \tag{2.5}$$

If we define the sum of two tensors $\mathbf{A}$ and $\mathbf{B}$ of $L(\mathscr{V})$ as

$$(\mathbf{A}+\mathbf{B})\mathbf{v}:=\mathbf{A}\mathbf{v}+\mathbf{B}\mathbf{v}\quad\text{for all }\mathbf{v}\in\mathscr{V},$$

and the multiplication of a tensor $\mathbf{A}$ by a scalar $\alpha\in\mathbb{R}$ as

$$(\alpha\mathbf{A})\mathbf{v}=\alpha(\mathbf{A}\mathbf{v})\quad\text{for all }\mathbf{v}\in\mathscr{V},$$

then $L(\mathscr{V})$ is itself a linear space, whose null element is the **null tensor** defined by

$$\mathbf{0}\mathbf{v}=\mathbf{0}\quad\text{for all }\mathbf{v}\in\mathscr{V}.$$

We apply the term **identity** to the tensor $\mathbf{I}$ defined by

$$\mathbf{I}\mathbf{v}=\mathbf{v}\quad\text{for all }\mathbf{v}\in\mathscr{V}.$$

We shall see in the following section how $L(\mathscr{V})$ can be endowed with a natural inner product.

A fourth-order tensor is a linear mapping of $L(\mathscr{V})$ into itself. We denote by $L^2(\mathscr{V})$ the set of all fourth-order tensors:

$$L^2(\mathscr{V}):=\{\mathbb{L}:L(\mathscr{V})\to L(\mathscr{V})\,|\,\mathbb{L}\text{ is linear}\}. \tag{2.6}$$

Such a notation is to be interpreted as follows:

$$L^2(\mathscr{V}):=L(L(\mathscr{V})).$$

$L^2(\mathscr{V})$ is a linear space, where the basic operations are defined just as in $L(\mathscr{V})$; the null element is the tensor that associates the null tensor of $L(\mathscr{V})$ with all others. The identity of $L^2(\mathscr{V})$ is the tensor $\mathbb{I}$ defined by

$$\mathbb{I}[\mathbf{L}] = \mathbf{L} \quad \text{for all } \mathbf{L} \in L(\mathscr{V}).$$

Let **a** and **b** be vectors of $\mathscr{V}$. We define the **tensor product** of **a** and **b** as the tensor $\mathbf{a} \otimes \mathbf{b} \in L(\mathscr{V})$ such that

$$(\mathbf{a} \otimes \mathbf{b})\mathbf{v} := (\mathbf{b} \cdot \mathbf{v})\mathbf{a} \quad \text{for all } \mathbf{v} \in \mathscr{V}. \tag{2.7}$$

If $e = (\mathbf{e}_1, \mathbf{e}_2, \mathbf{e}_3)$ is a basis of $\mathscr{V}$, then $e^2 := (\mathbf{e}_i \otimes \mathbf{e}_j \,|\, i,j = 1,2,3)$ is a basis of $L(\mathscr{V})$: the members of e^2 are linearly independent, and every tensor $\mathbf{L} \in L(\mathscr{V})$ can be written in the form

$$\mathbf{L} = \sum_{i,j=1}^{3} L_{ij} \mathbf{e}_i \otimes \mathbf{e}_j, \tag{2.8}$$

where

$$L_{ij} := \mathbf{e}_i \cdot \mathbf{L}\mathbf{e}_j, \quad \text{for } i,j = 1,2,3, \tag{2.9}$$

are the **Cartesian components** of **L** in the basis e^2. It is an immediate consequence of this fact that the dimension of $L(\mathscr{V})$ is 3^2. In the following section we shall also provide a basis for $L^2(\mathscr{V})$; it will then be clear that the dimension of $L^2(\mathscr{V})$ is 9^2.

Clearly, another basis of $L(\mathscr{V})$ could be generated from e^2 by taking linear combinations of its members so as to obtain precisely 9 linearly independent tensors. Here is a list of such tensors to be used in the following section:

$$\mathbf{E}_1 := \frac{1}{\sqrt{2}}(\mathbf{e}_2 \otimes \mathbf{e}_3 + \mathbf{e}_3 \otimes \mathbf{e}_2), \quad \mathbf{E}_2 := \frac{1}{\sqrt{2}}(\mathbf{e}_1 \otimes \mathbf{e}_3 + \mathbf{e}_3 \otimes \mathbf{e}_1),$$

$$\mathbf{E}_3 := \frac{1}{\sqrt{2}}(\mathbf{e}_1 \otimes \mathbf{e}_2 + \mathbf{e}_2 \otimes \mathbf{e}_1), \quad \mathbf{E}_4 := \frac{1}{\sqrt{2}}(\mathbf{e}_1 \otimes \mathbf{e}_1 - \mathbf{e}_2 \otimes \mathbf{e}_2),$$

$$\mathbf{E}_5 := \mathbf{e}_3 \otimes \mathbf{e}_3, \quad \mathbf{E}_6 := \frac{1}{\sqrt{2}}(\mathbf{e}_1 \otimes \mathbf{e}_1 + \mathbf{e}_2 \otimes \mathbf{e}_2),$$

$$\mathbf{E}_7 := \frac{1}{\sqrt{2}}(\mathbf{e}_3 \otimes \mathbf{e}_2 - \mathbf{e}_2 \otimes \mathbf{e}_3),$$

$$\mathbf{E}_8 := \frac{1}{\sqrt{2}}(\mathbf{e}_1 \otimes \mathbf{e}_3 - \mathbf{e}_3 \otimes \mathbf{e}_1), \quad \mathbf{E}_9 := \frac{1}{\sqrt{2}}(\mathbf{e}_2 \otimes \mathbf{e}_1 - \mathbf{e}_1 \otimes \mathbf{e}_2). \tag{2.10}$$

We shall occasionally employ the notation

$$E := (\mathbf{E}_i \mid i = 1, \ldots, 9). \tag{2.11}$$

The product of two tensors **A** and **B** of $L(\mathscr{V})$ is the tensor defined by

$$(\mathbf{AB})\mathbf{v} := \mathbf{A}(\mathbf{B}\mathbf{v}) \quad \text{for all } \mathbf{v} \in \mathscr{V}. \tag{2.12}$$

The product of two tensors $\mathbb{A}$ and $\mathbb{B}$ of $L^2(\mathscr{V})$ is defined likewise:

$$(\mathbb{A}\mathbb{B})[\mathbf{L}] := \mathbb{A}(\mathbb{B}[\mathbf{L}]) \quad \text{for all } \mathbf{L} \in L(\mathscr{V}).$$

It is an easy exercise to check that

$$(\mathbf{a} \otimes \mathbf{b})(\mathbf{c} \otimes \mathbf{d}) = (\mathbf{b} \cdot \mathbf{c})\mathbf{a} \otimes \mathbf{d} \quad \text{for all } \mathbf{a}, \mathbf{b}, \mathbf{c}, \mathbf{d} \in \mathscr{V},$$

and

$$\mathbf{A}(\mathbf{a} \otimes \mathbf{b}) = (\mathbf{A}\mathbf{a}) \otimes \mathbf{b}, \quad \text{for all } \mathbf{a} \in \mathscr{V} \text{ and } \mathbf{A} \in L(\mathscr{V}).$$

2.1.3 Third-order tensors

A third-order tensor can be regarded either as a linear mapping of $\mathscr{V}$ into $L(\mathscr{V})$, *i.e.* as a member of the space $L(\mathscr{V}, L(\mathscr{V}))$, or as a linear mapping of $L(\mathscr{V})$ into $\mathscr{V}$, *i.e.* as a member of the space $L(L(\mathscr{V}), \mathscr{V})$. Let **a**, **b** and **c** be vectors of $\mathscr{V}$; the **tensor product** $\mathbf{a} \otimes \mathbf{b} \otimes \mathbf{c}$ is an elementary third-order tensor: when it operates on $\mathscr{V}$, it is defined by

$$(\mathbf{a} \otimes \mathbf{b} \otimes \mathbf{c})[\mathbf{v}] := (\mathbf{v} \cdot \mathbf{c})(\mathbf{a} \otimes \mathbf{b}) \quad \text{for all } \mathbf{v} \in \mathscr{V},$$

whereas, when it operates on $L(\mathscr{V})$, it is defined by

$$(\mathbf{a} \otimes \mathbf{b} \otimes \mathbf{c})[\mathbf{L}] := (\mathbf{b} \cdot \mathbf{L}\mathbf{c})\mathbf{a} \quad \text{for all } \mathbf{L} \in L(\mathscr{V}).$$

Let $e = (\mathbf{e}_1, \mathbf{e}_2, \mathbf{e}_3)$ be a basis of $\mathscr{V}$. Then $e^3 := (\mathbf{e}_i \otimes \mathbf{e}_j \otimes \mathbf{e}_k \mid i, j, k = 1, 2, 3)$ is a basis of both $L(\mathscr{V}, L(\mathscr{V}))$ and $L(L(\mathscr{V}), \mathscr{V})$, and so their dimension is 3^3. Each third-order tensor L can be represented by the formula

$$\mathsf{L} = \sum_{i,j,k=1}^{3} L_{ijk} \mathbf{e}_i \otimes \mathbf{e}_j \otimes \mathbf{e}_k, \tag{2.13}$$

where L_{ijk}, for $i, j, k = 1, 2, 3$, are the **Cartesian components** of L in e^3 defined by

$$L_{ijk} := \mathbf{e}_i \cdot \mathsf{L}[\mathbf{e}_j \otimes \mathbf{e}_k] \quad \text{for } i, j, k \in \{1, 2, 3\}. \tag{2.14}$$

Another definition of the Cartesian components of I, equivalent to (2.14), will be given in Section 2.2.

2.2 Algebra

Here we go into further details of some of the linear spaces introduced in the preceding section, namely $\mathscr{V}$, $L(\mathscr{V})$, and $L^2(\mathscr{V})$. We single out several issues that are relevant to some or to all of these spaces.

2.2.1 Transposition

The following theorem states a fundamental fact, which holds in similar forms for all linear inner-product spaces.

Theorem 2.1. Let $\mathbf{L}$ be a tensor of $L(\mathscr{V})$. There is precisely one tensor of $L(\mathscr{V})$, which is denoted by $\mathbf{L}^T$, such that

$$\mathbf{u}\cdot\mathbf{L}\mathbf{v} = \mathbf{L}^T\mathbf{u}\cdot\mathbf{v} \quad \text{for all } \mathbf{u}, \mathbf{v} \in \mathscr{V}. \tag{2.15}$$

The tensor $\mathbf{L}^T$ is called the **transpose** of $\mathbf{L}$.

It is an easy exercise to prove that

$$(\mathbf{a}\otimes\mathbf{b})^T = \mathbf{b}\otimes\mathbf{a} \quad \text{for all } \mathbf{a}, \mathbf{b} \in \mathscr{V}, \tag{2.16}$$

and also that

$$(\mathbf{A}^T)^T = \mathbf{A},$$
$$(\mathbf{A}+\mathbf{B})^T = \mathbf{A}^T + \mathbf{B}^T,$$
$$(\mathbf{A}\mathbf{B})^T = \mathbf{B}^T\mathbf{A}^T, \quad \text{for all } \mathbf{A}, \mathbf{B} \in L(\mathscr{V}),$$

and

$$(\mathbf{a}\otimes\mathbf{b})\mathbf{A} = \mathbf{a}\otimes\mathbf{A}^T\mathbf{b} \quad \text{for all } \mathbf{a}, \mathbf{b} \in \mathscr{V} \text{ and } \mathbf{A} \in L(\mathscr{V}).$$

Definition 2.2. A tensor $\mathbf{L} \in L(\mathscr{V})$ is **symmetric** if

$$\mathbf{L}^T = \mathbf{L},$$

whereas it is **skew** if

$$\mathbf{L}^T = -\mathbf{L}.$$

The set of all symmetric second-order tensors is a linear subspace of $L(\mathscr{V})$, so is the set of all skew second-order tensors. We occasionally denote the former subspace by $Sym(\mathcal{V})$ and the latter by $Skw(\mathcal{V})$. Recalling the basis $(\mathbf{E}_i | i = 1, \ldots, 9)$, whose members are listed in (2.10), we easily see that $(\mathbf{E}_i | i = 1, \ldots, 6)$ is a basis for $Sym(\mathcal{V})$ while $(\mathbf{E}_i | i = 7, 8, 9)$ is a basis for $Skw(\mathcal{V})$. Thus, these two subspaces have dimension 6 and 3, respectively.

Any tensor $\mathbf{L} \in L(\mathcal{V})$ which is neither symmetric nor skew can be written as the sum of two tensors, one symmetric and the other skew, as is shown below:

$$\mathbf{L} = \mathbf{S} + \mathbf{W}, \quad \text{where} \quad \mathbf{S} := \tfrac{1}{2}(\mathbf{L} + \mathbf{L}^T) \quad \text{and} \quad \mathbf{W} := \tfrac{1}{2}(\mathbf{L} - \mathbf{L}^T). \tag{2.17}$$

Clearly, in (2.17) $\mathbf{S}$ is symmetric and $\mathbf{W}$ is skew: they are called the **symmetric part** and the **skew part** of $\mathbf{L}$, respectively.

To state the analogue of Theorem 2.1 for the tensors of $L^2(\mathcal{V})$, we need to endow $L(\mathcal{V})$ with an inner product. The transposition already defined for second-order tensors allows us to do so with the aid of a new concept, that of trace.

Theorem 2.2. There is precisely one linear mapping of $L(\mathcal{V})$ into $\mathbb{R}$ such that its value on $\mathbf{a} \otimes \mathbf{b}$ is $\mathbf{a} \cdot \mathbf{b}$ for all $\mathbf{a}, \mathbf{b} \in \mathcal{V}$.

We apply the term **trace** to such a linear mapping and we denote it by tr. It possesses the following properties:

$$\begin{aligned} &\operatorname{tr} \mathbf{A}^T = \operatorname{tr} \mathbf{A} && \text{for all } \mathbf{A} \in L(\mathcal{V}), \\ &\operatorname{tr} \mathbf{I} = 3 && \text{where } \mathbf{I} \text{ is the identity of } L(\mathcal{V}), \\ &\operatorname{tr} \mathbf{W} = 0 && \text{whenever } \mathbf{W} \text{ is skew.} \end{aligned}$$

If L_{ij}, for $i, j = 1, 2, 3$, are the Cartesian components of a tensor $\mathbf{L}$ as defined in (2.9) above, then

$$\operatorname{tr} \mathbf{L} = \sum_{i=1}^{3} L_{ii}.$$

It is easily seen that the mapping of $L(\mathcal{V}) \times L(\mathcal{V})$ into $\mathbb{R}$ defined by

$$(\mathbf{A}, \mathbf{B}) \mapsto \mathbf{A} \cdot \mathbf{B} := \operatorname{tr}(\mathbf{A}^T \mathbf{B}) \tag{2.18}$$

is an **inner product** in $L(\mathcal{V})$. It is such that

$$\mathbf{A}^T \cdot \mathbf{B}^T = \mathbf{A} \cdot \mathbf{B} \quad \text{for all } \mathbf{A}, \mathbf{B} \in L(\mathcal{V});$$

$$\mathbf{A} \cdot \mathbf{L} = \mathbf{A} \cdot \mathbf{S},$$

if **A** is symmetric and **L** is as in (2.17);

$$\mathbf{B}\cdot\mathbf{L}=\mathbf{B}\cdot\mathbf{W}$$

if **B** is skew and **L** is as in (2.17);

$$\mathbf{A}\cdot\mathbf{B}=0$$

if **A** is symmetric and **B** is skew. Furthermore, by (2.18),

$$\operatorname{tr}\mathbf{A}=\mathbf{I}\cdot\mathbf{A}\quad\text{for all } \mathbf{A}\in L(\mathscr{V}),$$

and also

$$\mathbf{AB}\cdot\mathbf{C}=\mathbf{A}\cdot\mathbf{CB}^T=\mathbf{B}\cdot\mathbf{A}^T\mathbf{C}\quad\text{for all } \mathbf{A},\mathbf{B},\mathbf{C}\in L(\mathscr{V}),$$
$$(\mathbf{a}\otimes\mathbf{b})\cdot(\mathbf{c}\otimes\mathbf{d})=(\mathbf{a}\cdot\mathbf{c})(\mathbf{b}\cdot\mathbf{d})\quad\text{for all } \mathbf{a},\mathbf{b},\mathbf{c},\mathbf{d}\in\mathscr{V}.$$

Thus, the basis e^2 of $L(\mathscr{V})$ introduced in the preceding section is orthonormal and so also is the basis E (*cf.* (2.10) and (2.11)).

We note in passing that formula (2.13) above, which expresses the third-order tensor $\mathbb{I}$, is still valid if the Cartesian components of $\mathbb{I}$ are defined by

$$L_{ijk}:=(\mathbf{e}_i\otimes\mathbf{e}_j)\cdot\mathbb{I}[\mathbf{e}_k]\quad\text{for } i,j,k\in\{1,2,3\}.$$

What makes the difference between this definition and (2.14) is the space on which $\mathbb{I}$ operates: here the inner product must be taken between second-order tensors.

The **magnitude** of a tensor $\mathbf{A}\in L(\mathscr{V})$ is defined in the usual way in terms of the inner product (2.18):

$$|\mathbf{A}|:=\sqrt{\mathbf{A}\cdot\mathbf{A}}=\sqrt{\operatorname{tr}(\mathbf{A}^T\mathbf{A})}.$$

If A_{ij} are the Cartesian components of **A** as in (2.9), then

$$|\mathbf{A}|^2=\sum_{i,j=1}^{3}A_{ij}^2.$$

We are now in a position to state the theorem about transposition in $L^2(\mathscr{V})$.

Theorem 2.3. For every tensor $\mathbb{L}\in L^2(\mathscr{V})$ there is precisely one tensor $\mathbb{L}^T\in L^2(\mathscr{V})$ such that

$$\mathbf{A}\cdot\mathbb{L}[\mathbf{B}]=\mathbb{L}^T[\mathbf{A}]\cdot\mathbf{B}\quad\text{for all } \mathbf{A},\mathbf{B}\in L(\mathscr{V}).\qquad(2.19)$$

The tensor $\mathbb{L}^T$ is called the **transpose** of $\mathbb{L}$. Furthermore, we can repeat Definition 2.2 *verbatim*, changing only $\mathbf{L}$ into $\mathbb{L}$; likewise, we can read (2.17) as the decomposition of $\mathbb{L}$ into a symmetric part and a skew part. The following rules are mere transliterations of those valid in $L(\mathscr{V})$:

$$(\mathbb{A}^T)^T = \mathbb{A},$$
$$(\mathbb{A}+\mathbb{B})^T = \mathbb{A}^T + \mathbb{B}^T,$$
$$(\mathbb{A}\mathbb{B})^T = \mathbb{B}^T\mathbb{A}^T \quad \text{for all } \mathbb{A},\mathbb{B}\in L^2(\mathscr{V}).$$

The **tensor product** $\mathbf{A}\otimes\mathbf{B}$ of two elements of $L(\mathscr{V})$ is defined as the element of $L^2(\mathscr{V})$ such that

$$(\mathbf{A}\otimes\mathbf{B})[\mathbf{L}] := (\mathbf{B}\cdot\mathbf{L})\mathbf{A} \quad \text{for all } \mathbf{L}\in L(\mathscr{V}).$$

It is an easy exercise to prove that

$$(\mathbf{A}\otimes\mathbf{B})^T = \mathbf{B}\otimes\mathbf{A} \quad \text{for all } \mathbf{A},\mathbf{B}\in L(\mathscr{V}).$$

Let a basis $e = (\mathbf{e}_1, \mathbf{e}_2, \mathbf{e}_3)$ be given in $\mathscr{V}$. We learned in the preceding section that $e^2 = (\mathbf{e}_i\otimes\mathbf{e}_j \mid i,j = 1,2,3)$ is then a basis for $L(\mathscr{V})$. In the same way we introduce a basis also in $L^2(\mathscr{V})$; it is defined by

$$e^4 := ((\mathbf{e}_i\otimes\mathbf{e}_j)\otimes(\mathbf{e}_h\otimes\mathbf{e}_k) \mid i,j,h,k = 1,2,3).$$

Recalling the definition (2.11) of E, we easily obtain another basis of $L^2(\mathscr{V})$, that is

$$E^2 := (\mathbf{E}_i\otimes\mathbf{E}_j \mid i,j = 1,\ldots,9).$$

Thus, every tensor $\mathbb{L}\in L^2(\mathscr{V})$ can be given the form

$$\mathbb{L} = \sum_{i,j=1}^{9} \lambda_{ij}\mathbf{E}_i\otimes\mathbf{E}_j, \tag{2.20}$$

where

$$\lambda_{ij} := \mathbf{E}_i\cdot\mathbb{L}[\mathbf{E}_j].$$

It is sometimes convenient to write (2.20) in a different way:

$$\mathbb{L} = \sum_{h=1}^{9^2} \lambda_h\mathbb{E}_h,$$

where

$$\lambda_{j+9(i-1)} := \lambda_{ij} \quad \text{and} \quad \mathbb{E}_{j+9(i-1)} := \mathbf{E}_i\otimes\mathbf{E}_j \quad \text{for all } i,j = 1,\ldots,9.$$

Let $\mathscr{U}$ be a linear subspace of $\mathscr{V}$. We apply the term the **orthogonal complement** of $\mathscr{U}$ to the linear subspace of $\mathscr{V}$ defined by

$$\mathscr{U}^{\perp} := \{\mathbf{v} \in \mathscr{V} \mid \mathbf{v} \cdot \mathbf{u} = 0 \quad \text{for all } \mathbf{u} \in \mathscr{U}\}.$$

The subspaces $\mathscr{U}$ and $\mathscr{U}^{\perp}$ **decompose** $\mathscr{V}$, that is for every $\mathbf{v} \in \mathscr{V}$ there is precisely one vector $\mathbf{u}_1 \in \mathscr{U}$ and one vector $\mathbf{u}_2 \in \mathscr{U}^{\perp}$ such that

$$\mathbf{v} = \mathbf{u}_1 + \mathbf{u}_2.$$

Let $\mathbf{L}$ be given in $L(\mathscr{V})$. We say that $\mathscr{U}$ is **invariant** under $\mathbf{L}$ if

$$\mathbf{L}\mathbf{u} \in \mathscr{U} \quad \text{for all } \mathbf{u} \in \mathscr{U},$$

that is, if the restriction of $\mathbf{L}$ to $\mathscr{U}$ is a tensor on $\mathscr{U}$:

$$\mathbf{L}|_{\mathscr{U}} \in L(\mathscr{U}),$$

where $L(\mathscr{U})$ is the space of all linear mappings of $\mathscr{U}$ into itself.

Theorem 2.4. If $\mathscr{U}$ is a linear subspace of $\mathscr{V}$ invariant under $\mathbf{L} \in L(\mathscr{V})$, then $\mathscr{U}^{\perp}$ is invariant under $\mathbf{L}^T$.

A similar theorem holds also for the invariant subspaces of $L(\mathscr{V})$, as it does for the invariant subspaces of all inner-product linear spaces.

Another theorem on $L(\mathscr{V})$ will often be employed in the following:

Decomposition Theorem of Second-order Tensors. $Sym(\mathscr{V})$ and $Skw(\mathscr{V})$, the linear subspaces of $L(\mathscr{V})$ comprising all symmetric and all skew second-order tensors, are such that

$$(Sym(\mathscr{V}))^{\perp} = Skw(\mathscr{V}) \quad \text{and} \quad (Skw(\mathscr{V}))^{\perp} = Sym(\mathscr{V}),$$

and so they decompose $L(\mathscr{V})$.

2.2.2 Spectral decomposition

Definition 2.3. Let $\mathbf{L}$ be given in $L(\mathscr{V})$. We say that $\mathbf{v} \in \mathscr{V} \setminus \{\mathbf{0}\}$ is an eigenvector of $\mathbf{L}$ and $\lambda \in \mathbb{R}$ the corresponding eigenvalue if

$$\mathbf{L}\mathbf{v} = \lambda \mathbf{v}.$$

The set of all eigenvalues of $\mathbf{L}$, which may well be empty, is called the **spectrum** of $\mathbf{L}$. Clearly, if λ belongs to the spectrum of $\mathbf{L}$ and $\mathbf{v}$ is an eigenvector corresponding to λ, then so is $\mu\mathbf{v}$, for all $\mu \neq 0$. Furthermore, if $\mathbf{v}_1$ and $\mathbf{v}_2$ are eigenvectors of $\mathbf{L}$ corresponding to the same eigenvalue λ, then so is $\mu_1\mathbf{v}_1 + \mu_2\mathbf{v}_2$ for all $\mu_1, \mu_2 \in \mathbb{R}$ that do not vanish simultaneously. Thus, all eigenvectors of $\mathbf{L}$ corresponding to a given eigenvalue constitute a linear subspace of $\mathscr{V}$, which is sometimes called a **proper space** of $\mathbf{L}$.

Theorem 2.5. If $\mathbf{L} \in L(\mathscr{V})$ is symmetric then there is a basis $(\mathbf{e}_1, \mathbf{e}_2, \mathbf{e}_3)$ of $\mathscr{V}$ whose elements are all eigenvectors of $\mathbf{L}$.

A consequence of this theorem that deserves notice is the following representation formula for $\mathbf{L}$:

$$\mathbf{L} = \lambda_1 \mathbf{e}_1 \otimes \mathbf{e}_1 + \lambda_2 \mathbf{e}_2 \otimes \mathbf{e}_2 + \lambda_3 \mathbf{e}_3 \otimes \mathbf{e}_3, \tag{2.21}$$

where λ_1, λ_2 and λ_3 are the eigenvalues of $\mathbf{L}$ corresponding to $\mathbf{e}_1, \mathbf{e}_2$ and $\mathbf{e}_3$. Thus, there is at least one basis of $\mathscr{V}$ in which the Cartesian components of a symmetric tensor of $L(\mathscr{V})$ are the entries of a diagonal matrix.

Another way of stating Theorem 2.5 is the following.

Spectral Theorem. The proper spaces of a symmetric tensor of $L(\mathscr{V})$ decompose $\mathscr{V}$.

A definition akin to Definition 2.3 could be easily laid down also for the **proper tensors** of a tensor in $L^2(\mathscr{V})$ and the corresponding **proper values.** The exact analogue of Theorem 2.5 would then follow, leading to the representation formula for all symmetric tensors of $L^2(\mathscr{V})$

$$\mathbb{L} = \sum_{i=1}^{9} \lambda_i \mathbf{L}_i \otimes \mathbf{L}_i, \tag{2.22}$$

where λ_i and $\mathbf{L}_i$, for $i = 1, \ldots, 9$, are the proper values of $\mathbb{L}$ and the corresponding proper tensors.

2.2.3 Duality

A **linear form** on $\mathscr{V}$ is a linear mapping of $\mathscr{V}$ into $\mathbb{R}$. The set of all linear forms on $\mathscr{V}$ is called the **dual space** of $\mathscr{V}$ and it is denoted by $\mathscr{V}^*$:

$$\mathscr{V}^* := \{\gamma : \mathscr{V} \to \mathbb{R} \,|\, \gamma \text{ is linear}\}.$$

$\mathscr{V}^*$ can easily be endowed with the structure of a linear space on $\mathbb{R}$ if addition and scalar multiplication of forms are defined through the addition and the multiplication of their values. Thus, if γ_1 and γ_2 belong to $\mathscr{V}^*$ and λ is a scalar

$$(\gamma_1 + \gamma_2)(\mathbf{v}) := \gamma_1(\mathbf{v}) + \gamma_2(\mathbf{v})$$

and

$$(\lambda\gamma_1)(\mathbf{v}) := \lambda\gamma_1(\mathbf{v}) \quad \text{for all } \mathbf{v}\in\mathscr{V}.$$

A fundamental theorem of linear algebra asserts that $\mathscr{V}^*$ and $\mathscr{V}$ can be identified with one another.

Representation Theorem of Linear Forms. For any given γ in $\mathscr{V}^*$ there is precisely one $\mathbf{v}$ in $\mathscr{V}$ such that

$$\gamma(\mathbf{u}) = \mathbf{v}\cdot\mathbf{u} \quad \text{for all } \mathbf{u}\in\mathscr{V}.$$

A linear form on $\mathscr{V}$ is thus represented through a vector of $\mathscr{V}$. We now see how second-order tensors can similarly represent bilinear forms. A **bilinear form** ω on $\mathscr{V}$ is a mapping of $\mathscr{V}\times\mathscr{V}$ into $\mathbb{R}$ such that both $\omega(\mathbf{u},\cdot)$ and $\omega(\cdot,\mathbf{u})$ are linear forms for all $\mathbf{u}\in\mathscr{V}$.

Representation Theorem of Bilinear Forms. For any given bilinear form ω on $\mathscr{V}$ there is precisely one tensor $\mathbf{L}$ in $L(\mathscr{V})$ such that

$$\omega(\mathbf{u},\mathbf{v}) = \mathbf{u}\cdot\mathbf{L}\mathbf{v} \quad \text{for all} \quad \mathbf{u},\mathbf{v}\in\mathscr{V}.$$

It follows immediately from (2.15) that if $\mathbf{L}\in L(\mathscr{V})$ represents the bilinear form ω then $\mathbf{L}^T$ represents the form ω^T, often called the **transpose** of ω, whose value on every pair of vectors equals that taken by ω on the pair of exchanged vectors:

$$\omega^T(\mathbf{u},\mathbf{v}) := \omega(\mathbf{v},\mathbf{u}) \quad \text{for all } \mathbf{u},\mathbf{v}\in\mathscr{V}.$$

2.2.4 Orientation of $\mathscr{V}$

We have already established that the linear space of all skew tensors of $L(\mathscr{V})$ has dimension 3, just as the dimension of $\mathscr{V}$. Hence, there is a one-to-one correspondence between second-order skew tensors and vectors. We now embody such a correspondence.

Theorem 2.6. If $\mathbf{W}\in L(\mathscr{V})\setminus\{\mathbf{0}\}$ is skew, then its spectrum is $\{0\}$ and its proper space has dimension 1.

Thus, the unique proper space of a skew tensor $\mathbf{W}$ is its null space, which is often referred to as the **axis** of $\mathbf{W}$:

$$\mathscr{A}(\mathbf{W}) := \{\mathbf{v}\in\mathscr{V} \mid \mathbf{W}\mathbf{v} = \mathbf{0}\}.$$

Let a skew tensor $\mathbf{W}$ of $L(\mathscr{V})$ other than $\mathbf{0}$ be given. There are precisely two opposite vectors $\mathbf{w}$ of $\mathscr{A}(\mathbf{W})$ such that

$$\mathbf{w}\cdot\mathbf{w}=\tfrac{1}{2}\mathbf{W}\cdot\mathbf{W}. \tag{2.23}$$

Choosing one of these vectors amounts to fixing one of two possible **orientations** of $\mathscr{V}$. There is no intrinsic way to make such a choice. We denote by $\mathbf{w}(\mathbf{W})$ the element of $\mathscr{V}$ that is thus associated to $\mathbf{W}$ and we call it the **axial vector** of $\mathbf{W}$. Clearly, changing the orientation of $\mathscr{V}$ changes $\mathbf{w}(\mathbf{W})$ into its opposite. Hereafter one orientation of $\mathscr{V}$ is prescribed once and for all. If $\mathbf{v}$ is any element of $\mathscr{V}$ different from $\mathbf{0}$ there is precisely one skew tensor $\mathbf{W}$ of $L(\mathscr{V})$ whose axial vector is $\mathbf{v}$:

$$\mathbf{v}=\mathbf{w}(\mathbf{W});$$

we say that $\mathbf{W}$ is the **skew tensor associated** with $\mathbf{v}$ and sometimes we denote it by $\mathbf{W}(\mathbf{v})$.

So far we have left aside the case when $\mathbf{W}=\mathbf{0}$. We set

$$\mathbf{w}(\mathbf{0})=\mathbf{0} \quad\text{and}\quad \mathbf{W}(\mathbf{0})=\mathbf{0}.$$

Thus, both $\mathbf{w}(\cdot)$ and $\mathbf{W}(\cdot)$ are linear isomorphisms, one the inverse of the other.

Let $\mathbf{W}_1$ and $\mathbf{W}_2$ be skew tensors and let $\mathbf{w}_1$ and $\mathbf{w}_2$ be the corresponding axial vectors. It can be proved that

$$\mathbf{w}_1\cdot\mathbf{w}_2=\tfrac{1}{2}\mathbf{W}_1\cdot\mathbf{W}_2,$$

a formula which generalizes (2.23).

Let $\mathbf{a}$ and $\mathbf{b}$ be any two vectors of $\mathscr{V}$. We call the **exterior product** of $\mathbf{a}$ and $\mathbf{b}$ the vector defined by

$$\mathbf{a}\wedge\mathbf{b}:=\mathbf{W}(\mathbf{a})\mathbf{b}, \tag{2.24}$$

that is, the vector that we obtain by operating on $\mathbf{b}$ by the skew tensor associated with $\mathbf{a}$.[1] It is easily seen that the product defined by (2.24) is linear in both operands. Moreover, $\mathbf{a}\wedge\mathbf{b}$ is orthogonal to both $\mathbf{a}$ and $\mathbf{b}$, for all $\mathbf{a},\mathbf{b}\in\mathscr{V}$, and $\mathbf{a}\wedge\mathbf{b}$ vanishes when $\mathbf{a}$ or $\mathbf{b}$ vanishes and also when $\mathbf{a}$ and $\mathbf{b}$ are parallel. The exterior product of two vectors satisfies the identity:

$$\mathbf{a}\wedge\mathbf{b}=\mathbf{w}(\mathbf{b}\otimes\mathbf{a}-\mathbf{a}\otimes\mathbf{b}) \quad \text{for all } \mathbf{a},\mathbf{b}\in\mathscr{V}, \tag{2.25}$$

---

[1] Often, in the English and German literature the notation $\mathbf{a}\times\mathbf{b}$ is used instead of $\mathbf{a}\wedge\mathbf{b}$ and the terms **cross-product** or **vector product** are used instead of exterior product. In this case, the term exterior product and the notation $\mathbf{a}\wedge\mathbf{b}$ are used for the skew tensor $\mathbf{a}\otimes\mathbf{b}-\mathbf{b}\otimes\mathbf{a}$. Here we follow the usage common in both French and Italian literature.

whence it follows immediately that

$$\mathbf{b} \wedge \mathbf{a} = -\mathbf{a} \wedge \mathbf{b} \quad \text{for all } \mathbf{a}, \mathbf{b} \in \mathscr{V}. \tag{2.26}$$

Here is a useful identity, which is a consequence of (2.24) and (2.25):

$$\mathbf{a} \wedge (\mathbf{b} \wedge \mathbf{c}) = (\mathbf{a} \cdot \mathbf{c})\mathbf{b} - (\mathbf{a} \cdot \mathbf{b})\mathbf{c} \quad \text{for all } \mathbf{a}, \mathbf{b}, \mathbf{c} \in \mathscr{V}. \tag{2.27}$$

To prove (2.27) we set

$$\mathbf{W} = \mathbf{c} \otimes \mathbf{b} - \mathbf{b} \otimes \mathbf{c},$$

so that, by (2.25), the axial vector of $\mathbf{W}$ is $\mathbf{b} \wedge \mathbf{c}$. Hence

$$\mathbf{W}\mathbf{a} = (\mathbf{b} \cdot \mathbf{a})\mathbf{c} - (\mathbf{c} \cdot \mathbf{a})\mathbf{b},$$

while, by (2.24),

$$\mathbf{W}\mathbf{a} = (\mathbf{b} \wedge \mathbf{c}) \wedge \mathbf{a}.$$

Then (2.27) follows easily.

Let $e = (\mathbf{e}_1, \mathbf{e}_2, \mathbf{e}_3)$ and $e' = (\mathbf{e}'_1, \mathbf{e}'_2, \mathbf{e}'_3)$ be two bases of $\mathscr{V}$. We say that e and e' are **equally oriented** if

$$\mathbf{e}_1 \wedge \mathbf{e}_2 \cdot \mathbf{e}_3 = \mathbf{e}'_1 \wedge \mathbf{e}'_2 \cdot \mathbf{e}'_3.$$

A basis of $\mathscr{V}$, say e, may have two opposite **orientations** depending on whether

$$\mathbf{e}_1 \wedge \mathbf{e}_2 \cdot \mathbf{e}_3 = 1 \quad \text{or} \quad \mathbf{e}_1 \wedge \mathbf{e}_2 \cdot \mathbf{e}_3 = -1;$$

correspondingly, we say that it is **positively** or **negatively oriented.** Clearly, the orientation of a basis is closely related to the orientation of $\mathscr{V}$, because so is the definition of exterior product. When we change the orientation of $\mathscr{V}$, the orientation of a given basis is changed into its opposite.

All the bases employed throughout this book are positively oriented. Thus, in view of (2.25), the skew tensors associated with $\mathbf{e}_1, \mathbf{e}_2$ and $\mathbf{e}_3$ are, respectively

$$\mathbf{W}(\mathbf{e}_1) = \mathbf{e}_3 \otimes \mathbf{e}_2 - \mathbf{e}_2 \otimes \mathbf{e}_3, \quad \mathbf{W}(\mathbf{e}_2) = \mathbf{e}_1 \otimes \mathbf{e}_3 - \mathbf{e}_3 \otimes \mathbf{e}_1,$$
$$\mathbf{W}(\mathbf{e}_3) = \mathbf{e}_2 \otimes \mathbf{e}_1 - \mathbf{e}_1 \otimes \mathbf{e}_2.$$

A quick glance at (2.10) shows that

$$\mathbf{W}(\mathbf{e}_i) = \sqrt{2}\mathbf{E}_{i+6} \quad \text{for} \quad i = 1, 2, 3.$$

Let the basis $e = (\mathbf{e}_1, \mathbf{e}_2, \mathbf{e}_3)$ of $\mathscr{V}$ be given. If W_{ij} are the Cartesian components in e of a skew tensor $\mathbf{W}$ and w_i those of its axial vector

$\mathbf{w}(\mathbf{W})$, they are related through the equations

$$W_{ij} = \sum_{h=1}^{3} \varepsilon_{ihj} w_h, \tag{2.28}$$

$$w_i = \frac{1}{2} \sum_{h,j=1}^{3} \varepsilon_{hij} W_{hj}, \tag{2.29}$$

where ε_{ijh}, for $i, j, h = 1, 2, 3$, are the components of the **Ricci alternator**:

$$\varepsilon_{ijh} := \begin{cases} 1 & \text{if } (i,j,h) \text{ is an even permutation of } (1,2,3), \\ 0 & \text{if } (i,j,h) \text{ is not a permutation of } (1,2,3), \\ -1 & \text{if } (i,j,h) \text{ is an odd permutation of } (1,2,3). \end{cases}$$

The reader might care to take the trouble of rewriting formulae (2.24)–(2.27) using Cartesian components. In doing so he is advised to make use of the identity

$$\sum_{i=1}^{3} \varepsilon_{ijh} \varepsilon_{ilm} = \delta_{jl} \delta_{hm} - \delta_{jm} \delta_{hl} \quad \text{for all } j, h, l, m \in \{1, 2, 3\}.$$

For example, (2.24) becomes

$$(\mathbf{a} \wedge \mathbf{b})_i = \sum_{j,h=1}^{3} \varepsilon_{ijh} a_j b_h \quad \text{for all } i \in \{1, 2, 3\}, \tag{2.30}$$

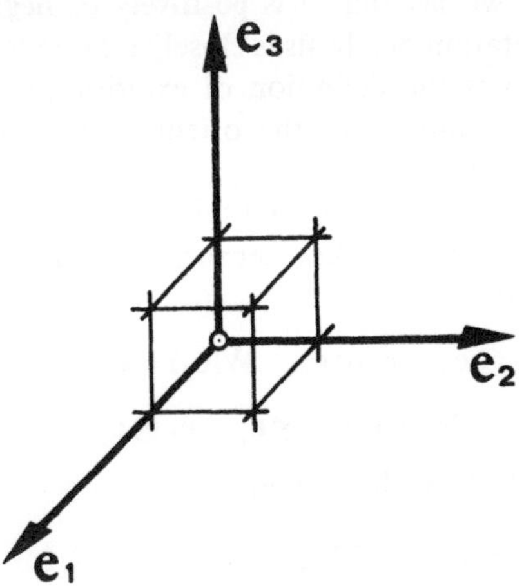

Figure 2.3

where a_i, b_i, and $(\mathbf{a} \wedge \mathbf{b})_i$, for $i = 1, 2, 3$, are the Cartesian components in e of $\mathbf{a}$, $\mathbf{b}$, and $(\mathbf{a} \wedge \mathbf{b})$.

Hereafter we take the orientation of $\mathscr{V}$ to be chosen so that the basis sketched in Figure 2.3 be positively oriented.

2.2.5 Determinant of a tensor

The reader will certainly be familiar with the determinant of a matrix. Here we first define intrinsically the determinant of a tensor and then we see that it agrees with the determinant of the matrix of its Cartesian components in all bases of $\mathscr{V}$ (*cf.* Section 53 of Halmos (1958)).

We say that a mapping $\alpha: \mathscr{V} \times \mathscr{V} \times \mathscr{V} \to \mathbb{R}$ is a **skew trilinear form** if

$$\alpha(\mathbf{u}, \mathbf{v}, \cdot), \ \alpha(\mathbf{u}, \cdot, \mathbf{v}), \text{ and } \alpha(\cdot, \mathbf{u}, \mathbf{v})$$

are all linear forms on $\mathscr{V}$ for all $\mathbf{u}, \mathbf{v} \in \mathscr{V}$, and if exchanging any two arguments of α reverses its sign, that is if

$$\alpha(\mathbf{u}, \mathbf{v}, \mathbf{w}) = -\alpha(\mathbf{v}, \mathbf{u}, \mathbf{w}) = -\alpha(\mathbf{u}, \mathbf{w}, \mathbf{v}) = -\alpha(\mathbf{w}, \mathbf{v}, \mathbf{u}) \quad \text{for all } \mathbf{u}, \mathbf{v}, \mathbf{w} \in \mathscr{V}. \tag{2.31}$$

It follows immediately from the above definition that every skew trilinear form vanishes whenever it is evaluated on a triple of linearly dependent vectors. By contrast, if a skew trilinear form does not vanish on a triple of vectors, then they are linearly independent.

It is easy to check that the set of all skew trilinear forms is a linear space if addition and scalar multiplication of forms are defined through the addition and the multiplication by a scalar of their values. The null element is the **null form**, that is the form that vanishes identically. We denote by $\mathscr{T}$ such a space.

Theorem 2.7. The dimension of $\mathscr{T}$ is 1.

Thus, given two forms of $\mathscr{T}$, α_1 and α_2, with α_1 other than the null form, there exists $\lambda \in \mathbb{R}$ such that

$$\alpha_2(\mathbf{u}, \mathbf{v}, \mathbf{w}) = \lambda \alpha_1(\mathbf{u}, \mathbf{v}, \mathbf{w}) \quad \text{for all } \mathbf{u}, \mathbf{v}, \mathbf{w} \in \mathscr{V}. \tag{2.32}$$

Let α be any form of $\mathscr{T}$ different from the null form and let $\mathbf{L}$ be a tensor of $L(\mathscr{V})$. Another form of $\mathscr{T}$ can be induced by $\mathbf{L}$ from α; it is defined by

$$\alpha_{\mathbf{L}}(\mathbf{u}, \mathbf{v}, \mathbf{w}) := \alpha(\mathbf{L}\mathbf{u}, \mathbf{L}\mathbf{v}, \mathbf{L}\mathbf{w}) \quad \text{for all } \mathbf{u}, \mathbf{v}, \mathbf{w} \in \mathscr{V}.$$

By (2.32), the forms $\alpha_{\mathbf{L}}$ and α must be proportional: so there exists $\lambda_{\mathbf{L}} \in \mathbb{R}$, which depends only on $\mathbf{L}$, such that[2]

$$\alpha_{\mathbf{L}}(\mathbf{u}, \mathbf{v}, \mathbf{w}) = \lambda_{\mathbf{L}} \alpha(\mathbf{u}, \mathbf{v}, \mathbf{w}) \quad \text{for all } \mathbf{u}, \mathbf{v}, \mathbf{w} \in \mathscr{V}. \tag{2.33}$$

We call $\lambda_{\mathbf{L}}$ the **determinant** of $\mathbf{L}$ and we denote it by $\det \mathbf{L}$.

Definition 2.4. For every $\mathbf{L} \in L(\mathscr{V})$ $\det \mathbf{L} \in \mathbb{R}$ is characterized by

$$\alpha(\mathbf{L}\mathbf{u}, \mathbf{L}\mathbf{v}, \mathbf{L}\mathbf{w}) = (\det \mathbf{L})\alpha(\mathbf{u}, \mathbf{v}, \mathbf{w}) \quad \text{for all } \mathbf{u}, \mathbf{v}, \mathbf{w} \in \mathscr{V} \tag{2.34}$$

and for every $\alpha \in \mathscr{T}$.

There are a few consequences of this definition that deserve note:

$$\begin{aligned} &\det \mathbf{0} = 0, \quad \det \mathbf{I} = 1, \\ &\det(\lambda \mathbf{L}) = \lambda^3 \det \mathbf{L} \quad \text{for all } \mathbf{L} \in L(\mathscr{V}) \text{ and } \lambda \in \mathbb{R}, \\ &\det(\mathbf{A}\mathbf{B}) = (\det \mathbf{A})(\det \mathbf{B}) \quad \text{for all } \mathbf{A}, \mathbf{B} \in L(\mathscr{V}). \end{aligned} \tag{2.35}$$

Another consequence, which can be drawn with little more labour, is

$$\det \mathbf{A}^T = \det \mathbf{A} \quad \text{for all } \mathbf{A} \in L(\mathscr{V}). \tag{2.36}$$

The reader will easily deduce from (2.34) that if L_{ij} are the Cartesian components of $\mathbf{L} \in L(\mathscr{V})$ in a basis of $\mathscr{V}$, then

$$\det \mathbf{L} = \sum_{\pi \in \mathscr{S}_3} \varepsilon_{\pi(1),\pi(2),\pi(3)} L_{1,\pi(1)} L_{2,\pi(2)} L_{3,\pi(3)},$$

where $\mathscr{S}_3$ is the set of all permutations of $\{1, 2, 3\}$. This shows that the matrices whose entries are the Cartesian components of $\mathbf{L}$ in different bases all share the same determinant.

A tensor $\mathbf{L}$ in $L(\mathscr{V})$ is said to be **invertible** if there is a tensor $\mathbf{L}^{-1}$ in $L(\mathscr{V})$, called the **inverse** of $\mathbf{L}$, such that

$$\mathbf{L}^{-1}\mathbf{L} = \mathbf{L}\mathbf{L}^{-1} = \mathbf{I}. \tag{2.37}$$

If a tensor is invertible, then its inverse is unique. A characteristic property of all invertible tensors in $L(\mathscr{V})$ is to map triples of linearly independent vectors into triples of vectors that are still linearly independent. Hence, by Definition 2.4 a tensor is invertible if, and only if, its determinant does not vanish. Thus, putting together $(2.35)_4$ and (2.37), we see that for an invertible tensor $\mathbf{L}$

$$\det \mathbf{L}^{-1} = \frac{1}{\det \mathbf{L}}.$$

---

[2] It is an easy exercise to prove that $\lambda_{\mathbf{L}}$ does not depend on the particular form α.

It is easy to check that the mapping $\beta: \mathscr{V} \times \mathscr{V} \times \mathscr{V} \to \mathbb{R}$ defined by

$$\beta(\mathbf{u}, \mathbf{v}, \mathbf{w}) := \mathbf{u} \wedge \mathbf{v} \cdot \mathbf{w} \tag{2.38}$$

belongs to $\mathscr{T}$. Hence, by Definition 2.4, any two equally oriented bases of $\mathscr{V}$ are related to each other by a tensor of positive determinant. It also follows from (2.38) and (2.34) that

$$\mathbf{Lu} \wedge \mathbf{Lv} \cdot \mathbf{Lw} = (\det \mathbf{L}) \mathbf{u} \wedge \mathbf{v} \cdot \mathbf{w} \quad \text{for all } \mathbf{u}, \mathbf{v}, \mathbf{w} \in \mathscr{V}. \tag{2.39}$$

There is a geometric interpretation of the determinant of a tensor which is worth noting. Let $\mathbf{u}$, $\mathbf{v}$, and $\mathbf{w}$ be linearly independent vectors of $\mathscr{V}$ and let o be given in $\mathscr{E}$. Then $|\mathbf{u} \wedge \mathbf{v} \cdot \mathbf{w}|$ is the volume of the parallelepiped $\mathscr{P}$ of $\mathscr{E}$ whose edges are parallel to one or the other of the following segments

$$\{p \in \mathscr{E} \,|\, p - o = t\mathbf{u},\ t \in [0, 1]\},$$
$$\{p \in \mathscr{E} \,|\, p - o = t\mathbf{v},\ t \in [0, 1]\},$$
$$\{p \in \mathscr{E} \,|\, p - o = t\mathbf{w},\ t \in [0, 1]\}.$$

Likewise, if $\mathbf{L}$ is an invertible tensor, $|\mathbf{Lu} \wedge \mathbf{Lv} \cdot \mathbf{Lw}|$ is the volume of

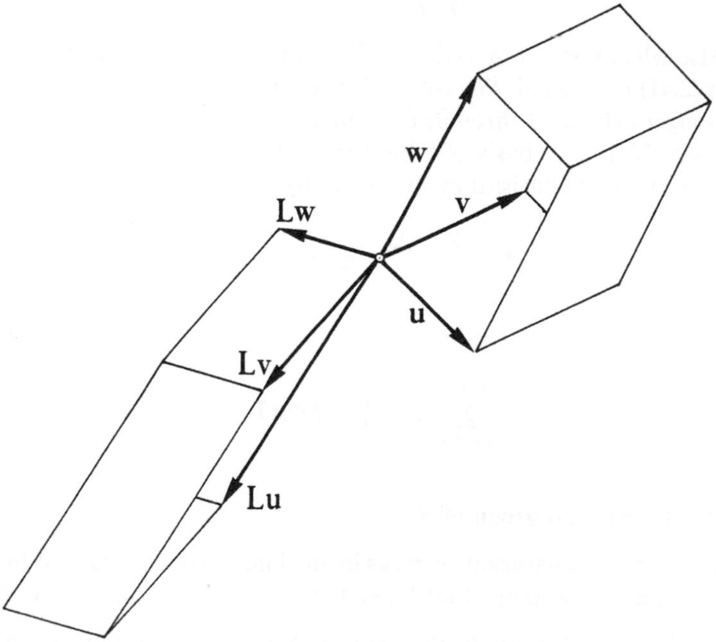

Figure 2.4

the parallelepiped of $\mathscr{E}$ whose edges are obtained from those of $\mathscr{P}$ through the action of $\mathbf{L}$ (see Figure 2.4). By (2.39), $|\det \mathbf{L}|$ can be interpreted as a volume dilation factor. When $\mathbf{L}$ is not invertible and $\mathbf{u}$, $\mathbf{v}$, $\mathbf{w}$ are still linearly independent, (2.39) tells us that $\mathbf{L}$ squeezes $\mathscr{P}$ into a flat region of $\mathscr{E}$.

Let $\mathbf{L} \in L(\mathscr{V})$ be invertible. We call the tensor in $L(\mathscr{V})$ defined by

$$\mathbf{L}^* := (\det \mathbf{L})(\mathbf{L}^{-1})^T \tag{2.40}$$

the **adjugate** of $\mathbf{L}$. The tensor $\mathbf{L}^*$ has a remarkable property, which is a consequence of (2.39), that is

$$\mathbf{L}\mathbf{u} \wedge \mathbf{L}\mathbf{v} = \mathbf{L}^*(\mathbf{u} \wedge \mathbf{v}) \quad \text{for all } \mathbf{u}, \mathbf{v} \in \mathscr{V}. \tag{2.41}$$

To prove (2.41), we write (2.39) in the form

$$\mathbf{L}^T(\mathbf{L}\mathbf{u} \wedge \mathbf{L}\mathbf{v}) \cdot \mathbf{w} = (\det \mathbf{L})\mathbf{u} \wedge \mathbf{v} \cdot \mathbf{w}$$

for any given $\mathbf{u}, \mathbf{v}$ and all $\mathbf{w} \in \mathscr{V}$, whence

$$(\mathbf{L}\mathbf{u} \wedge \mathbf{L}\mathbf{v}) = (\det \mathbf{L})(\mathbf{L}^T)^{-1}(\mathbf{u} \wedge \mathbf{v}) \quad \text{for all } \mathbf{u}, \mathbf{v} \in \mathscr{V}.$$

The desired result then follows, since

$$(\mathbf{L}^T)^{-1} = (\mathbf{L}^{-1})^T.$$

Also when $\mathbf{L}$ is not invertible there is precisely one tensor $\mathbf{L}^*$ such that (2.41) is satisfied. We still call it the adjugate of $\mathbf{L}$, though it is no longer expressed through formula (2.40). The Cartesian components of $\mathbf{L}^*$ in any basis of $\mathscr{V}$ are the **cofactors** of the components of $\mathbf{L}$ in the same basis; they are given by the following formula:

$$L^*_{lm} = \frac{1}{2} \sum_{i,h,j,k=1}^{3} \varepsilon_{lih}\varepsilon_{mjk} L_{ij} L_{hk}.$$

It is an easy exercise to see that

$$\sum_{l,m=1}^{3} L_{lm} L^*_{lm} = 3 \det \mathbf{L}.$$

2.2.6 Orthogonal group of $\mathscr{V}$

A question arises naturally in $\mathscr{V}$, as in any inner-product space: Which are the linear mappings that leave the inner product unchanged?

Definition 2.5. We apply the term **orthogonal** to the tensors that belong to

$$O(\mathscr{V}) := \{\mathbf{Q} \in L(\mathscr{V}) \mid \mathbf{Qu} \cdot \mathbf{Qv} = \mathbf{u} \cdot \mathbf{v} \quad \text{for all } \mathbf{u}, \mathbf{v} \in \mathscr{V}\}. \tag{2.42}$$

By the Representation Theorem of Bilinear Forms, a tensor $\mathbf{Q} \in L(\mathscr{V})$ is orthogonal if, and only if, it satisfies

$$\mathbf{Q}^T\mathbf{Q} = \mathbf{I}. \tag{2.43}$$

It follows from (2.36) and (2.43) that

$$(\det \mathbf{Q})^2 = 1 \quad \text{for all } \mathbf{Q} \in O(\mathscr{V}). \tag{2.44}$$

Thus, every orthogonal tensor is invertible and, by (2.43), its inverse and its transpose coincide. $O(\mathscr{V})$ is a group with respect to the product defined by (2.12); the neutral element of $O(\mathscr{V})$ is the identity tensor $\mathbf{I}$. Moreover, (2.40) implies that

$$\mathbf{Q}^* = (\det \mathbf{Q})\mathbf{Q}. \tag{2.45}$$

Combining (2.45) and (2.44), we learn that the adjugate of an orthogonal tensor $\mathbf{Q}$ is either equal or opposite to $\mathbf{Q}$.

Among all orthogonal tensors some are special: those relating equally oriented bases of $\mathscr{V}$.

Definition 2.6. We apply the term **special orthogonal** or **rotation** to each tensor of

$$SO(\mathscr{V}) := \{\mathbf{Q} \in O(\mathscr{V}) \mid \det \mathbf{Q} = 1\}.$$

$SO(\mathscr{V})$ is indeed a proper subgroup of $O(\mathscr{V})$ and, moreover, it plays a peculiar role in $O(\mathscr{V})$.

Theorem 2.8. Let $\mathbf{Q}$ be given in $O(\mathscr{V})$. Then either $\mathbf{Q} \in SO(\mathscr{V})$ or $(-1)\mathbf{Q} \in SO(\mathscr{V})$.

A member of $O(\mathscr{V})$ which does not belong to $SO(\mathscr{V})$ is often called an **improper orthogonal** tensor.

When a rotation $\mathbf{Q} \in SO(\mathscr{V})$ leaves an axis unchanged, it is rather natural to give $\mathbf{Q}$ a special name.

Definition 2.7. Let $\mathbf{e} \in \mathbb{S}^2$ be given. We call each tensor of

$$SO(\mathbf{e}, \mathscr{V}) := \{\mathbf{Q} \in SO(\mathscr{V}) \mid \mathbf{Qe} = \mathbf{e}\}.$$

a **rotation about e**. $SO(\mathbf{e}, \mathscr{V})$ is a proper subgroup of $SO(\mathscr{V})$, for any $\mathbf{e} \in \mathbb{S}^2$.

The following theorem is originally due to Euler.

Theorem 2.9. If $\mathbf{Q}\in SO(\mathscr{V})$, then there is an $\mathbf{e}\in\mathbb{S}^2$ such that $\mathbf{Q}\in SO(\mathbf{e},\mathscr{V})$.

We could also state this theorem as follows: a rotation always has an eigenvector with eigenvalue equal to 1.

In view of Theorem 2.9, we shall be able to represent all elements of $SO(\mathscr{V})$ if we learn how to represent all elements of $SO(\mathbf{e},\mathscr{V})$, for a given $\mathbf{e}\in\mathbb{S}^2$.

Theorem 2.10. Let $\mathbf{e}\in\mathbb{S}^2$ be given. If $\mathbf{Q}\in SO(\mathbf{e},\mathscr{V})$, then there is a $\vartheta\in[0,2\pi[$ such that

$$\mathbf{Q}=\mathbf{I}+\sin\vartheta\mathbf{W}(\mathbf{e})+(1-\cos\vartheta)\mathbf{W}(\mathbf{e})^2, \tag{2.46}$$

where $\mathbf{W}(\mathbf{e})$ is the skew tensor associated with $\mathbf{e}$.

To give this theorem a geometric interpretation, one should realize that $\mathbf{W}(\mathbf{e})^2$, which is clearly symmetric, is indeed the opposite of the projection onto the plane orthogonal to $\mathbf{e}$:

$$\mathbf{W}(\mathbf{e})^2=-\mathbf{P}(\mathbf{e}),$$

where

$$\mathbf{P}(\mathbf{e}):=\mathbf{I}-\mathbf{e}\otimes\mathbf{e}. \tag{2.47}$$

Besides, the parameter ϑ which appears in (2.46) can be interpreted as an angle of rotation; Fig. 2.5 represents it. A simple geometric proof of Theorem 2.10 could indeed be built upon Fig. 2.5. We leave such a proof to the reader as an exercise.

A remarkable algebraic consequence of (2.46) is that all members of $SO(\mathbf{e},\mathscr{V})$ commute with $\mathbf{W}(\mathbf{e})$:

$$\mathbf{QW}(\mathbf{e})=\mathbf{W}(\mathbf{e})\mathbf{Q}\quad\text{for all}\quad\mathbf{Q}\in SO(\mathbf{e},\mathscr{V}).$$

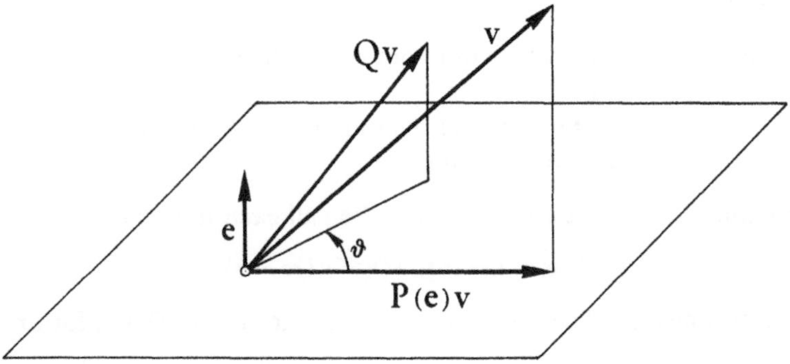

Figure 2.5

2.3 Analysis

In this section we recall some definitions and theorems from different areas of analysis. Together with the algebraic structures outlined in the preceding section, they will complete the collection of elementary mathematical tools employed in this book.

2.3.1 Curves, fields and deformations

We have already learned in Section 2.1 that a metric, and hence a topology, is naturally induced in the Euclidean space $\mathscr{E}$ from the inner product of $\mathscr{V}$ (*cf.* equation (2.4)). Let $(p_n \in \mathscr{E} \mid n \in \mathbb{N})$ be a sequence of points in $\mathscr{E}$. We say that it **converges** to a point $p \in \mathscr{E}$ if

$$\lim_{n \to \infty} d(p_n, p) = \lim_{n \to \infty} |p_n - p| = 0,$$

and we write that

$$p_n \to p \quad \text{as } n \to \infty.$$

Likewise, a topology can be induced in $\mathscr{V}$ and $L(\mathscr{V})$. Two vectors are close to each other if the length of the difference between them is close to zero: correspondingly, two tensors are close to each other if the magnitude of the difference between them is close to zero. Thus, if there are a sequence of vectors $(\mathbf{a}_n \in \mathscr{V} \mid \mathbf{n} \in \mathbb{N})$ and a sequence of tensors $(\mathbf{A}_n \in L(\mathscr{V}) \mid n \in \mathbb{N})$ such that

$$\lim_{n \to \infty} |\mathbf{a}_n - \mathbf{a}| = 0 \quad \text{and} \quad \lim_{n \to \infty} |\mathbf{A}_n - \mathbf{A}| = 0,$$

for some $\mathbf{a} \in \mathscr{V}$ and $\mathbf{A} \in L(\mathscr{V})$, we say that the former sequence **converges** to $\mathbf{a}$ and the latter to $\mathbf{A}$; we customarily employ the notation

$$\mathbf{a}_n \to \mathbf{a} \quad \text{and} \quad \mathbf{A}_n \to \mathbf{A} \quad \text{as } n \to \infty.$$

Standard definitions such as boundary and closure of a set, boundedness and compactness, *etc.* could easily be rephrased in $\mathscr{E}, \mathscr{V}$ and $L(\mathscr{V})$ with the aid of the topologies just defined thereon. These notions would extend to $\mathscr{E}, \mathscr{V}$ and $L(\mathscr{V})$ notions which are elementary in $\mathbb{R}$ and need not be recalled here.

Let $[a, b]$ be an interval of $\mathbb{R}$. We say that a mapping $p: [a, b] \to \mathscr{E}$ is **continuous** at $t \in [a, b]$ if, given any sequence $(t_n \in [a, b] \mid n \in \mathbb{N})$

converging to t in $\mathbb{R}$, the sequence $(p_n \in \mathcal{E} \mid n \in \mathbb{N})$ defined by

$$p_n := p(t_n) \quad \text{for all } n \in \mathbb{N}$$

converges to $p(t)$ in $\mathcal{E}$. We say that the mapping $t \mapsto p(t)$ is a **curve** in $\mathcal{E}$ if it is continuous at every $t \in [a, b]$.

Likewise, we say that a mapping $\mathbf{v}: [a, b] \to \mathcal{V}$ is **continuous** at $t \in [a, b]$ if, given any sequence $(t_n \in [a, b] \mid n \in \mathbb{N})$ converging to t in $\mathbb{R}$, the sequence $(\mathbf{v}_n := \mathbf{v}(t_n) \mid n \in \mathbb{N})$ converges to $\mathbf{v}(t)$ in $\mathcal{V}$. We also say that the mapping $t \mapsto \mathbf{v}(t)$ is a **curve** in $\mathcal{V}$ if it is continuous at every $t \in [a, b]$.

If $p: [a, b] \to \mathcal{E}$ is a curve and o is any given point in $\mathcal{E}$, then the mapping into $\mathcal{V}$ defined by

$$\mathbf{v}(t) := p(t) - o \quad \text{for all } t \in [a, b]$$

is clearly a curve in $\mathcal{V}$.

The reader should now be able to state the appropriate definition for curves in $L(\mathcal{V})$.

Let $\mathcal{B}$ be an open set in $\mathcal{E}$ and $\mathbf{f}$ a mapping from $\mathcal{B}$ into $\mathcal{V}$. We say that $\mathbf{f}$ is **continuous** at $p \in \mathcal{B}$ if for any given sequence $(p_n \in \mathcal{B} \mid n \in \mathbb{N})$ converging to p in $\mathcal{E}$ the sequence $(\mathbf{v}_n := \mathbf{f}(p_n) \mid n \in \mathbb{N})$ converges to $\mathbf{f}(p)$ in $\mathcal{V}$. We say that $\mathbf{f}: \mathcal{B} \to \mathcal{V}$ is a **vector field** on $\mathcal{B}$ if it is continuous at every $p \in \mathcal{B}$.

Consider a scalar-valued mapping φ on $\mathcal{B}$. The definition of continuity at a point of $\mathcal{B}$ can be given also for such a mapping by the use of corresponding converging sequences in $\mathcal{B}$ and $\mathbb{R}$. We say that φ is a **scalar field** on $\mathcal{B}$ if it is continuous at every point therein.

I invite the reader to write down the definition of a **tensor field** on $\mathcal{B}$, following the same path that has led us to define scalar and vector fields above.

I briefly mention here also other mappings on $\mathcal{B}$, which are neither curves nor fields. To be precise, they are mappings into $\mathcal{E}$ which are one-to-one and continuous with respect to the topology of $\mathcal{E}$. They will be called **deformations** of $\mathcal{B}$ in $\mathcal{E}$.

2.3.2 Differentiation

We now extend to $\mathcal{V}$ and $L(\mathcal{V})$ the use of the order symbol o, which in $\mathbb{R}$ is certainly well known to the reader. This will open the way to the differentiation of curves, fields and deformations.

Let $\mathbf{v}: [a, b] \to \mathcal{V}$ be a curve and let g be a function of $[a, b]$ into $\mathbb{R}$. We say that $\mathbf{v}$ is of **order** o with respect to g at $t_0 \in [a, b]$, and we write

$$\mathbf{v}(t) = o(g(t)) \quad \text{as } t \to t_0,$$

if

$$\lim_{t \to t_0} \frac{|\mathbf{v}(t)|}{|g(t)|} = 0.$$

The reader who is not already familiar with this topic should write down the analogous definition of the order symbol o for curves in $L(\mathscr{V})$.

Likewise, if both **h** and **g** are mappings of a subset $\mathscr{U}$ of $\mathscr{V}$ into $\mathscr{V}$, we say that **h** is of **order** o (or **negligible**) with respect to **g** at $\mathbf{u}_0 \in \mathscr{U}$, and we write

$$\mathbf{h}(\mathbf{u}) = o(\mathbf{g}(\mathbf{u})) \quad \text{as } \mathbf{u} \to \mathbf{u}_0,$$

if

$$\lim_{\mathbf{u} \to \mathbf{u}_0} \frac{|\mathbf{h}(\mathbf{u})|}{|\mathbf{g}(\mathbf{u})|} = 0.$$

The definition of continuity for curves in $\mathscr{V}$ can easily be rephrased in terms of the order symbol o. Let $\mathbf{v}$ be a mapping of the interval $[a, b]$ into $\mathscr{V}$. For a given $t_0 \in [a, b]$ we define the mapping $\mathbf{w}\colon [a, b] \to \mathscr{V}$ as follows

$$\mathbf{w}(t) := \mathbf{v}(t) - \mathbf{v}(t_0) \quad \text{for all } t \in [a, b].$$

The mapping $\mathbf{v}$ is continuous at t_0 if $\mathbf{w}$ is of order o at t_0 with respect to the function $g \equiv 1$ on $[a, b]$. This is also expressed by the formula

$$\mathbf{v}(t) - \mathbf{v}(t_0) = o(1) \quad \text{as } t \to t_0. \tag{2.48}$$

Let $\mathbf{v}\colon [a, b] \to \mathscr{V}$ be a curve. We say that $\mathbf{v}$ is **differentiable** at $t_0 \in]a, b[$ if there is a $\mathbf{w} \in \mathscr{V}$ such that

$$\mathbf{v}(t) - \mathbf{v}(t_0) = (t - t_0)\mathbf{w} + o(t - t_0) \quad \text{as } t \to t_0 \tag{2.49}$$

(here the order symbol o operates on the function $t \mapsto t - t_0$ defined on $[a, b]$). We call $\mathbf{w}$ the **derivative** of $\mathbf{v}$ at t_0. If $\mathbf{v}$ is differentiable at every $t \in]a, b[$, then the derivative $\mathbf{w}$ is defined everywhere in $]a, b[$, and we denote it by $\dot{\mathbf{v}}$:

$$\mathbf{w}(t) =: \dot{\mathbf{v}}(t).$$

If $\dot{\mathbf{v}}$ turns out to be continuous everywhere in $]a, b[$, we say that the curve $\mathbf{v}$ is of class C^1.

The reader can now easily arrive at the definition of the classes of smoother curves that are customarily denoted by C^n, for $n \geqslant 1$.

It is plain that these classes are ordered with respect to inclusion:

$$C^{n+1} \subset C^n \quad \text{for all } n \geqslant 1. \tag{2.50}$$

Sometimes the mappings here called curves are said to be of class C^0 then (2.50) holds for all $n \in \mathbb{N}$.

The differentiation of a curve in $\mathscr{E}$ is essentially defined in the same way as for a curve in $\mathscr{V}$. We say that the curve $p: [a, b] \to \mathscr{E}$ is **differentiable** at $t_0 \in]a, b[$, if there is a $\mathbf{w} \in \mathscr{V}$ such that

$$p(t) - p(t_0) = (t - t_0)\mathbf{w} + o(t - t_0) \quad \text{as } t \to t_0. \tag{2.51}$$

If the curve p is differentiable at every $t \in]a, b[$, then its derivative is a mapping into $\mathscr{V}$, which is denoted by $\dot{p}$. When $\dot{p}$ itself is a curve in $\mathscr{V}$, we say that the curve p is of class C^1. Curves of higher regularity are defined likewise.

The reader can now easily work out the appropriate definition of derivative for curves in $L(\mathscr{V})$, as well as a hierarchy of smoothness among them.[3]

I record here for later use some elementary rules for the differentiation of curves in $\mathscr{V}$ and $L(\mathscr{V})$. All of them could easily be proved by the use of (2.49) and the parallel formula for the derivative of a curve in $L(\mathscr{V})$. Below, α is a scalar-valued function, $\mathbf{u}$ and $\mathbf{v}$ are curves in $\mathscr{V}$, while $\mathbf{L}$ and $\mathbf{M}$ are curves in $L(\mathscr{V})$; all of them are defined on $[a, b]$ and are of class C^1 thereon; their derivative is denoted by a superimposed dot:

$$\begin{aligned}
(\alpha\mathbf{u})^{\cdot} &= \dot{\alpha}\mathbf{u} + \alpha\dot{\mathbf{u}}, \\
(\mathbf{u}\cdot\mathbf{v})^{\cdot} &= \dot{\mathbf{u}}\cdot\mathbf{v} + \mathbf{u}\cdot\dot{\mathbf{v}}, \\
(\mathbf{u} \wedge \mathbf{v})^{\cdot} &= \dot{\mathbf{u}} \wedge \mathbf{v} + \mathbf{u} \wedge \dot{\mathbf{v}}, \\
(\mathbf{L}\mathbf{u})^{\cdot} &= \dot{\mathbf{L}}\mathbf{u} + \mathbf{L}\dot{\mathbf{u}}, \\
(\mathbf{u} \otimes \mathbf{v})^{\cdot} &= \dot{\mathbf{u}} \otimes \mathbf{v} + \mathbf{u} \otimes \dot{\mathbf{v}}, \\
(\alpha\mathbf{L})^{\cdot} &= \dot{\alpha}\mathbf{L} + \alpha\dot{\mathbf{L}}, \\
(\mathbf{L}\mathbf{M})^{\cdot} &= \dot{\mathbf{L}}\mathbf{M} + \mathbf{L}\dot{\mathbf{M}}, \\
(\mathbf{L}^T)^{\cdot} &= (\dot{\mathbf{L}})^T, \\
(\mathbf{L}^{-1})^{\cdot} &= -\mathbf{L}^{-1}\dot{\mathbf{L}}\mathbf{L}^{-1}, \\
(\det \mathbf{L})^{\cdot} &= \mathbf{L}^* \cdot \dot{\mathbf{L}}.
\end{aligned}$$

---

[3] Both in this section and in the preceding one, our development could have been made shorter if we had employed an abstract inner-product space $\mathscr{W}$ to illustrate the properties that $\mathscr{V}$, $L(\mathscr{V})$, and $L^2(\mathscr{V})$ have in common. I prefer to give a detailed account of each of these spaces to be fair to the reader who is not much inclined towards abstraction.

In the last two formulae $\mathbf{L}(t)$ is assumed to be invertible for all $t\in[a,b]$ (*cf.* (2.40) for the definition of $\mathbf{L}^*$).

We now turn our attention to differentiation of vector and scalar fields. Our development here will largely parallel that on differentiable curves above.

Let $\mathbf{f}:\mathscr{B}\to\mathscr{V}$ be a vector field on an open set $\mathscr{B}$ of $\mathscr{E}$. We say that $\mathbf{f}$ is **differentiable** at $p_0\in\mathscr{B}$ if there is a tensor $\mathbf{F}\in L(\mathscr{V})$ such that

$$\mathbf{f}(p_0+\mathbf{u})-\mathbf{f}(p_0)=\mathbf{F}\mathbf{u}+o(\mathbf{u})\quad\text{as } \mathbf{u}\to\mathbf{0}.$$

We call $\mathbf{F}$ the **gradient** of $\mathbf{f}$ at p_0. If $\mathbf{f}$ is differentiable everywhere in $\mathscr{B}$, its gradient is a tensor-valued mapping on $\mathscr{B}$, which we denote by $\nabla\mathbf{f}$:

$$\mathbf{f}(p+\mathbf{u})-\mathbf{f}(p)=(\nabla\mathbf{f}(p))\mathbf{u}+o(\mathbf{u})\quad\text{as } \mathbf{u}\to\mathbf{0} \tag{2.52}$$

for all $p\in\mathscr{B}$. If $\nabla\mathbf{f}$ is a field on $\mathscr{B}$, we say that $\mathbf{f}$ is of class C^1. Smoother vector fields are defined by the use of tensors of order higher than 2, as is done shortly below.

Let $\mathbf{f}$ be a vector field of class C^1 on $\mathscr{B}$. The **divergence** of $\mathbf{f}$ is the scalar field defined on $\mathscr{B}$ by

$$\operatorname{div}\mathbf{f}:=\operatorname{tr}\nabla\mathbf{f}. \tag{2.53}$$

Let $\mathbf{W}_\mathbf{f}(p)$ be the skew part of $\nabla\mathbf{f}(p)$ and $\mathbf{w}_\mathbf{f}(p)$ the axial vector of $\mathbf{W}_\mathbf{f}(p)$, for all $p\in\mathscr{B}$. The **curl** of $\mathbf{f}$ is the vector field defined on $\mathscr{B}$ by

$$\operatorname{curl}\mathbf{f}:=2\mathbf{w}_\mathbf{f}. \tag{2.54}$$

It follows from (2.23) that

$$|\operatorname{curl}\mathbf{f}|^2=2|\mathbf{W}_\mathbf{f}|^2. \tag{2.55}$$

When $\mathbf{f}$ happens to be of class C^2 on $\mathscr{B}$, (2.52) is replaced by the formula

$$\mathbf{f}(p+\mathbf{u})-\mathbf{f}(p)=(\nabla\mathbf{f}(p))\mathbf{u}+\tfrac{1}{2}(\nabla^2\mathbf{f}(p))[\mathbf{u}\otimes\mathbf{u}]+o(|\mathbf{u}|\mathbf{u})\quad\text{as}\quad \mathbf{u}\to\mathbf{0},$$

for all $p\in\mathscr{B}$, where $\nabla^2\mathbf{f}(p)$ is the **second gradient** of $\mathbf{f}$ at p, a third-order tensor which vanishes on all skew second-order tensors.

A rather telling application of (2.53) and (2.54) arises from the kinematics of rigid bodies. If we regard $\mathscr{B}$ as the region in space occupied by a rigid body, the velocity field $\mathbf{v}$ on $\mathscr{B}$ is traditionally represented in the form

$$\mathbf{v}(p)=\mathbf{v}(p_0)+\boldsymbol{\omega}\wedge(p-p_0)\quad\text{for all } p\in\mathscr{B}, \tag{2.56}$$

where p_0 is any given point of $\mathscr{B}$, and $\boldsymbol{\omega}$ is the **angular velocity** of the body. By (2.24), (2.56) and the following equation are equivalent:

$$\mathbf{v}(p) = \mathbf{v}(p_0) + \mathbf{A}(p - p_0) \quad \text{for all } p \in \mathscr{B}, \tag{2.57}$$

where $\mathbf{A}$, the **spin tensor**, is the skew tensor associated with $\boldsymbol{\omega}$. It readily follows from (2.57) that

$$\nabla \mathbf{v} = \mathbf{A},$$

and so, by (2.53) and (2.54),

$$\operatorname{div} \mathbf{v} = 0 \quad \text{and} \quad \boldsymbol{\omega} = \tfrac{1}{2} \operatorname{curl} \mathbf{v}.$$

Also when the velocity field is not that of a rigid body, its curl still expresses a local measure of vorticity.

Making use of (2.53) we can also define the divergence of a tensor field $\mathbf{L}: \mathscr{B} \to L(\mathscr{V})$ of class C^1. Let $\mathbf{u}$ be any given vector in $\mathscr{V}$. The mapping $\mathbf{f}: \mathscr{B} \to \mathscr{V}$ defined by

$$\mathbf{f}(p) := (\mathbf{L}^T(p))\mathbf{u} \quad \text{for all } p \in \mathscr{B}$$

is clearly a vector field of class C^1 on $\mathscr{B}$, whose divergence defines at every $p \in \mathscr{B}$ a linear form in $\mathbf{u}$. Thus, by the Representation Theorem of Linear Forms of the preceding section, there is a vector field, which we call the **divergence** of $\mathbf{L}$ and denote by $\operatorname{div} \mathbf{L}$, such that

$$\mathbf{u} \cdot \operatorname{div} \mathbf{L} := \operatorname{div}(\mathbf{L}^T \mathbf{u}) \quad \text{for all } \mathbf{u} \in \mathscr{V}. \tag{2.58}$$

Let $\mathbf{W}: \mathscr{B} \to Skw(\mathscr{V})$ be a tensor field of class C^1, and let $\mathbf{w}(p)$ be the axial vector of $\mathbf{W}(p)$ for all $p \in \mathscr{B}$. The mapping $p \mapsto \mathbf{w}(p)$ defines a vector field of class C^1 on $\mathscr{B}$; it satisfies the equation

$$\operatorname{div} \mathbf{W} = -\operatorname{curl} \mathbf{w}.$$

Let a scalar field φ on $\mathscr{B}$ be given. We say that φ is **differentiable** at $p_0 \in \mathscr{B}$ if there is a vector $\mathbf{w}$ such that

$$\varphi(p_0 + \mathbf{u}) - \varphi(p_0) = \mathbf{w} \cdot \mathbf{u} + o(|\mathbf{u}|) \quad \text{as } |\mathbf{u}| \to 0;$$

$\mathbf{w}$ is called the **gradient** of φ at p_0. If φ is differentiable everywhere in $\mathscr{B}$, its gradient is a vector-valued mapping on $\mathscr{B}$, which we denote by $\nabla\varphi$:

$$\varphi(p + \mathbf{u}) - \varphi(p) = \nabla\varphi(p) \cdot \mathbf{u} + o(|\mathbf{u}|) \quad \text{as } |\mathbf{u}| \to 0, \tag{2.59}$$

for all $p \in \mathscr{B}$. If $\nabla\varphi$ is a vector field, then we say that φ is of class C^1. When this is the case, $\nabla\varphi$ may happen to be differentiable in $\mathscr{B}$;

its gradient is then the **second gradient** of φ, which we denote by $\nabla^2\varphi$, and the following formula holds in $\mathscr{B}$:

$$\varphi(p+\mathbf{u})-\varphi(p)=\nabla\varphi(p)\cdot\mathbf{u}+\tfrac{1}{2}\mathbf{u}\cdot(\nabla^2\varphi(p))\mathbf{u}+o(|\mathbf{u}|^2) \quad \text{as } |\mathbf{u}|\to 0. \tag{2.60}$$

It is not difficult to show that $\nabla^2\varphi(p)$ is a symmetric tensor for all $p\in\mathscr{B}$.

If $\nabla^2\varphi$ is a tensor field on $\mathscr{B}$, then φ is of class C^2 and its **Laplacian** is defined by

$$\Delta\varphi:=\operatorname{tr}\nabla^2\varphi=\operatorname{div}\nabla\varphi. \tag{2.61}$$

Let $f:\mathscr{B}\to\mathscr{E}$ be a deformation of $\mathscr{B}$. We say that f is **differentiable** at $p_0\in\mathscr{B}$ if there is a tensor $\mathbf{F}_0$, called the gradient of f at p_0, such that

$$f(p_0+\mathbf{u})-f(p_0)=\mathbf{F}_0\mathbf{u}+o(\mathbf{u}) \quad \text{as } \mathbf{u}\to\mathbf{0}. \tag{2.62}$$

If f is differentiable everywhere in $\mathscr{B}$, the gradient of f is a tensor-valued mapping on $\mathscr{B}$, which we denote by ∇f. Then (2.62) reads as follows

$$f(p+\mathbf{u})-f(p)=(\nabla f(p))\mathbf{u}+o(\mathbf{u}) \quad \text{as } \mathbf{u}\to\mathbf{0}, \tag{2.63}$$

for all $p\in\mathscr{B}$. When ∇f is continuous in $\mathscr{B}$ the deformation f is said to be of class C^1.

Let $\varphi, \mathbf{f}$ and $\mathbf{L}$ be a scalar field, a vector field and a tensor field respectively, and let all of them be of class C^1 on the same open set $\mathscr{B}$ in $\mathscr{E}$. The following differentiation rules are easy consequences of (2.52), (2.53) and (2.58):

$$\nabla(\varphi\mathbf{f})=\mathbf{f}\otimes\nabla\varphi+\varphi\nabla\mathbf{f},$$
$$\operatorname{div}(\varphi\mathbf{f})=\mathbf{f}\cdot\nabla\varphi+\varphi\operatorname{div}\mathbf{f},$$
$$\operatorname{div}(\mathbf{L}\mathbf{f})=(\operatorname{div}\mathbf{L}^T)\cdot\mathbf{f}+\operatorname{tr}(\mathbf{L}\nabla\mathbf{f}).$$

Furthermore, if $\mathbf{g}$ is a vector field of class C^1 on $\mathscr{B}$, making use also of (2.59), we easily prove that

$$\nabla(\mathbf{f}\cdot\mathbf{g})=(\nabla\mathbf{f})^T\mathbf{g}+(\nabla\mathbf{g})^T\mathbf{f}. \tag{2.64}$$

One consequence of (2.64) will often be employed in this book. Let a continuous mapping $\mathbf{n}:\mathscr{B}\to\mathbb{S}^2$ be given. We may regard it as a vector field on $\mathscr{B}$ such that

$$\mathbf{n}\cdot\mathbf{n}=1.$$

If such a field is of class C^1, differentiating both sides of the latter equation with the aid of (2.64), we get

$$(\nabla\mathbf{n})^T\mathbf{n} = \mathbf{0}. \tag{2.65}$$

Now we turn our attention to two instances of the chain rule; the symbol $\circ$ denotes composition in both formulae below:

$$\nabla(\psi\circ f) = (\nabla f)^T((\nabla\psi)\circ f),$$

where f is a deformation of class C^1 of $\mathscr{B}$ and ψ is a scalar field of class C^1 on $f(\mathscr{B})$;

$$\nabla(p\circ\varphi) = (\dot{p}\circ\varphi)\otimes\nabla\varphi,$$

where p is a curve of class C^1 on $\mathbb{R}$ into $\mathscr{E}$ and φ is a scalar field as above. The reader will find it an instructive exercise to prove these equations, making use of (2.63), (2.59), (2.61) and Theorem 2.1 of Section 2.2.1.

2.3.3 Co-ordinates

So far we have dealt with curves, fields and deformations, as it were, in the abstract. When specific computations are needed, the use of an appropriate system of co-ordinates is of the essence. We shall confine attention here to scalar and vector fields of class C^1 and we shall learn how to express their gradients in different co-ordinates. Rather than simply recording a list of formulae, I attempt to illustrate a method, working out explicitly a few typical examples.

Cartesian co-ordinates, being the most familiar ones, come first. Let a point o in $\mathscr{E}$ and a basis $e = (\mathbf{e}_1, \mathbf{e}_2, \mathbf{e}_3)$ of $\mathscr{V}$ be given. We call $(o, \mathbf{e}_1, \mathbf{e}_2, \mathbf{e}_3)$ a **frame** of $\mathscr{E}$ with origin o. For any given point $p\in\mathscr{E}$ there are three scalars, x_1, x_2 and x_3, such that

$$p - o = x_1\mathbf{e}_1 + x_2\mathbf{e}_2 + x_3\mathbf{e}_3; \tag{2.66}$$

they are the Cartesian co-ordinates of p.

Let $\mathscr{B}\subset\mathscr{E}$ be an open set and φ a scalar field of class C^1 on $\mathscr{B}$. A subset B of $\mathbb{R}^3$ then corresponds to $\mathscr{B}$ through (2.66): it is defined by

$$B := \left\{(x_1, x_2, x_3)\in\mathbb{R}^3 \,|\, o + \sum_{i=1}^{3} x_i\mathbf{e}_i \in\mathscr{B}\right\}.$$

Likewise, a scalar-valued mapping $\tilde{\varphi}$ on B corresponds to φ:

$$\varphi\left(o + \sum_{i=1}^{3} x_i\mathbf{e}_i\right) =: \tilde{\varphi}(x_1, x_2, x_3) \quad \text{for all } (x_1, x_2, x_3)\in B.$$

It is easily seen from (2.59) that the Cartesian components of $\nabla\varphi$ in e are

$$(\nabla\varphi)_i = \tilde{\varphi}_{,i} \quad \text{for } i = 1, 2, 3, \tag{2.67}$$

where the comma denotes the partial derivative with respect to a space co-ordinate:

$$\tilde{\varphi}_{,i} := \frac{\partial \tilde{\varphi}}{\partial x_i} \quad \text{for } i = 1, 2, 3.$$

I tolerate a common abuse of notation allowing $\tilde{\varphi}$ to be identified with φ, because confusion is unlikely to arise. Thus, (2.67) becomes

$$(\nabla\varphi)_i = \varphi_{,i} \quad \text{for } i = 1, 2, 3,$$

hence

$$\nabla\varphi = \sum_{i=1}^{3} \varphi_{,i}\mathbf{e}_i.$$

Similarly, if $\mathbf{f}$ is a vector field of class C^1 on $\mathscr{B}$, we denote by f_i, with $i = 1, 2, 3$, its Cartesian components in e; they are indeed scalar fields of class C^1 on $\mathscr{B}$, as are the Cartesian components in e of $\nabla\mathbf{f}$, which in the above notation read as

$$(\nabla\mathbf{f})_{ij} = f_{i,j} \quad \text{for } i, j = 1, 2, 3,$$

whence

$$\nabla\mathbf{f} = \sum_{i,j=1}^{3} f_{i,j}\mathbf{e}_i \otimes \mathbf{e}_j. \tag{2.68}$$

By (2.53) and (2.54), with the aid of (2.29), we arrive at

$$\operatorname{div}\mathbf{f} = \sum_{i=1}^{3} f_{i,i},$$

$$\operatorname{curl}\mathbf{f} = \sum_{i,j,h=1}^{3} \varepsilon_{ijh} f_{h,j}\mathbf{e}_i.$$

If $\mathbf{f}$ is of class C^2 on $\mathscr{B}$, then f_i, for each $i \in \{1, 2, 3\}$, is a scalar field of class C^2 and

$$\nabla^2\mathbf{f} = \sum_{i,j,k=1}^{3} f_{i,jk}\mathbf{e}_i \otimes \mathbf{e}_j \otimes \mathbf{e}_k,$$

where

$$f_{i,jk} := \frac{\partial^2 f_i}{\partial x_j \partial x_k}.$$

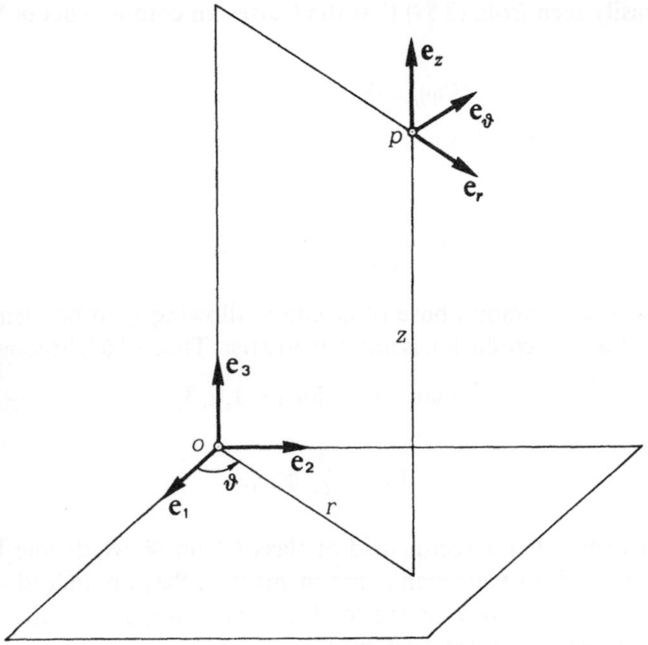

Figure 2.6

Clearly,

$$f_{i,jk} = f_{i,kj} \quad \text{for all } i, j, k \in \{1, 2, 3\}.$$

We now turn our attention to **cylindrical co-ordinates**. Here a **movable frame** $(p, \mathbf{e}_r, \mathbf{e}_\vartheta, \mathbf{e}_z)$, whose origin is the arbitrary point $p \in \mathscr{E}$, is to be added to the frame introduced above (see Fig. 2.6). Cylindrical and Cartesian co-ordinates are related through the following formulae

$$x_1 = r\cos\vartheta, \quad x_2 = r\sin\vartheta, \quad x_3 = z, \tag{2.69}$$

where $r > 0$ and $0 \leqslant \vartheta < 2\pi$; correspondingly, the unit vectors of the movable frame and those of e are related through

$$\mathbf{e}_r = \cos\vartheta\mathbf{e}_1 + \sin\vartheta\mathbf{e}_2, \quad \mathbf{e}_\vartheta = -\sin\vartheta\mathbf{e}_1 + \cos\vartheta\mathbf{e}_2, \quad \mathbf{e}_z = \mathbf{e}_3. \tag{2.70}$$

It then follows from (2.66) that

$$p - o = r\mathbf{e}_r + z\mathbf{e}_z. \tag{2.71}$$

Let φ be a field of class C^1 as above and $t \mapsto p(t)$ any curve of class C^1 in $\mathscr{B}$. If we denote by $\dot{\varphi}$ the derivative of the scalar-valued function defined by $t \mapsto \varphi(p(t))$, then by the use of (2.59) we easily arrive at

$$\dot{\varphi} = \nabla\varphi \cdot \dot{p}, \tag{2.72}$$

where $\dot{p}$ is the derivative of p.

If φ is expressed in cylindrical co-ordinates, then

$$\dot{\varphi} = \varphi_{,r}\dot{r} + \varphi_{,\vartheta}\dot{\vartheta} + \varphi_{,z}\dot{z}, \tag{2.73}$$

where $\dot{r}$, $\dot{\vartheta}$, and $\dot{z}$ are the derivatives of the functions which describe the curve $t \mapsto p(t)$ in cylindrical co-ordinates, and (2.71) implies that

$$\dot{p} = \dot{r}\mathbf{e}_r + r\dot{\vartheta}\mathbf{e}_\vartheta + \dot{z}\mathbf{e}_z, \tag{2.74}$$

since, by (2.70),

$$\dot{\mathbf{e}}_r = \dot{\vartheta}\mathbf{e}_\vartheta. \tag{2.75}$$

Making use of both (2.74) and (2.73) in (2.72), since the curve p is arbitrary we easily arrive at the formula that expresses $\nabla\varphi$ in the movable frame:

$$\nabla\varphi = \varphi_{,r}\mathbf{e}_r + \frac{1}{r}\varphi_{,\vartheta}\mathbf{e}_\vartheta + \varphi_{,z}\mathbf{e}_z. \tag{2.76}$$

Likewise, if $\mathbf{f}$ is a vector field of class C^1, equation (2.52) implies that

$$\dot{\mathbf{f}} = (\nabla\mathbf{f})\dot{p}. \tag{2.77}$$

This formula is to be applied first to the unit vectors of the movable frame away from the z-axis. Comparing (2.77) and (2.75), with the aid of (2.74), we easily see that

$$\nabla\mathbf{e}_r = \frac{1}{r}\mathbf{e}_\vartheta \otimes \mathbf{e}_\vartheta. \tag{2.78}$$

Similarly, since by (2.70)

$$\dot{\mathbf{e}}_\vartheta = -\dot{\vartheta}\mathbf{e}_r,$$

we also arrive at

$$\nabla\mathbf{e}_\vartheta = -\frac{1}{r}\mathbf{e}_r \otimes \mathbf{e}_\vartheta. \tag{2.79}$$

Finally, $\nabla\mathbf{e}_z = \mathbf{0}$ because $\mathbf{e}_z$ is constant.

Let f_r, f_ϑ and f_z be the components of $\mathbf{f}$ in the movable frame, so that

$$\mathbf{f} = f_r \mathbf{e}_r + f_\vartheta \mathbf{e}_\vartheta + f_z \mathbf{e}_z.$$

Thus, by (2.78), (2.79) and the differentiation rules recalled above, we get

$$\nabla \mathbf{f} = \frac{1}{r} f_r \mathbf{e}_\vartheta \otimes \mathbf{e}_\vartheta + \mathbf{e}_r \otimes \nabla f_r - \frac{1}{r} f_\vartheta \mathbf{e}_r \otimes \mathbf{e}_\vartheta + \mathbf{e}_\vartheta \otimes \nabla f_\vartheta + \mathbf{e}_z \otimes \nabla f_z.$$

Hence, if we regard f_r, f_ϑ and f_z as functions of r, ϑ and z, by (2.76) we obtain the formula that expresses $\nabla \mathbf{f}$ in the movable frame:

$$\begin{aligned}\nabla \mathbf{f} = {} & f_{r,r} \mathbf{e}_r \otimes \mathbf{e}_r + \frac{1}{r}(f_{r,\vartheta} - f_\vartheta) \mathbf{e}_r \otimes \mathbf{e}_\vartheta + f_{r,z} \mathbf{e}_r \otimes \mathbf{e}_z \\ & + f_{\vartheta,r} \mathbf{e}_\vartheta \otimes \mathbf{e}_r + \frac{1}{r}(f_{\vartheta,\vartheta} + f_r) \mathbf{e}_\vartheta \otimes \mathbf{e}_\vartheta + f_{\vartheta,z} \mathbf{e}_\vartheta \otimes \mathbf{e}_z \\ & + f_{z,r} \mathbf{e}_z \otimes \mathbf{e}_r + \frac{1}{r} f_{z,\vartheta} \mathbf{e}_z \otimes \mathbf{e}_\vartheta + f_{z,z} \mathbf{e}_z \otimes \mathbf{e}_z,\end{aligned} \quad (2.80)$$

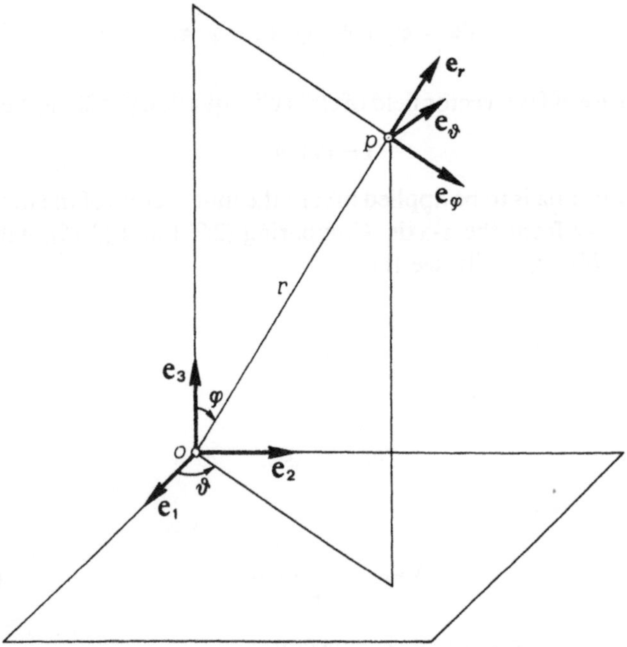

Figure 2.7

and hence, with the aid of (2.53), (2.54), and (2.25),

$$\operatorname{div}\mathbf{f} = f_{r,r} + \frac{1}{r}(f_{\vartheta,\vartheta} + f_r) + f_{z,z}, \tag{2.81}$$

$$\operatorname{curl}\mathbf{f} = \left(\frac{1}{r}f_{z,\vartheta} - f_{\vartheta,z}\right)\mathbf{e}_r + (f_{r,z} - f_{z,r})\mathbf{e}_\vartheta + \left(f_{\vartheta,r} + \frac{1}{r}(f_\vartheta - f_{r,\vartheta})\right)\mathbf{e}_z. \tag{2.82}$$

The last co-ordinates in $\mathscr{E}$ that we consider here are **spherical co-ordinates**. The movable frame $(p, \mathbf{e}_r, \mathbf{e}_\varphi, \mathbf{e}_\vartheta)$ of these co-ordinates is illustrated in Fig. 2.7 together with the frame $(o, \mathbf{e}_1, \mathbf{e}_2, \mathbf{e}_3)$. Cartesian and spherical co-ordinates are related through

$$x_1 = r\sin\varphi\cos\vartheta, \quad x_2 = r\sin\varphi\sin\vartheta, \quad x_3 = r\cos\varphi,$$

where $r > 0$, $0 \leqslant \varphi \leqslant \pi$, and $0 \leqslant \vartheta < 2\pi$. Away from o, the unit vectors of the movable frame are expressed in e by the formulae below:

$$\begin{aligned}
\mathbf{e}_r &= \sin\varphi\cos\vartheta\,\mathbf{e}_1 + \sin\varphi\sin\vartheta\,\mathbf{e}_2 + \cos\varphi\,\mathbf{e}_3,\\
\mathbf{e}_\varphi &= \cos\varphi\cos\vartheta\,\mathbf{e}_1 + \cos\varphi\sin\vartheta\,\mathbf{e}_2 - \sin\varphi\,\mathbf{e}_3,\\
\mathbf{e}_\vartheta &= -\sin\vartheta\,\mathbf{e}_1 + \cos\vartheta\,\mathbf{e}_2.
\end{aligned} \tag{2.83}$$

Equation (2.71) is replaced here by

$$p - o = r\mathbf{e}_r.$$

Hence, with the aid of (2.83), we arrive at the following equations

$$\nabla\mathbf{e}_r = \frac{1}{r}(\mathbf{e}_\varphi \otimes \mathbf{e}_\varphi + \mathbf{e}_\vartheta \otimes \mathbf{e}_\vartheta),$$

$$\nabla\mathbf{e}_\varphi = -\frac{1}{r}(\mathbf{e}_r \otimes \mathbf{e}_\varphi - \cot\varphi\,\mathbf{e}_\vartheta \otimes \mathbf{e}_\vartheta),$$

$$\nabla\mathbf{e}_\vartheta = -\frac{1}{r}(\mathbf{e}_r \otimes \mathbf{e}_\vartheta + \cot\varphi\,\mathbf{e}_\varphi \otimes \mathbf{e}_\vartheta).$$

Let $\psi: \mathscr{B} \to \mathbb{R}$ and $\mathbf{f}: \mathscr{B} \to \mathscr{V}$ be fields on an open subset of $\mathscr{E}$. If ψ is regarded as a function of the co-ordinates r, φ and ϑ, and if f_r, f_φ and f_ϑ denote the components of $\mathbf{f}$ in the movable frame, proceeding along the same lines described above, the reader will easily arrive at the formulae

$$\nabla\psi = \psi_{,r}\mathbf{e}_r + \frac{1}{r}\psi_{,\varphi}\mathbf{e}_\varphi + \frac{1}{r\sin\varphi}\psi_{,\vartheta}\mathbf{e}_\vartheta, \tag{2.84}$$

$$\begin{aligned}\nabla\mathbf{f} = {} & f_{r,r}\mathbf{e}_r\otimes\mathbf{e}_r + \frac{1}{r}(f_{r,\varphi} - f_\varphi)\mathbf{e}_r\otimes\mathbf{e}_\varphi \\ & + \frac{1}{r}\left(\frac{1}{\sin\varphi}f_{r,\vartheta} - f_\vartheta\right)\mathbf{e}_r\otimes\mathbf{e}_\vartheta + f_{\varphi,r}\mathbf{e}_\varphi\otimes\mathbf{e}_r \\ & + \frac{1}{r}(f_r + f_{\varphi,\varphi})\mathbf{e}_\varphi\otimes\mathbf{e}_\varphi + \frac{1}{r}\left(\frac{1}{\sin\varphi}f_{\varphi,\vartheta} - \cot\varphi f_\vartheta\right)\mathbf{e}_\varphi\otimes\mathbf{e}_\vartheta \\ & + f_{\vartheta,r}\mathbf{e}_\vartheta\otimes\mathbf{e}_r + \frac{1}{r}f_{\vartheta,\varphi}\mathbf{e}_\vartheta\otimes\mathbf{e}_\varphi \\ & + \frac{1}{r}\left(f_r + \cot\varphi f_\varphi + \frac{1}{\sin\varphi}f_{\vartheta,\vartheta}\right)\mathbf{e}_\vartheta\otimes\mathbf{e}_\vartheta. \end{aligned} \tag{2.85}$$

It then follows from the latter equation that

$$\operatorname{div}\mathbf{f} = f_{r,r} + \frac{2}{r}f_r + \frac{1}{r}(\cot\varphi f_\varphi + f_{\varphi,\varphi}) + \frac{1}{r\sin\varphi}f_{\vartheta,\vartheta} \tag{2.86}$$

and

$$\begin{aligned}\operatorname{curl}\mathbf{f} = {} & \frac{1}{r}\left(f_{\vartheta,\varphi} + \cot\varphi f_\vartheta - \frac{1}{\sin\varphi}f_{\varphi,\vartheta}\right)\mathbf{e}_r \\ & + \left(\frac{1}{r\sin\varphi}f_{r,\vartheta} - \frac{1}{r}f_\vartheta - f_{\vartheta,r}\right)\mathbf{e}_\varphi \\ & + \left(f_{\varphi,r} + \frac{1}{r}f_\varphi - \frac{1}{r}f_{r,\varphi}\right)\mathbf{e}_\vartheta. \end{aligned} \tag{2.87}$$

2.3.4 Measures

This is not a book on analysis, and so it is not appropriate to give an exhaustive account of all the analytical concepts employed here, especially of the most subtle ones, such as those related to measure theory. For a quick and direct introduction to this subject I refer the reader to Chapter 2 of Vol'pert and Hudjaev (1985) and to Chapter 1 of Ziemer (1989). Nonetheless, I cannot avoid mentioning a few elementary facts in this connection.

The **volume-measure** in $\mathscr{E}$ is taken here to be **Lebesgue measure** and it is denoted by v.

Definition 2.8. **Lebesgue measure** is a set function on $\mathscr{E}$ whose value on parallelepipeds equals their volume classically defined, and which is a regular measure[4] on each ball of $\mathscr{E}$.

When we consider subsets of $\mathscr{E}$ which are less bulky than parallelepipeds, the prototypes of Lebesgue measurable sets, their volume may well vanish. Thus the need arises to define in $\mathscr{E}$ measures of lower dimension which make precise the intuitive ideas of *area* and *length*. It has recently become clear that **Hausdorff measure** fills this need better than others. The definition of this measure is too technical to be reproduced here, but one of its features deserves notice: for any given subset $\mathscr{B}$ of $\mathscr{E}$ there is one Hausdorff measure $H^\gamma(\mathscr{B})$ for each scalar $\gamma \geqslant 0$. $H^\gamma(\mathscr{B})$ is called the **γ-dimensional Hausdorff measure** of $\mathscr{B}$. A special notation is used here for H^γ when $\gamma = 2$ or $\gamma = 1$:

$$H^2 =: a \quad \text{and} \quad H^1 =: \ell.$$

We call a **area-measure** and ℓ **length-measure.**

$H^\gamma(\mathscr{B})$ may happen to be infinite or zero for most values of γ. Here are two theorems that tell us how H^γ depends on γ.

Theorem 2.11. $H^3(\mathscr{B}) = v(\mathscr{B})$ for all $\mathscr{B} \subset \mathscr{E}$.

That is to say that 3-dimensional Hausdorff measure and Lebesgue measure coincide.

Theorem 2.12. Let a subset $\mathscr{B}$ of $\mathscr{E}$ be given. There is a scalar $d_{\mathscr{B}} \geqslant 0$ such that

$$H^\gamma(\mathscr{B}) = 0 \quad \text{for all } \gamma > d_{\mathscr{B}},$$
$$H^\gamma(\mathscr{B}) = \infty \quad \text{for all } \gamma < d_{\mathscr{B}}.$$

We call $d_{\mathscr{B}}$ the **Hausdorff dimension** of $\mathscr{B}$.

2.3.5 Regions

The only regions in space we have encountered so far are open subsets of $\mathscr{E}$, but among them sets could be imagined which are not likely to be occupied by material bodies. This book aims to treat the mathematical theory of liquid crystals as a branch of modern continuum mechanics, and so it is sensible to adopt here also the

---

[4] Regular measures are defined *e.g.* on p. 76 of Vol'pert and Hudjaev (1985).

mathematical structures devised to describe the sets fit to be occupied by continuous bodies and their subbodies.

As was pointed out in Noll (1986), such a class of fit regions should meet certain requirements which ensure that what is usually done in continuum physics can be justified in precise mathematical terms. First, the set of all fit subregions of a given fit region should satisfy the axioms of a material universe as described in the Appendix to Noll (1973) or in Sections I.2 and I.3 of Truesdell (1991). Second, the class of all fit regions should be invariant under **transplacements**, which should include adjustments to fit regions of smooth **diffeomorphisms** from $\mathscr{E}$ into itself. Third, each fit region should have a surface-like boundary for which a form of the Gauss–Green Theorem should be valid.

It is explained in Noll and Virga (1988) why naïve classes of sets fail to fulfil the above requirements; an account is also given there of previous attempts to find the smallest class of fit regions.

The proposal of Noll and Virga (1988) rests upon deep aspects of the theory of sets of **finite perimeter**. I state below the appropriate definitions and theorems for readers who are already familiar with this theory. I refer to the textbooks for complete justification. (See *e.g.* Section II.1 of Truesdell (1991), Chapter 5 of Ziemer (1989), or Chapter 5 of Vol'pert and Hudjaev (1985).)

It is interesting that the surface-like boundary of a fit region need not be the topological boundary, but it is the **reduced boundary**, on which an outer normal $\mathbf{v}$ can be defined.

Definition 2.9. A subset $\mathscr{D}$ of $\mathscr{E}$ is a **fit region** if it

(a) is bounded,
(b) is regularly open,[5]
(c) has finite perimeter,
(d) has a boundary of volume-measure zero.

Hereafter in this section a fit region $\mathscr{D}$ is assumed to be given in $\mathscr{E}$.

Theorem 2.13. The reduced boundary $\partial^*\mathscr{D}$ of $\mathscr{D}$ has finite area-measure, and the outer normal $\mathbf{v}: \partial^*\mathscr{D} \to \mathbb{S}^2$ is area-integrable on $\partial^*\mathscr{D}$.

Integral–Gradient Theorem 2.14. Let φ be a scalar field of class C^1

[5] A subset of $\mathscr{E}$ is regularly open if it is equal to the interior of its closure.

on $\mathscr{D}$, continuous up to the closure of $\mathscr{D}$. We have that

$$\int_{\mathscr{D}} \nabla\varphi \, dv = \int_{\partial^*\mathscr{D}} \varphi \mathbf{v} \, da.$$

A similar theorem holds also for vector fields; from this we draw the following conclusion.

Divergence Theorem 2.15. Let $\mathbf{f}$ be a vector field of class C^1 on $\mathscr{D}$, continuous up to the closure of $\mathscr{D}$. We have

$$\int_{\mathscr{D}} \operatorname{div} \mathbf{f} \, dv = \int_{\partial^*\mathscr{D}} \mathbf{f} \cdot \mathbf{v} \, da.$$

The last two theorems are classic when $\mathscr{D}$ is a region of $\mathscr{E}$ with piecewise C^1-boundary.

Though everyone would agree that a subset of $\mathscr{E}$ with a piecewise smooth boundary can be defined unambiguously, when challenged to make this concept precise, different readers might give non-equivalent definitions (see *e.g.* Appendix II to Fraenkel (1979)). Noll and Virga (1990), laying the foundations of a theory for **edge interactions** between material bodies which extends and completes the classical theory of **surface interactions**, needed to introduce a precise definition of regions in space with piecewise C^2-boundary. These are called **regular regions** and are carefully described in Section 1 of their paper (see, in particular, Definition 1).

They will often be employed in this book. It would be a little clumsy to give here every detail of their definition, but a few qualitative facts must be recorded, with the observation that they stand on firm footing, though no justification is presented here.

Let a regular region $\mathscr{R}$ be given in $\mathscr{E}$. There is at least one finite partition P of the closure of $\mathscr{R}$ such that $\mathscr{R}$ itself belongs to P as well as **sides, edges** and **vertices** of its boundary. P_2 is the subpartition of P of all sides, P_1 is that of all edges, and P_0 that of all vertices. The partition P possesses two further properties. First, for each edge in P_1 there are precisely two sides in P_2 **adjacent** to it. Second, for each vertex in P_0 there are precisely two edges in P_1 adjacent to it relative to one side in P_2.

The sides of a regular region should be thought of as **regular orientable surfaces** in $\mathscr{E}$, to be considered next in more detail.

I finally note that regular regions are fit regions and that their definition seems to be equivalent to that given by Kellogg (1929) on p. 112.

2.3.6 Regular orientable surfaces

A surface in space seems to be a highly intuitive concept, but making it precise requires some skill. Here I follow in part Section 2 of Noll and Virga (1990) and unpublished lecture notes by Noll, revised in 1987, which are to become Chapter 3 of the sequel to Noll (1987). (See also Section 3 of Gurtin and Murdoch (1974–75).)

Definition 2.10. A surface $\mathscr{S}$ is a bounded connected C^2-manifold of dimension 2 embedded in $\mathscr{E}$.[6]

A 2-dimensional linear space is associated with every point p of a surface $\mathscr{S}$; it is the **tangent space** of $\mathscr{S}$ at p, which we denote by $\mathscr{T}_{\mathscr{S}}(p)$.

Both scalar and vector fields are easily defined on $\mathscr{S}$. An adequate definition of differentiation can also be introduced for them. If a scalar field φ on $\mathscr{S}$ is differentiable, then the gradient of φ at each $p\in\mathscr{S}$ is a linear mapping of $\mathscr{T}_{\mathscr{S}}(p)$ into $\mathbb{R}$. Likewise, if a vector field $\mathbf{f}$ on $\mathscr{S}$ is differentiable, then the gradient of $\mathbf{f}$ at each $p\in\mathscr{S}$ is a linear mapping of $\mathscr{T}_{\mathscr{S}}(p)$ into $\mathscr{V}$. We say that φ and $\mathbf{f}$ are of class C^1 whenever their gradients are continuous on $\mathscr{S}$. One can always define on a surface $\mathscr{S}$ a mapping $\mathbf{v}$ into $\mathbb{S}^2$ such that

$$\mathbf{v}(p)\in(\mathscr{T}_{\mathscr{S}}(p))^{\perp} \quad \text{for all } p\in\mathscr{S}.$$

Every such mapping is called a **normal** of $\mathscr{S}$. When there is a normal of $\mathscr{S}$ which is everywhere continuous on $\mathscr{S}$, then we say that it endows $\mathscr{S}$ with an **orientation**, and hence $\mathscr{S}$ is **orientable**.

An orientable surface possesses two orientations; one has the normal opposite to the other. The normal of each orientation is a vector field of class C^1 on $\mathscr{S}$. Hereafter one of these normals is taken as given and will be denoted by $\mathbf{v}\colon \mathscr{S}\to\mathbb{S}^2$.

A special role will be played below by the tensor field defined on $\mathscr{S}$ by $\mathbf{E}_{\mathscr{S}}:=\mathbf{P}\circ\mathbf{v}$ (*cf.* equation (2.47)), that is

$$\mathbf{E}_{\mathscr{S}}(p):=\mathbf{I}-\mathbf{v}(p)\otimes\mathbf{v}(p) \quad \text{for all } p\in\mathscr{S}.$$

---

[6] For the definition of C^2-manifolds embedded in $\mathscr{E}$ the reader is referred *e.g.* to pp. 21–25 of Lang (1985).

We assume that there is a unique continuous extension of $\mathbf{E}_{\mathscr{S}}$ to the closure of $\mathscr{S}$. We shall also denote this extension by $\mathbf{E}_{\mathscr{S}}$.

The set difference between the closure of $\mathscr{S}$ and $\mathscr{S}$ itself is termed the **border** of $\mathscr{S}$; we denote it by $\mathrm{d}\mathscr{S}$, Clearly, $\mathrm{d}\mathscr{S}$ differs from $\partial\mathscr{S}$, the topological boundary of $\mathscr{S}$, which would be the same as the closure of $\mathscr{S}$.

Let a finite partition P of the closure of $\mathscr{S}$ be given. Every bounded connected C^2-manifold of dimension 1 in P is called an **edge**, and every manifold of dimension 0 in P is called a **vertex**. Two edges of $\mathscr{S}$ are said to be **adjacent** to a vertex of $\mathscr{S}$ if the only member of the latter belongs to the closure of both edges.

Definition 2.11. We say that an orientable surface $\mathscr{S}$ is **regular** if there is a finite partition P of $\mathscr{S}$ whose members are $\mathscr{S}$ itself and the edges and vertices of its border such that for every vertex of P there are precisely two edges of P adjacent to it. We call P a **regular partition** of $\mathscr{S}$.

A regular orientable surface $\mathscr{S}$ admits infinitely many regular partitions; new ones can always be obtained by subdivision of the border of $\mathscr{S}$. Of course, there are always regular partitions with a minimum number of pieces, but there may be more than one such partition.

Figure 2.8 shows a flat regular surface with this property. A regular partition for the surface must have at least five pieces (the surface itself, two edges and two vertices). There are infinitely many regular partitions with five pieces because one of the vertices can be put into infinitely many positions. The surface illustrated in Figure 2.8 is a good example also for another reason: it shows that at a vertex of a regular partition the tangent to $\mathrm{d}\mathscr{S}$ need not suffer a jump.

Definition 2.12. Let $\mathscr{S}$ be a regular orientable surface. We call the union of all edges of $\mathscr{S}$ the **reduced border** of $\mathscr{S}$, and we denote it by $\mathrm{d}^*\mathscr{S}$.

It should be noted that the edges of $\mathscr{S}$ whose union is $\mathrm{d}^*\mathscr{S}$ need not belong to the same regular partition of $\mathscr{S}$. Thus, for example, the reduced border of the surface in Figure 2.8 differs from its border only by one vertex.

At each point of the reduced border of a regular orientable surface $\mathscr{S}$ there is precisely one unit vector pointing outwards and

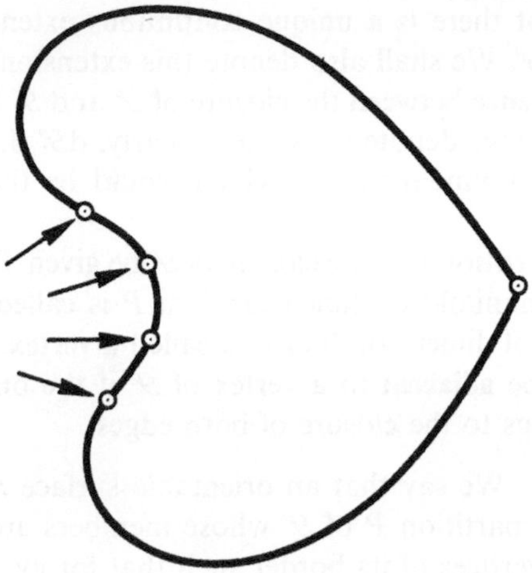

Figure 2.8

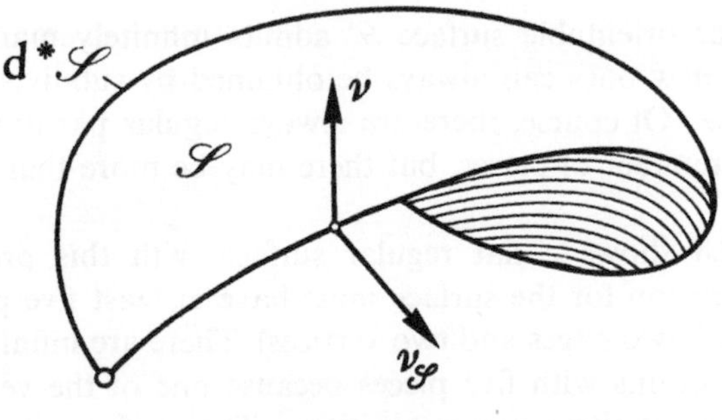

Figure 2.9

orthogonal to both $\mathbf{v}$ and the tangent to $d^*\mathscr{S}$ at that point (see Figure 2.9); a vector field is thus defined on $d^*\mathscr{S}$:

$$\mathbf{v}_{\mathscr{S}}: d^*\mathscr{S} \to \mathbb{S}^2.$$

Hereafter in this section $\mathscr{S}$ will denote a regular orientable surface.

Let $\mathbf{f}$ be a vector field of class C^1 on $\mathscr{S}$. We call the **surface-gradient** of $\mathbf{f}$ the mapping $\nabla_s\mathbf{f}: \mathscr{S} \to L(\mathscr{V})$ that assigns to each point $p \in \mathscr{S}$ the tensor

$$\nabla_s\mathbf{f}(p) := \nabla\mathbf{f}(p)\mathbf{E}_{\mathscr{S}}(p).$$

It should be noted that

$$(\nabla_s\mathbf{f}(p))\mathbf{v}(p) = \mathbf{0} \quad \text{for all } p \in \mathscr{S}.$$

Let $\mathbf{e}$ be any unit vector lying in $\mathscr{T}_{\mathscr{S}}(p)$; the **surface derivative** of $\mathbf{f}$ along $\mathbf{e}$ at p is the vector defined by

$$\nabla_{\mathbf{e}}\mathbf{f}(p) = (\nabla_s \mathbf{f}(p))\mathbf{e}.$$

Like $\nabla\mathbf{f}$, $\nabla_s\mathbf{f}$ can also be extended up to the border of $\mathscr{S}$; we shall denote such an extension by $\nabla_s\mathbf{f}$ also.

Let $\varphi:\mathscr{S}\to\mathbb{R}$ and $\mathbf{f}:\mathscr{S}\to\mathscr{V}$ be fields of class C^1. It is an easy exercise to show that

$$\nabla_s(\varphi\mathbf{f}) = \varphi\nabla_s\mathbf{f} + (\mathbf{f}\otimes\nabla\varphi)\mathbf{E}_{\mathscr{S}}. \tag{2.88}$$

The **surface divergence** of $\mathbf{f}$ is the field $\operatorname{div}_s\mathbf{f}$: $\mathscr{S}\to\mathbb{R}$ defined by

$$(\operatorname{div}_s\mathbf{f})(p) = \operatorname{tr}(\nabla_s\mathbf{f}(p)) \quad \text{for all } p\in\mathscr{S}. \tag{2.89}$$

Likewise, if $\mathbf{L}$ is a tensor field of class C^1 on $\mathscr{S}$, then its **surface divergence** is the field $\operatorname{div}_s\mathbf{L}:\mathscr{S}\to\mathscr{V}$ characterized by

$$\mathbf{u}\cdot\operatorname{div}_s\mathbf{L}(p) := \operatorname{div}_s(\mathbf{L}^T\mathbf{u})(p) \quad \text{for all } p\in\mathscr{S} \text{ and } \mathbf{u}\in\mathscr{V}.$$

Finally, a vector field $\mathbf{f}$ on $\mathscr{S}$ is called **tangential** if $\mathbf{f}(p)\in\mathscr{T}_{\mathscr{S}}(p)$ for all $p\in\mathscr{S}$; a tensor field $\mathbf{L}$ on $\mathscr{S}$ is called **tangential** if $\mathbf{L}(p)\mathbf{v}(p) = \mathbf{0}$ for all $p\in\mathscr{S}$.

Surface-Divergence Theorem 2.16. Let $\bar{\mathscr{S}}$ be the closure of $\mathscr{S}$. For every field $\mathbf{f}:\bar{\mathscr{S}}\to\mathscr{V}$ that is of class C^1 and tangential we have

$$\int_{\mathscr{S}} \operatorname{div}_s\mathbf{f}\,da = \int_{\mathrm{d}^*\mathscr{S}} (\mathbf{f}\cdot\mathbf{v}_{\mathscr{S}})\,d\ell.$$

Similarly, for every field $\mathbf{L}:\bar{\mathscr{S}}\to L(\mathscr{V})$ that is of class C^1 and tangential we have

$$\int_{\mathscr{S}} \operatorname{div}_s\mathbf{L}\,da = \int_{\mathrm{d}^*\mathscr{S}} (\mathbf{L}\mathbf{v}_{\mathscr{S}})\,d\ell.$$

The case when the vector field $\mathbf{f}$ on $\mathscr{S}$ is the normal $\mathbf{v}$ deserves special attention. The tensor field defined on $\mathscr{S}$ by

$$\mathbf{L}_{\mathscr{S}} := \nabla_s\mathbf{v}$$

is important. It turns out that $\mathbf{L}_{\mathscr{S}}(p)$ is a symmetric tensor for all $p\in\mathscr{S}$; $(\mathscr{T}_{\mathscr{S}}(p))^{\perp}$ is its proper space with eigenvalue 0; the other eigenvalues of $\mathbf{L}_{\mathscr{S}}(p)$ are called **principal curvatures** of $\mathscr{S}$ at p. The restriction to $\mathscr{T}_{\mathscr{S}}(p)$ of the bilinear form represented by $\mathbf{L}_{\mathscr{S}}(p)$ is called the **second fundamental form** of $\mathscr{S}$ at p. A scalar field is defined on $\mathscr{S}$ by

$$h_{\mathscr{S}} := \tfrac{1}{2}\operatorname{tr}\mathbf{L}_{\mathscr{S}};$$

it gives the **mean curvature** at each point of $\mathscr{S}$. Clearly,

$$2h_{\mathscr{S}} = \operatorname{div}_s \mathbf{v}.$$

The product of the two principal curvatures of $\mathscr{S}$ at a point p is called the **Gaussian curvature** of $\mathscr{S}$ at p and it is denoted by $k_{\mathscr{S}}(p)$.

Let $\mathbf{u}$ be a vector field on $\bar{\mathscr{S}}$ of class C^1; if $\mathbf{u}_s$ is the tangential field defined on $\mathscr{S}$ by

$$\mathbf{u}_s(p) := \mathbf{E}_{\mathscr{S}}(p)\mathbf{u}(p) \quad \text{for all } p \in \mathscr{S},$$

we can write $\mathbf{u}$ in the form

$$\mathbf{u} = \mathbf{u}_s + (\mathbf{u}\cdot\mathbf{v})\mathbf{v}.$$

Thus, by (2.88) and (2.89), we arrive at

$$\operatorname{div}_s \mathbf{u} = \operatorname{div}_s \mathbf{u}_s + 2(\mathbf{u}\cdot\mathbf{v})h_{\mathscr{S}},$$

whence, by the **Surface-Divergence Theorem**, we have

$$\int_{\mathscr{S}} (\operatorname{div}_s \mathbf{u} - 2(\mathbf{u}\cdot\mathbf{v})h_{\mathscr{S}})\,da = \int_{\mathrm{d}^*\mathscr{S}} (\mathbf{u}_s\cdot\mathbf{v}_{\mathscr{S}})\,d\ell. \tag{2.90}$$

2.4 Isotropic tensors

Intuitively, a material property of a body is isotropic when it is not affected by any rotation of the body. To make this idea more precise, suppose that at a point of the body the material response $\mathbf{r}$ to an external action $\mathbf{a}$ is given by

$$\mathbf{r} = \mathbf{L}\mathbf{a}, \tag{2.91}$$

where $\mathbf{r}$ and $\mathbf{a}$ are both taken as vectors in $\mathscr{V}$ and $\mathbf{L} \in L(\mathscr{V})$. Continuum physics offers many examples of linear constitutive relations like (2.91). For instance, it is customary in classical electrostatics to take $\mathbf{r}$ and $\mathbf{a}$ in (2.91) as the electric displacement $\mathbf{d}$ and the electric field $\mathbf{e}$, respectively: in such a case $\mathbf{L}$ is the **dielectric tensor**. Another example is the linear constitutive relation between the magnetization $\mathbf{m}$ and the magnetic field $\mathbf{h}$:

$$\mathbf{m} = \mathbf{L}\mathbf{h},$$

where $\mathbf{L}$ is now to be interpreted as the **magnetic susceptibility tensor**.

The reader may have noted that our notation here departs from tradition: it is indeed a well established custom to denote by capital letters the electric and magnetic quantities recalled above. Such a usage, however, would clash with that followed throughout this book of employing capital letters only for tensors. Since this book is not primarily concerned with electromagnetism, I have taken the liberty of employing on occasion unusual notations for electric and magnetic quantities.

Let $\mathbf{Q} \in SO(\mathscr{V})$ be any rotation. We define

$$\mathbf{a}^* := \mathbf{Q}\mathbf{a} \quad \text{and} \quad \mathbf{r}^* := \mathbf{Q}\mathbf{r}.$$

The tensor $\mathbf{L}$ in (2.91) represents an **isotropic** material property of the body if $\mathbf{r}^*$ is still related to $\mathbf{a}^*$ through (2.91), that is if

$$\mathbf{Q}\mathbf{r} = \mathbf{L}\mathbf{Q}\mathbf{a} \quad \text{for all } \mathbf{Q} \in SO(\mathscr{V}), \tag{2.92}$$

whenever (2.91) holds. Inserting (2.91) into (2.92), since $\mathbf{a}$ is arbitrary, we arrive at

$$\mathbf{Q}\mathbf{L}\mathbf{Q}^T = \mathbf{L} \quad \text{for all } \mathbf{Q} \in SO(\mathscr{V}). \tag{2.93}$$

Definition 2.13. We say that the tensor $\mathbf{L}$ is **isotropic** whenever it satisfies (2.93).

Generally, the definition of isotropic tensor differs from this in that all members of $O(\mathscr{V})$ are allowed in (2.93) (cf. *e.g.* Section 11 of Truesdell and Noll (1965), who call the tensors that satisfy (2.93) **hemitropic**). Here, when cholesteric liquid crystals are at issue, improper orthogonal tensors should be treated with care: they reverse the orientation of $\mathscr{V}$ and also the chirality of the molecules.

We shall employ the following lemma to represent all isotropic tensors.

Lemma 2.17. A tensor $\mathbf{L} \in L(\mathscr{V})$ commutes with all skew tensors of $L(\mathscr{V})$ if, and only if

$$\mathbf{L} = \alpha \mathbf{I}, \tag{2.94}$$

where α is a scalar.

Proof. Clearly each multiple of the identity commutes with any tensor, be it skew or not. Thus, to prove the lemma it will suffice to show that every tensor that commutes with all skew tensors is of the form (2.94).

Let $\mathbf{L} \in L(\mathscr{V})$ be symmetric. Suppose that

$$\mathbf{L}\mathbf{W} = \mathbf{W}\mathbf{L} \tag{2.95}$$

for all skew tensors $\mathbf{W}$. If $\mathbf{e}$ is a eigenvector of $\mathbf{L}$ with eigenvalue λ, then it follows from (2.95) that

$$\mathbf{L}(\mathbf{W}\mathbf{e}) = \lambda\mathbf{W}\mathbf{e}. \tag{2.96}$$

Thus, $\mathbf{W}\mathbf{e}$ is also, for all skew tensors $\mathbf{W}$, an eigenvector of $\mathbf{L}$ with eigenvalue λ. Let $\mathbf{e}_1$ and $\mathbf{e}_2$ be the eigenvectors of $\mathbf{L}$ orthogonal to $\mathbf{e}$. If we first take $\mathbf{W}$ as the skew tensor with axial vector $\mathbf{e}_1$ and then as the skew tensor with axial vector $\mathbf{e}_2$, we see that all the eigenvectors of $\mathbf{L}$ have λ as eigenvalue. This shows that $\mathbf{L}$ is a multiple of the identity.

Now let $\mathbf{L}$ be a skew tensor that commutes with another skew tensor $\mathbf{W}$, different from zero. If $\mathbf{k}$ is the axial vector of $\mathbf{L}$ and $\mathbf{w}$ that of $\mathbf{W}$, we have from (2.95) that

$$\mathbf{k} \wedge (\mathbf{w} \wedge \mathbf{v}) = \mathbf{w} \wedge (\mathbf{k} \wedge \mathbf{v}) \quad \text{for all } \mathbf{v} \in \mathscr{V},$$

that is, by (2.27),

$$(\mathbf{k}\cdot\mathbf{v})\mathbf{w} = (\mathbf{w}\cdot\mathbf{v})\mathbf{k}, \tag{2.97}$$

which implies that either $\mathbf{k}$ is parallel to $\mathbf{w}$ or $\mathbf{k}$ is the null vector. Thus, a skew tensor commutes with *all* skew tensors only if it vanishes.

If $\mathbf{L}$ is neither symmetric nor skew, we can write it as $\mathbf{L} = \mathbf{L}_1 + \mathbf{L}_2$, where $\mathbf{L}_1$ is symmetric and $\mathbf{L}_2$ is skew. Since when $\mathbf{W}$ is skew the tensor

$$\mathbf{L}_i\mathbf{W} - \mathbf{W}\mathbf{L}_i$$

is symmetric for $i = 1$, and skew for $i = 2$, we see from (2.95) that $\mathbf{L}$ commutes with all skew tensors whenever both $\mathbf{L}_1$ and $\mathbf{L}_2$ do. This completes the proof of the lemma.

The role of the lemma just proved is made clear by the following theorem.

Theorem 2.18. A tensor $\mathbf{L} \in L(\mathscr{V})$ is isotropic if, and only if, it commutes with all skew tensors of $L(\mathscr{V})$.

Proof. It is clear from the lemma above that a tensor $\mathbf{L} \in L(\mathscr{V})$ which commutes with all skew tensors of $L(\mathscr{V})$ satisfies (2.93), since it must be a multiple of the identity.

Let $\mathbf{L}$ satisfy (2.93). By Theorems 2.9 and 2.10 of Section 2.2.6, we can write $\mathbf{Q}$ in (2.93) as

$$\mathbf{Q} = \mathbf{I} + \sin\vartheta\mathbf{W} + (1 - \cos\vartheta)\mathbf{W}^2, \tag{2.98}$$

where $\vartheta \in [0, 2\pi[$ and $\mathbf{W}$ is any skew tensor. Let $\mathbf{W}$ be given. Inserting (2.98) into (2.93), we get an equation whose left-hand side is a smooth function of ϑ. Evaluating its first derivative at $\vartheta = 0$, we arrive at

$$\mathbf{WL} - \mathbf{LW} = \mathbf{0}.$$

Since $\mathbf{W}$ can be chosen arbitrarily among all skew tensors, the proof of the theorem is complete.

Combining the lemma and the theorem above, we conclude that all isotropic tensors are proportional to the identity.[7] Thus, for isotropic materials

$$\mathbf{d} = \varepsilon \mathbf{e} \quad \text{and} \quad \mathbf{m} = \chi \mathbf{h}, \tag{2.99}$$

where ε and χ are, respectively, the **dielectric constant** and the **magnetic susceptibility**. The reader will see in the next section how these constitutive relations change for materials, such as liquid crystals, which exhibit a preferred direction which destroys isotropy.

Remark. Had we allowed all members of the full orthogonal group in (2.93), the theorem above would still have remained true. A requirement of invariance under the members of $SO(\mathscr{V})$ should *a priori* be less restrictive than the same requirement under all members of $O(\mathscr{V})$, since the latter group is richer in tensors than the former. Here, however, the tensors that are isotropic according to the above definition, being multiples of the identity, happen to obey (2.93) for all $\mathbf{Q} \in O(\mathscr{V})$ also.

When in this book we seek explicit formulae which represent the effect on tensors, vectors or scalars of various invariance requirements like (2.93), usually we first determine the formula in question as if the requirement were valid only for a subgroup of $SO(\mathscr{V})$ and then we obtain the formula suitable for the whole of $SO(\mathscr{V})$ (or for the whole of $O(\mathscr{V})$, when the need arises) by removing the terms that would not comply with the invariance under the members of the larger group.

In this section we have seen an example of this method, in which extending the invariance requirement from a group to a larger one does not affect the representation formula. In the proof of Theorem 2.28 in Section 2.6 we shall see another example of the same sort, less trivial than this.

---

[7] This conclusion is valid also when the dimension of the underlying space $\mathscr{V}$ is other than 3, but not 2.

2.5 Transversely isotropic tensors

In this section we introduce a type of invariance which is suited for liquid crystals. It is less restrictive than isotropy and it is enjoyed by all materials which exhibit a preferred direction: it is called **transverse isotropy**.

Definition 2.14. Let $\mathbf{n}$ be any given element of $\mathbb{S}^2$; we define the following collections of tensors:

$$L(\mathbf{n},\mathscr{V}) := \{\mathbf{L}\in L(\mathscr{V})\,|\,\mathbf{L}^T\mathbf{n}=\mathbf{0}\},$$
$$L^2(\mathbf{n},\mathscr{V}) := \{\mathbb{L}: L(\mathbf{n},\mathscr{V})\rightarrow L(\mathbf{n},\mathscr{V})\,|\,\mathbb{L}\quad\text{is linear}\}.$$

We also recall from Section 2.2 the definition of $SO(\mathbf{n},\mathscr{V})$:

$$SO(\mathbf{n},\mathscr{V}) := \{\mathbf{Q}\in SO(\mathscr{V})\,|\,\mathbf{Q}\mathbf{n}=\mathbf{n}\}.$$

Remark 1. $L(\mathbf{n},\mathscr{V})$ is a subspace of $L(\mathscr{V})$. The inner product defined in $L(\mathscr{V})$ by (2.18) induces an inner product in $L(\mathbf{n},\mathscr{V})$. A similar statement applies to the spaces $L^2(\mathscr{V})$ and $L^2(\mathbf{n},\mathscr{V})$.

Throughout this section we take $\mathbf{n}$ as a given element of $\mathbb{S}^2$.

We let $\mathbf{P}(\mathbf{n})$ and $\mathbf{W}(\mathbf{n})$ be the tensors of $L(\mathbf{n},\mathscr{V})$ defined, respectively, by

$$\mathbf{P}(\mathbf{n}) := \mathbf{I}-\mathbf{n}\otimes\mathbf{n}, \tag{2.100}$$

where $\mathbf{I}$ is the identity tensor, and

$$\mathbf{W}(\mathbf{n})\mathbf{v} := \mathbf{n}\wedge\mathbf{v}\quad\text{for all } \mathbf{v}\in\mathscr{V}; \tag{2.101}$$

(*cf.* equations (2.24) and (2.47) above).

It follows from (2.100) that

$$\mathbf{P}(\mathbf{Q}\mathbf{n})=\mathbf{Q}\mathbf{P}(\mathbf{n})\mathbf{Q}^T\quad\text{for all } \mathbf{n}\in\mathbb{S}^2 \text{ and } \mathbf{Q}\in O(\mathscr{V}). \tag{2.102}$$

On the other hand, $\mathbf{W}(\mathbf{n})$ is the skew tensor associated with $\mathbf{n}$ and obeys a transformation rule slightly different from (2.102) when $\mathbf{n}$ is changed into $\mathbf{Q}\mathbf{n}$ with $\mathbf{Q}\in O(\mathscr{V})$. We now deduce such a rule. It follows from (2.101) and (2.102) that

$$\mathbf{W}(\mathbf{Q}\mathbf{n})\mathbf{Q}\mathbf{v}=\mathbf{Q}\mathbf{n}\wedge\mathbf{Q}\mathbf{v}=\mathbf{Q}^*(\mathbf{n}\wedge\mathbf{v})$$

for all $\mathbf{n}\in\mathbb{S}^2, \mathbf{v}\in\mathscr{V}$ and $\mathbf{Q}\in O(\mathscr{V})$, where (*cf.* (2.45))

$$\mathbf{Q}^* := (\det\mathbf{Q})\mathbf{Q}.$$

Thus, for given $\mathbf{n}\in\mathbb{S}^2$ and $\mathbf{Q}\in O(\mathscr{V})$, applying (2.101) again, we arrive at

$$\mathbf{W}(\mathbf{Qn})\mathbf{Qv} = (\det \mathbf{Q})\mathbf{QW}(\mathbf{n})\mathbf{v} \quad \text{for all } \mathbf{v}\in\mathscr{V},$$

whence

$$\mathbf{W}(\mathbf{Qn})\mathbf{Q} = (\det \mathbf{Q})\mathbf{QW}(\mathbf{n}).$$

Multiplying both sides of this equation by $\mathbf{Q}^T$ on the right, we get

$$\mathbf{W}(\mathbf{Qn}) = (\det \mathbf{Q})\mathbf{QW}(\mathbf{n})\mathbf{Q}^T \quad \text{for all } \mathbf{n}\in\mathbb{S}^2 \text{ and } \mathbf{Q}\in O(\mathscr{V}). \tag{2.103}$$

Of course, (2.102) and (2.103) have precisely the same form for all $\mathbf{Q}\in SO(\mathscr{V})$.

Following the pattern set up in the preceding section, we make two further definitions.

Definition 2.15. A tensor $\mathbf{K}\in L(\mathscr{V})$ is **transversely isotropic** about $\mathbf{n}$ if it satisfies

$$\mathbf{QKQ}^T = \mathbf{K} \quad \text{for all } \mathbf{Q}\in SO(\mathbf{n},\mathscr{V}). \tag{2.104}$$

Definition 2.16. A tensor $\mathbb{K}\in L^2(\mathbf{n},\mathscr{V})$ is **transversely isotropic** about $\mathbf{n}$ if it satisfies

$$\mathbb{K}[\mathbf{QNQ}^T] = \mathbf{Q}\mathbb{K}[\mathbf{N}]\mathbf{Q}^T \quad \text{for all } \mathbf{N}\in L(\mathbf{n},\mathscr{V}) \text{ and } \mathbf{Q}\in SO(\mathbf{n}\mathscr{V}). \tag{2.105}$$

Remark 2. The reader should note that Definition 2.16 differs from the classical definition of transversely isotropic elastic tensors, since these latter operate on a subspace of $L(\mathscr{V})$ different from $L(\mathbf{n},\mathscr{V})$. Here the analysis carried out in Podio-Guidugli and Virga (1987) is to be slightly recast.

We now seek in $L(\mathscr{V})$ the tensors that satisfy (2.104) and in $L^2(\mathscr{V})$ the symmetric tensors that satisfy (2.105).

Lemma 2.19. A tensor $\mathbf{L}\in L(\mathscr{V})$ commutes with $\mathbf{W}(\mathbf{n})$ if, and only if, there are three scalars α_1, α_2 and α_3 such that

$$\mathbf{L} = \alpha_1 \mathbf{n}\otimes\mathbf{n} + \alpha_2\mathbf{P}(\mathbf{n}) + \alpha_3\mathbf{W}(\mathbf{n}). \tag{2.106}$$

Proof. It is easily seen that every tensor of the form (2.106) commutes with $\mathbf{W}(\mathbf{n})$ for all values of the scalars α_1, α_2 and α_3.

We suppose now that $\mathbf{L}$ commutes with $\mathbf{W}(\mathbf{n})$:

$$\mathbf{W}(\mathbf{n})\mathbf{L} - \mathbf{LW}(\mathbf{n}) = \mathbf{0}. \tag{2.107}$$

In addition, let $\mathbf{L}$ be symmetric. By the Spectral Theorem there is a basis $(\mathbf{e}_1, \mathbf{e}_2, \mathbf{e}_3)$ of $\mathscr{V}$ whose members are eigenvectors of $\mathbf{L}$; we denote by λ_1, λ_2 and λ_3 the corresponding eigenvalues. It follows from (2.107) that

$$\mathbf{L}(\mathbf{W}(\mathbf{n})\mathbf{e}_i) = \lambda_i \mathbf{W}(\mathbf{n})\mathbf{e}_i \quad \text{for } i = 1, 2, 3.$$

Thus, for any given i, the vector

$$\mathbf{w}_i := \mathbf{W}(\mathbf{n})\mathbf{e}_i = \mathbf{n} \wedge \mathbf{e}_i$$

either vanishes or belongs to the same proper space of $\mathbf{L}$ as $\mathbf{e}_i$. We conclude that all symmetric tensors $\mathbf{L}$ that commute with $\mathbf{W}(\mathbf{n})$ must be such that

$$\mathbf{L} = \alpha_1 \mathbf{n} \otimes \mathbf{n} + \alpha_2 \mathbf{P}(\mathbf{n}). \tag{2.108}$$

We now let $\mathbf{L}$ be skew. Hence, there is a vector $\mathbf{k} \in \mathscr{V}$ such that

$$\mathbf{L}\mathbf{v} = \mathbf{k} \wedge \mathbf{v} \quad \text{for all } \mathbf{v} \in \mathscr{V}.$$

It follows from (2.107) that

$$\mathbf{n} \wedge (\mathbf{k} \wedge \mathbf{v}) = \mathbf{k} \wedge (\mathbf{n} \wedge \mathbf{v}) \quad \text{for all } \mathbf{v} \in \mathscr{V}. \tag{2.109}$$

Setting $\mathbf{v} = \mathbf{n}$ in (2.109), we get

$$\mathbf{k} = (\mathbf{k} \cdot \mathbf{n})\mathbf{n},$$

and so there is a scalar α_3 such that

$$\mathbf{L} = \alpha_3 \mathbf{W}(\mathbf{n}). \tag{2.110}$$

If $\mathbf{L}$ is neither symmetric nor skew, it can be written as $\mathbf{L} = \mathbf{S} + \mathbf{W}$, where $\mathbf{S}$ is symmetric and $\mathbf{W}$ is skew. It is easily seen that

$$\mathbf{W}(\mathbf{n})\mathbf{S} - \mathbf{S}\mathbf{W}(\mathbf{n})$$

is the symmetric part of

$$\mathbf{W}(\mathbf{n})\mathbf{L} - \mathbf{L}\mathbf{W}(\mathbf{n}),$$

while

$$\mathbf{W}(\mathbf{n})\mathbf{W} - \mathbf{W}\mathbf{W}(\mathbf{n})$$

is its skew part. Thus, (2.107) splits into the two equations that we have already solved.

The lemma we have just proved will play a central role in the following theorems.

Theorem 2.20. A tensor $\mathbf{K}\in L(\mathscr{V})$ is transversely isotropic about $\mathbf{n}$ if, and only if, it commutes with $\mathbf{W}(\mathbf{n})$.

Proof. First, recall from Section 2.2 that all members of $SO(\mathbf{n},\mathscr{V})$ are represented by the formula

$$\mathbf{Q}(\vartheta)=\mathbf{I}+\sin\vartheta\,\mathbf{W}(\mathbf{n})+(1-\cos\vartheta)\mathbf{W}^2(\mathbf{n}), \tag{2.111}$$

where ϑ ranges in $[0,2\pi[$ (*cf.* equation (2.46)). Hence, if $\mathbf{K}\in L(\mathscr{V})$ commutes with $\mathbf{W}(\mathbf{n})$, it also commutes with $\mathbf{Q}(\vartheta)$ for all $\vartheta\in[0,2\pi[$, and so (2.104) is satisfied since

$$\mathbf{Q}(\vartheta)\mathbf{K}\mathbf{Q}^T(\vartheta)=\mathbf{K}\mathbf{Q}(\vartheta)\mathbf{Q}^T(\vartheta)=\mathbf{K}\quad\text{for all }\vartheta\in[0,2\pi[.$$

Suppose, conversely, that $\mathbf{K}\in L(\mathscr{V})$ is transversely isotropic about $\mathbf{n}$. Inserting (2.111) into the left-hand side of (2.104), we get a smooth tensor-valued function of ϑ. Equation (2.104) requires such a function to be constant. It is easily seen that its first derivative at $\vartheta=0$ vanishes whenever $\mathbf{K}$ commutes with $\mathbf{W}(\mathbf{n})$, and so the proof of the theorem is complete.

Combining this theorem and the lemma above, we learn that (2.106) is the representation formula for all tensors of $L(\mathscr{V})$ transversely isotropic about $\mathbf{n}$. To see an application of this formula we consider a material with a preferred direction represented by $\mathbf{n}\in\mathbb{S}^2$. If such a material is subject to an electric field $\mathbf{e}$, it experiences an electric displacement $\mathbf{d}$ which is related to $\mathbf{e}$ as in Section 2.4:

$$\mathbf{d}=\mathbf{L}\mathbf{e}, \tag{2.112}$$

where the dielectric tensor $\mathbf{L}$ is now transversely isotropic about $\mathbf{n}$. Taking $\mathbf{L}$ to be symmetric, from (2.106) and (2.112) we arrive at

$$\mathbf{d}=\varepsilon_{\parallel}(\mathbf{e}\cdot\mathbf{n})\mathbf{n}+\varepsilon_{\perp}(\mathbf{e}-(\mathbf{e}\cdot\mathbf{n})\mathbf{n}), \tag{2.113}$$

where $\varepsilon_{\parallel}$ and $\varepsilon_{\perp}$ have taken the place of α_1 and α_2, respectively. When $\mathbf{e}$ is parallel to $\mathbf{n}$, equation (2.113) reduces to $(2.99)_1$ with $\varepsilon_{\parallel}$ instead of ε. Similarly, when $\mathbf{e}$ is orthogonal to $\mathbf{n}$, (2.113) reduces to $(2.99)_1$ with $\varepsilon_{\perp}$ instead of ε. Equation (2.113) is commonly written as

$$\mathbf{d}=\varepsilon_{\perp}\mathbf{e}+\varepsilon_a(\mathbf{e}\cdot\mathbf{n})\mathbf{n},$$

where $\varepsilon_a:=\varepsilon_{\parallel}-\varepsilon_{\perp}$ is the **dielectric anisotropy**. Of course, when ε_a vanishes the material becomes fully isotropic. A similar formula holds for the magnetization (*cf.* equation $(2.99)_2$):

$$\mathbf{m}=\chi_{\perp}\mathbf{h}+\chi_a(\mathbf{h}\cdot\mathbf{n})\mathbf{n},$$

where $\chi_a := \chi_\parallel - \chi_\perp$ is the **diamagnetic anisotropy** and the susceptibilities $\chi_\parallel$ and $\chi_\perp$ play precisely the same role as the dielectric constants $\varepsilon_\parallel$ and $\varepsilon_\perp$.

We now prove a representation formula for symmetric tensors of $L^2(\mathbf{n}, \mathscr{V})$ that are transversely isotropic about $\mathbf{n}$. Such a formula will be applied in the following section and will ultimately lead in Section 3.2 of Chapter 3 to the classical Frank's formula for the free energy density of both nematic and cholesteric liquid crystals. As above, we first need a preparatory lemma.

Lemma 2.21. Let $\mathbf{n} \in \mathbb{S}^2$ be given. A symmetric tensor $\mathbb{K} \in L^2(\mathbf{n}, \mathscr{V})$ commutes with the tensor $\mathbb{W} \in L^2(\mathbf{n}, \mathscr{V})$ defined by

$$\mathbb{W}[\mathbf{N}] := \mathbf{W}(\mathbf{n})\mathbf{N} - \mathbf{N}\mathbf{W}(\mathbf{n}) \quad \text{for all } \mathbf{N} \in L(\mathbf{n}, \mathscr{V}) \tag{2.114}$$

if, and only if, there are five scalars β_i, $i = 1, \dots, 5$, such that

$$\begin{aligned} \mathbb{K} = {} & \beta_1 \mathbf{W}(\mathbf{n}) \otimes \mathbf{W}(\mathbf{n}) + \beta_2 \mathbf{P}(\mathbf{n}) \otimes \mathbf{P}(\mathbf{n}) \\ & + \beta_3 (\mathbf{P}(\mathbf{n}) \otimes \mathbf{W}(\mathbf{n}) + \mathbf{W}(\mathbf{n}) \otimes \mathbf{P}(\mathbf{n})) + \beta_4 \mathbb{I}(\mathbf{n}) + \beta_5 \mathbb{P}(\mathbf{n}), \end{aligned} \tag{2.115}$$

where $\mathbb{I}(\mathbf{n})$ denotes the identity in $L^2(\mathbf{n}, \mathscr{V})$ and $\mathbb{P}(\mathbf{n})$ is the tensor of $L^2(\mathbf{n}, \mathscr{V})$ defined by

$$\mathbb{P}(\mathbf{n})[\mathbf{N}] := \mathbf{P}(\mathbf{n})\mathbf{N}\mathbf{P}(\mathbf{n}) \quad \text{for all } \mathbf{N} \in L(\mathbf{n}, \mathscr{V}). \tag{2.116}$$

Proof. We aim to find all members of the class

$$\mathscr{K} := \{\mathbb{K} \in L^2(\mathbf{n}, \mathscr{V}) \mid \mathbb{K}^T = \mathbb{K}, \mathbb{K}\mathbb{W} = \mathbb{W}\mathbb{K}\}.$$

It easily follows from (2.114) that

$$\mathbf{A} \cdot \mathbb{W}[\mathbf{B}] = -\mathbb{W}[\mathbf{B}] \cdot \mathbf{A} \quad \text{for all } \mathbf{A}, \mathbf{B} \in L(\mathbf{n}, \mathscr{V}),$$

and so $\mathbb{W}$ is skew.

Let $\mathscr{W}$ be the null space of $\mathbb{W}$. By (2.114), $\mathscr{W}$ consists of all tensors of $L(\mathbf{n}, \mathscr{V})$ that commute with $\mathbf{W}(\mathbf{n})$. Lemma 2.19 then implies that

$$\mathscr{W} = \operatorname{span}\{\mathbf{W}(\mathbf{n}), \mathbf{P}(\mathbf{n})\}.$$

Both $\mathscr{W}$ and $\mathscr{W}^\perp$, the orthogonal complement of $\mathscr{W}$ in $L(\mathbf{n}, \mathscr{V})$, are invariant under the tensors of $\mathscr{K}$, because these latter are symmetric. Thus, $\mathscr{K}$ is included in the class of all symmetric tensors of $L^2(\mathbf{n}, \mathscr{V})$ under which both $\mathscr{W}$ and $\mathscr{W}^\perp$ are invariant.

A symmetric tensor $\mathbb{K} \in L^2(\mathbf{n}, \mathscr{V})$ maps $\mathscr{W}$ into itself if, and only if, there are three scalars, β_1, β_2 and β_3, such that

$$\mathbb{K} = \beta_1 \mathbf{W}(\mathbf{n}) \otimes \mathbf{W}(\mathbf{n}) + \beta_2 \mathbf{P}(\mathbf{n}) \otimes \mathbf{P}(\mathbf{n}) + \beta_3(\mathbf{P}(\mathbf{n}) \otimes \mathbf{W}(\mathbf{n}) + \mathbf{W}(\mathbf{n}) \otimes \mathbf{P}(\mathbf{n})). \quad (2.117)$$

It is easily seen that all tensors of this form belong indeed to $\mathscr{K}$.

We now consider the symmetric tensors of $L^2(\mathbf{n}, \mathscr{V})$ under which $\mathscr{W}^\perp$ is invariant. Let $\mathbf{e}_1$ and $\mathbf{e}_2$ be unit vectors orthogonal to $\mathbf{n}$ such that

$$\mathbf{W}(\mathbf{n})\mathbf{e}_1 = \mathbf{e}_2 \quad \text{and} \quad \mathbf{W}(\mathbf{n})\mathbf{e}_2 = -\mathbf{e}_1. \quad (2.118)$$

The list of tensors $(\mathbf{W}(\mathbf{n}), \mathbf{P}(\mathbf{n}), \mathbf{E}_1, \mathbf{E}_2, \mathbf{E}_3, \mathbf{E}_4)$, where

$$\mathbf{E}_1 := \mathbf{e}_1 \otimes \mathbf{e}_2 + \mathbf{e}_2 \otimes \mathbf{e}_1, \mathbf{E}_2 := \mathbf{e}_1 \otimes \mathbf{e}_1 - \mathbf{e}_2 \otimes \mathbf{e}_2,$$
$$\mathbf{E}_3 := \mathbf{e}_1 \otimes \mathbf{n}, \mathbf{E}_4 := \mathbf{e}_2 \otimes \mathbf{n}, \quad (2.119)$$

is a basis of $L(\mathbf{n}, \mathscr{V})$, and so

$$\mathscr{W}^\perp = \operatorname{span}\{\mathbf{E}_1, \mathbf{E}_2, \mathbf{E}_3, \mathbf{E}_4\}.$$

All symmetric tensors of $L^2(\mathbf{n}, \mathscr{V})$ that map $\mathscr{W}^\perp$ into itself are represented by the formula

$$\mathbb{K} = \sum_{i,j=1}^{4} \gamma_{ij} \mathbf{E}_i \otimes \mathbf{E}_j \quad \text{with} \quad \gamma_{ij} = \gamma_{ji}. \quad (2.120)$$

To see which tensors of this form actually belong to $\mathscr{K}$ we first note that

$$\mathbb{W}[\mathbf{E}_1] = -2\mathbf{E}_2, \mathbb{W}[\mathbf{E}_2] = 2\mathbf{E}_1, \mathbb{W}[\mathbf{E}_3] = \mathbf{E}_4, \mathbb{W}[\mathbf{E}_4] = -\mathbf{E}_3, \quad (2.121)$$

and then we compute both $\mathbb{K}\mathbb{W}$ and $\mathbb{W}\mathbb{K}$, employing (2.120), (2.121) and the rule

$$\mathbb{W}(\mathbf{A} \otimes \mathbf{B}) = \mathbb{W}[\mathbf{A}] \otimes \mathbf{B},$$
$$(\mathbf{A} \otimes \mathbf{B})\mathbb{W} = -\mathbf{A} \otimes \mathbb{W}[\mathbf{B}] \quad \text{for all } \mathbf{A}, \mathbf{B} \in L(\mathbf{n}, \mathscr{V}).$$

It is a tedious but easy task to check that $\mathbb{K}\mathbb{W} = \mathbb{W}\mathbb{K}$ whenever

$$\mathbb{K} = \gamma_1 \mathbb{K}_1 + \gamma_2 \mathbb{K}_2, \quad (2.122)$$

where γ_1 and γ_2 are arbitrary scalars and

$$\mathbb{K}_1 := \mathbf{E}_1 \otimes \mathbf{E}_1 + \mathbf{E}_2 \otimes \mathbf{E}_2, \quad \mathbb{K}_2 := \mathbf{E}_3 \otimes \mathbf{E}_3 + \mathbf{E}_4 \otimes \mathbf{E}_4.$$

Combining (2.117) and (2.122), we obtain that

$$\mathscr{K} = \text{span}\{\mathbf{W(n)} \otimes \mathbf{W(n)}, \mathbf{P(n)} \otimes \mathbf{P(n)}, \mathbf{P(n)} \otimes \mathbf{W(n)} + \mathbf{W(n)} \otimes \mathbf{P(n)}, \mathbb{K}_1, \mathbb{K}_2\}.$$

To complete the proof of the lemma it remains to express both $\mathbb{K}_1$ and $\mathbb{K}_2$ in terms of $\mathbf{n}$ only. Since

$$|\mathbf{W(n)}| = |\mathbf{P(n)}| = |\mathbf{E}_1| = |\mathbf{E}_2| = 2 \quad \text{and} \quad |\mathbf{E}_3| = |\mathbf{E}_4| = 1,$$

we have that

$$\mathbb{I}(\mathbf{n}) = \tfrac{1}{2}(\mathbf{W(n)} \otimes \mathbf{W(n)} + \mathbf{P(n)} \otimes \mathbf{P(n)} + \mathbb{K}_1) + \mathbb{K}_2,$$

where $\mathbb{I}(\mathbf{n})$ is the identity in $L^2(\mathbf{n}, \mathscr{V})$. On the other hand, since

$$\mathbf{P(n)W(n)} = \mathbf{W(n)P(n)} = \mathbf{W(n)},$$

making use also of (2.118) and (2.119), we easily see that

$$\mathbb{P}(\mathbf{n}) = \tfrac{1}{2}(\mathbf{W(n)} \otimes \mathbf{W(n)} + \mathbf{P(n)} \otimes \mathbf{P(n)} + \mathbb{K}_1),$$

where $\mathbb{P}(\mathbf{n})$ is defined as in (2.116).

Theorem 2.22. Let $\mathbf{n} \in \mathbb{S}^2$ be given. A symmetric tensor of $L^2(\mathbf{n}, \mathscr{V})$ is transversely isotropic about $\mathbf{n}$ if, and only if, it commutes with the tensor $\mathbb{W} \in L^2(\mathbf{n}, \mathscr{V})$ defined by (2.114)

Proof. Let $\mathbb{K}$ be a symmetric tensor of $L^2(\mathbf{n}, \mathscr{V})$. If $\mathbb{K}$ commutes with $\mathbb{W}$, then by Lemma 2.21 it can be written as in (2.115). All tensors of this form are transversely isotropic about $\mathbf{n}$ in the sense of Definition 2.16, as the reader can easily verify, recalling that both $\mathbf{W(n)}$ and $\mathbf{P(n)}$ commute with all $\mathbf{Q} \in SO(\mathbf{n}, \mathscr{V})$.

Suppose now that $\mathbb{K}$ satisfies (2.105). Inserting (2.111) into (2.105) and evaluating at $\vartheta = 0$ the derivative with respect to ϑ of both sides of the equation thus obtained, we arrive at

$$(\mathbb{K}\mathbb{W})[\mathbf{N}] = (\mathbb{W}\mathbb{K})[\mathbf{N}] \quad \text{for all } \mathbf{N} \in L(\mathbf{n}, \mathscr{V}),$$

where $\mathbb{W}$ is the same as in (2.114).

By Lemma 2.21 and Theorem 2.22, we conclude that (2.115) represents all symmetric tensors of $L^2(\mathbf{n}, \mathscr{V})$ transversely isotropic about $\mathbf{n}$.

2.6 Isotropic functions

We collect here for future reference the representation formulae of some scalar-valued functions, called **isotropic**, whose values remain unchanged when the rotations of $SO(\mathscr{V})$ operate suitably on their arguments.[8]

To make our development more specific we introduce a notion of congruence in $\mathscr{V}$.

Definition 2.17. Two vectors of $\mathscr{V}$, $\mathbf{u}$ and $\mathbf{v}$, are **congruent** if there is a $\mathbf{Q} \in SO(\mathscr{V})$ such that

$$\mathbf{u} = \mathbf{Q}\mathbf{v}. \tag{2.123}$$

Similarly, two pairs of vectors, $(\mathbf{u}_1, \mathbf{u}_2)$ and $(\mathbf{v}_1, \mathbf{v}_2)$ are **congruent** if there is a $\mathbf{Q} \in SO(\mathscr{V})$ such that

$$\mathbf{u}_1 = \mathbf{Q}\mathbf{v}_1 \quad \text{and} \quad \mathbf{u}_2 = \mathbf{Q}\mathbf{v}_2. \tag{2.124}$$

Remark 1. The congruence thus defined establishes an equivalence among vectors and pairs of vectors. It should also be noted that the null vector of $\mathscr{V}$ is congruent only to itself.

Lemma 2.23. The vectors $\mathbf{u}$ and $\mathbf{v}$ of $\mathscr{V}$ are congruent if, and only if, they have equal length:

$$|\mathbf{u}| = |\mathbf{v}|. \tag{2.125}$$

Proof. Let $\mathbf{u}$ and $\mathbf{v}$ be given in $\mathscr{V}$. If (2.123) holds then clearly (2.125) does also.

If $|\mathbf{u}| = |\mathbf{v}| = 0$ then (2.123) holds for any $\mathbf{Q} \in SO(\mathscr{V})$. Suppose now that $|\mathbf{u}| = |\mathbf{v}| \neq 0$. If $\mathbf{u}$ and $\mathbf{v}$ are parallel then nothing remains to be proved, since (2.123) is satisfied with $\mathbf{Q}$ equal to the identity, when $\mathbf{u}$ and $\mathbf{v}$ share the same orientation, or with $\mathbf{Q}$ equal to a rotation of angle π about a direction orthogonal to both $\mathbf{u}$ and $\mathbf{v}$, when they have opposite orientations. Thus, we suppose that $\mathbf{u}$ and $\mathbf{v}$ are not parallel and let $\mathbf{e}$ be the unit vector defined by

$$\mathbf{e} := \frac{\mathbf{u} \wedge \mathbf{v}}{|\mathbf{u} \wedge \mathbf{v}|}.$$

It is easy to check that $\mathbf{u} = \mathbf{Q}_e\mathbf{v}, \mathbf{Q}_e$ being the element of $SO(\mathbf{e}, \mathscr{V})$ given by

$$\mathbf{Q}_e := \mathbf{I} + \sin \vartheta_e \mathbf{W}(\mathbf{e}) + (1 - \cos \vartheta_e)\mathbf{W}^2(\mathbf{e}),$$

where

$$\vartheta_e := \arccos\left(\frac{\mathbf{u} \cdot \mathbf{v}}{|\mathbf{u}||\mathbf{v}|}\right),$$

and $\mathbf{W}(\mathbf{e})$ is the skew tensor whose axial vector is $\mathbf{e}$.

---

8 Truesdell and Noll (1965) call these functions hemitropic (cf. p. 24 of their article).

Lemma 2.24. The pairs of vectors $(\mathbf{u}_1, \mathbf{u}_2)$ and $(\mathbf{v}_1, \mathbf{v}_2)$ are congruent if, and only if,

$$|\mathbf{u}_1| = |\mathbf{v}_1|, |\mathbf{u}_2| = |\mathbf{v}_2| \quad \text{and} \quad \mathbf{u}_1 \cdot \mathbf{u}_2 = \mathbf{v}_1 \cdot \mathbf{v}_2. \tag{2.126}$$

Proof. Let the vectors $\mathbf{u}_1, \mathbf{v}_1, \mathbf{u}_2$ and $\mathbf{v}_2$ be given in $\mathscr{V}$. If $\mathbf{u}_1 = \mathbf{u}_2 = \mathbf{0}$, then the lemma is an immediate consequence of the definition, since $(\mathbf{0}, \mathbf{0})$ is the only pair of vectors congruent to itself. If one vector of the pair $(\mathbf{u}_1, \mathbf{u}_2)$ vanishes, say $\mathbf{u}_1$, then the lemma follows from Lemma 2.23, since $\mathbf{v}_1$ must also vanish and $\mathbf{v}_2$ must be congruent to $\mathbf{u}_2$. Thus, we suppose that $\mathbf{u}_1$ and $\mathbf{u}_2$ are both different from the null vector.

It is clear from (2.124) that equations (2.126) are satisfied by all pairs of congruent vectors. Conversely, it follows from $(2.126)_{1,2}$ and Lemma 2.23 that there are $\mathbf{Q}_1$ and $\mathbf{Q}_2$ in $SO(\mathscr{V})$ such that

$$\mathbf{u}_1 = \mathbf{Q}_1 \mathbf{v}_1 \quad \text{and} \quad \mathbf{u}_2 = \mathbf{Q}_2 \mathbf{v}_2. \tag{2.127}$$

We set

$$\mathbf{v}_1^* := \mathbf{Q}_2^T \mathbf{Q}_1 \mathbf{v}_1, \tag{2.128}$$

so that the pairs $(\mathbf{u}_1, \mathbf{u}_2)$ and $(\mathbf{v}_1^*, \mathbf{v}_2)$ are congruent. To prove the lemma it remains to be seen that the pairs $(\mathbf{v}_1^*, \mathbf{v}_2)$ and $(\mathbf{v}_1, \mathbf{v}_2)$ are also congruent. It is a consequence of $(2.126)_3$, (2.127) and (2.128) that

$$\mathbf{v}_1^* \cdot \mathbf{v}_2 = \mathbf{v}_1 \cdot \mathbf{v}_2 \quad \text{and} \quad |\mathbf{v}_1^*| = |\mathbf{v}_1|.$$

Thus, there is a rotation about $\mathbf{v}_2$, possibly equal to the identity, which maps $\mathbf{v}_1$ into $\mathbf{v}_1^*$. This shows that $(\mathbf{v}_1^*, \mathbf{v}_2)$ and $(\mathbf{v}_1, \mathbf{v}_2)$ are congruent.

The lemmas above are essentially taken from Section 2 of Wang (1970) (*cf.* also Wang (1971)); I refer the reader to that paper for a systematic treatment of the issues touched on in this section.

Theorem 2.25. A function $\varphi: \mathscr{V} \to \mathbb{R}$ satisfies

$$\varphi(\mathbf{Q}\mathbf{v}) = \varphi(\mathbf{v}) \quad \text{for all } \mathbf{v} \in \mathscr{V} \text{ and } \mathbf{Q} \in SO(\mathscr{V}) \tag{2.129}$$

if, and only if, there is a function $\tilde{\varphi}: \mathbb{R}^+ \to \mathbb{R}$ such that

$$\varphi(\mathbf{v}) = \tilde{\varphi}(|\mathbf{v}|) \quad \text{for all } \mathbf{v} \in \mathscr{V}. \tag{2.130}$$

Proof. Let $\mathbf{v}$ be given in $\mathscr{V}$. Condition (2.129) requires the function φ to take the same value on all vectors congruent to $\mathbf{v}$. Lemma 2.23 characterizes all such vectors as those having the same length as $\mathbf{v}$.

Letting $\mathbf{v}$ have arbitrary length we easily complete the proof of the theorem.

The following theorem closely parallels Theorem 2.25. Its proof, which is omitted here, is entirely based upon Lemma 2.24.

Theorem 2.26. A function $\varphi:\mathscr{V}\times\mathscr{V}\to\mathbb{R}$ satisfies

$$\varphi(\mathbf{Q}\mathbf{v}_1,\mathbf{Q}\mathbf{v}_2)=\varphi(\mathbf{v}_1,\mathbf{v}_2)\quad\text{for all } \mathbf{v}_1,\mathbf{v}_2\in\mathscr{V} \text{ and } \mathbf{Q}\in SO(\mathscr{V}) \tag{2.131}$$

if, and only if, there is a function $\tilde{\varphi}:\mathbb{R}^+\times\mathbb{R}^+\times\mathbb{R}\to\mathbb{R}$ such that

$$\varphi(\mathbf{v}_1,\mathbf{v}_2)=\tilde{\varphi}(|\mathbf{v}_1|,|\mathbf{v}_2|,\mathbf{v}_1\cdot\mathbf{v}_2)\quad\text{for all } \mathbf{v}_1,\mathbf{v}_2\in\mathscr{V}. \tag{2.132}$$

Remark 2. Both formulae (2.130) and (2.132) are still valid if all elements of $O(\mathscr{V})$ are allowed to be in (2.129) and (2.131).

Both Theorems 2.25 and 2.26 are included as special cases in a theorem of Cauchy (1850), whose generality we do not embrace here. The reader is referred to p. 30 of Truesdell and Noll (1965) for a modern statement of this classical result which, unlike Cauchy's, applies also to vector spaces whose dimension is greater than 3. In this section we shall prove two other theorems which, though not as classical as the two above, are especially germane to the subject of this book.

The following theorem is not strictly necessary to our development, but it is proved here because it closely resembles Theorem 2.25 above.

Theorem 2.27. A mapping $\mathbf{f}:\mathscr{V}\to\mathscr{V}$ satisfies

$$\mathbf{f}(\mathbf{Q}\mathbf{v})=\mathbf{Q}\mathbf{f}(\mathbf{v})\quad\text{for all } \mathbf{v}\in\mathscr{V} \text{ and } \mathbf{Q}\in SO(\mathscr{V}) \tag{2.133}$$

if, and only if, there is a function $\tilde{\varphi}:\mathbb{R}^+\to\mathbb{R}$ such that

$$\mathbf{f}(\mathbf{v})=\tilde{\varphi}(|\mathbf{v}|)\mathbf{v}\quad\text{for all } \mathbf{v}\in\mathscr{V}. \tag{2.134}$$

Proof. The function $\mathbf{f}$ given by (2.134) clearly satisfies (2.133). If this condition is valid, we may select $\mathbf{v}$ in $\mathscr{V}\setminus\{\mathbf{0}\}$ and choose $\mathbf{Q}$ arbitrarily in $SO(\mathbf{e},\mathscr{V})$, where $\mathbf{e}:=\mathbf{v}/|\mathbf{v}|$, so that

$$\mathbf{f}(\mathbf{v})=\mathbf{Q}\mathbf{f}(\mathbf{v})\quad\text{for all } \mathbf{Q}\in SO(\mathbf{e},\mathscr{V}). \tag{2.135}$$

Equation (2.135) tells us that $\mathbf{f}(\mathbf{v})$ must be parallel to $\mathbf{v}$, and so letting $\mathbf{v}$ vary in $\mathscr{V}\setminus\{\mathbf{0}\}$, we conclude that there is a function $\varphi:\mathscr{V}\setminus\{\mathbf{0}\}\to\mathbb{R}$ such that

$$\mathbf{f}(\mathbf{v})=\varphi(\mathbf{v})\mathbf{v}\quad\text{for all } \mathbf{v}\in\mathscr{V}\setminus\{\mathbf{0}\}. \tag{2.136}$$

Inserting (2.136) into (2.133) we see that φ must satisfy $\varphi(\mathbf{Q}\mathbf{v})=\varphi(\mathbf{v})$

for all $\mathbf{v}\in\mathscr{V}\setminus\{\mathbf{0}\}$. Thus, by Theorem 2.25, we arrive at (2.134), provided $\mathbf{v}\neq\mathbf{0}$. Finally, (2.133) implies that $\mathbf{f}(\mathbf{0})=\mathbf{0}$. This completes the proof of the lemma, whichever value we assign to $\tilde{\varphi}(0)$.

At first glance the statement of the following theorem might seem unnecessarily complicated; it becomes clearer in the light of equation (2.65), if we interpret the variable $\mathbf{N}$ below as the tensor $\nabla\mathbf{n}$, for some smooth mapping $\mathbf{n}$ of a region in space into $\mathbb{S}^2$. The reader should recall the sets $L(\mathbf{n},\mathscr{V})$ and $L^2(\mathbf{n},\mathscr{V})$ defined at the beginning of Section 2.5.

Theorem 2.28. For any given $\mathbf{n}\in\mathbb{S}^2$ let $\varphi(\mathbf{n},\cdot)$ be the scalar-valued function on $L(\mathbf{n},\mathscr{V})$ defined by

$$\varphi(\mathbf{n},\mathbf{N}):=k(\mathbf{n})+\mathbf{K}(\mathbf{n})\cdot\mathbf{N}+\mathbf{N}\cdot\mathbb{K}(\mathbf{n})[\mathbf{N}], \tag{2.137}$$

where $k(\mathbf{n})$ is a scalar, $\mathbf{K}(\mathbf{n})$ is a tensor in $L(\mathbf{n},\mathscr{V})$, and $\mathbb{K}(\mathbf{n})$ is a symmetric tensor in $L^2(\mathbf{n},\mathscr{V})$. The function φ satisfies

$$\varphi(\mathbf{Qn},\mathbf{QNQ}^T)=\varphi(\mathbf{n},\mathbf{N})\quad\text{for all }\mathbf{Q}\in SO(\mathscr{V}) \tag{2.138}$$

if, and only if,

$$\begin{aligned}\varphi(\mathbf{n},\mathbf{N})=\alpha_0&+\alpha_2\mathbf{P}(\mathbf{n})\cdot\mathbf{N}+\alpha_3\mathbf{W}(\mathbf{n})\cdot\mathbf{N}+\beta_1(\mathbf{W}(\mathbf{n})\cdot\mathbf{N})^2\\&+\beta_2(\mathbf{P}(\mathbf{n})\cdot\mathbf{N})^2+\beta_3(\mathbf{W}(\mathbf{n})\cdot\mathbf{N})(\mathbf{P}(\mathbf{N})\cdot\mathbf{N})\\&+\beta_4\mathbf{N}\cdot\mathbf{N}+\beta_5\mathbf{P}(\mathbf{n})\mathbf{N}\cdot\mathbf{N}\mathbf{P}(\mathbf{n}),\end{aligned} \tag{2.139}$$

where $\alpha_0,\alpha_2,\alpha_3$ and β_i, $i=1,\dots,5$, are scalars and $\mathbf{P}(\mathbf{n})$ and $\mathbf{W}(\mathbf{n})$ are defined as in (2.100) and (2.101).

Proof. The function φ in (2.137) is defined on the domain

$$F(\mathscr{V}):=\{(\mathbf{n},\mathbf{N})\,|\,\mathbf{n}\in\mathbb{S}^2,\quad \mathbf{N}\in L(\mathbf{n},\mathscr{V})\};$$

k is a scalar-valued function on $\mathbb{S}^2$, and $\mathbf{K}$ and $\mathbb{K}$ are mappings of $\mathbb{S}^2$ into $L(\mathscr{V})$ and $L^2(\mathscr{V})$, respectively, such that

$$K(\mathbf{n})\in L(\mathbf{n},\mathscr{V}),\ \mathbb{K}(\mathbf{n})\in L^2(\mathbf{n},\mathscr{V}),\text{ and }\mathbb{K}(\mathbf{n})^T=\mathbb{K}(\mathbf{n})\text{ for all }\mathbf{n}\in\mathbb{S}^2.$$

We first note that if $\mathbf{N}\in L(\mathbf{n},\mathscr{V})$ for some $\mathbf{n}\in\mathbb{S}^2$, then $\mathbf{QNQ}^T\in L(\mathbf{Qn},\mathscr{V})$ for all $\mathbf{Q}\in SO(\mathscr{V})$, and so (2.138) is compatible with the definition of φ, which requires $\varphi(\mathbf{Qn},\cdot)$ to map $L(\mathbf{Qn},\mathscr{V})$ into $\mathbb{R}$ for all $\mathbf{Q}\in SO(\mathscr{V})$.

Let $\mathbf{n}$ be given in $\mathbb{S}^2$. Then (2.138) implies that

$$\varphi(\mathbf{n},\mathbf{QNQ}^T)=\varphi(\mathbf{n},\mathbf{N})\quad\text{for all }\mathbf{Q}\in SO(\mathbf{n},\mathscr{V}). \tag{2.140}$$

Inserting (2.137) into (2.140) and bringing together all terms with the same degree of homogeneity in $\mathbf{N}$, we conclude that $\mathbf{K}(\mathbf{n})$ and $\mathbb{K}(\mathbf{n})$ must satisfy

$$\mathbf{K}(\mathbf{n})\cdot\mathbf{QNQ}^T = \mathbf{K}(\mathbf{n})\cdot\mathbf{N},$$

$$\mathbf{QNQ}^T\cdot\mathbb{K}(\mathbf{n})[\mathbf{QNQ}^T] = \mathbf{N}\cdot\mathbb{K}(\mathbf{n})[\mathbf{N}],$$

for all $\mathbf{N}\in L(\mathbf{n},\mathscr{V})$ and $\mathbf{Q}\in SO(\mathbf{n},\mathscr{V})$. If in these equations we make use of (2.111), which represents all members of $SO(\mathbf{n},\mathscr{V})$ as the angle ϑ varies in $[0,2\pi[$, and then we evaluate at $\vartheta=0$ the derivative with respect to ϑ of the functions thus obtained, we find that

$$\mathbf{K}(\mathbf{n})\cdot(\mathbf{W}(\mathbf{n})\mathbf{N}-\mathbf{N}\mathbf{W}(\mathbf{n}))=0, \tag{2.141}$$

$$\mathbb{W}[\mathbf{N}]\cdot\mathbb{K}(\mathbf{n})[\mathbf{N}]+\mathbf{N}\cdot(\mathbb{K}(\mathbf{n})\mathbb{W})[\mathbf{N}]=0 \tag{2.142}$$

for all $\mathbf{N}\in L(\mathbf{n},\mathscr{V})$, where $\mathbb{W}$ is the tensor in $L^2(\mathbf{n},\mathscr{V})$ defined by (2.114). Since both $\mathbf{W}(\mathbf{n})$ and $\mathbb{W}$ are skew, (2.141) and (2.142) can be written, respectively, as

$$(\mathbf{K}(\mathbf{n})\mathbf{W}(\mathbf{n})-\mathbf{W}(\mathbf{n})\mathbf{K}(\mathbf{n}))\cdot\mathbf{N}=0,$$

$$\mathbf{N}\cdot(\mathbb{K}(\mathbf{n})\mathbb{W}-\mathbb{W}\mathbb{K}(\mathbf{n}))[\mathbf{N}]=0$$

for all $\mathbf{N}\in L(\mathbf{n},\mathscr{V})$. From the former equation we deduce at once that $\mathbf{K}(\mathbf{n})$ must commute with $\mathbf{W}(\mathbf{n})$, since $(\mathbf{K}(\mathbf{n})\mathbf{W}(\mathbf{n})-\mathbf{W}(\mathbf{n})\mathbf{K}(\mathbf{n}))$ belongs to $L(\mathbf{n},\mathscr{V})$. From the latter equation and the Spectral Theorem we deduce that $\mathbb{K}(\mathbf{n})$ must commute with $\mathbb{W}$, since $(\mathbb{K}(\mathbf{n})\mathbb{W}-\mathbb{W}\mathbb{K}(\mathbf{n}))$ is a symmetric tensor. Thus, by Lemmas 2.19 and 2.21 of Section 2.5, we conclude that $\mathbf{K}(\mathbf{n})$ and $\mathbb{K}(\mathbf{n})$ can be represented by formulae (2.106) and (2.115), respectively. More precisely, α_1 must vanish in the former, since $\mathbf{n}\otimes\mathbf{n}$ does not belong to $L(\mathbf{n},\mathscr{V})$ while both $\mathbf{P}(\mathbf{n})$ and $\mathbf{W}(\mathbf{n})$ do. Making use of (2.106) and (2.115) in (2.137), we easily prove that, given any $\mathbf{n}\in\mathbb{S}^2$, φ obeys (2.138) for all $\mathbf{Q}\in SO(\mathbf{n},\mathscr{V})$ only if $(\varphi(\mathbf{n},\mathbf{N})-k(\mathbf{n}))$ in (2.137) is just the same as $(\varphi(\mathbf{n},\mathbf{N})-\alpha_0)$ computed from (2.139).

To conclude the proof of the theorem it remains to be seen that

$$k(\mathbf{n})=\alpha_0 \quad \text{for all } \mathbf{n}\in\mathbb{S}^2, \tag{2.143}$$

and that the function φ in (2.139) indeed obeys (2.138) for all $\mathbf{Q}\in SO(\mathscr{V})$, not just for all $\mathbf{Q}\in SO(\mathbf{n},\mathscr{V})$.

If we insert (2.137) into (2.138) and restrict attention to the terms that do not depend on $\mathbf{N}$, we arrive at

$$k(\mathbf{Qn})=k(\mathbf{n}) \quad \text{for all } \mathbf{n}\in\mathbb{S}^2 \text{ and } \mathbf{Q}\in SO(\mathscr{V}),$$

whence, by Theorem 2.25, we see that k must be constant, since $|\mathbf{n}| = 1$.

Finally, employing (2.102) and (2.103) and using the properties of the inner product in $L(\mathscr{V})$ (*cf.* Section 2.2), we easily verify that the function given by (2.139) obeys (2.138).

Remark 3. If the function φ defined by (2.137) is required to satisfy (2.138) for all $\mathbf{Q} \in O(\mathscr{V})$, it must be of the form (2.139) with $\alpha_3 = 0$ and $\beta_3 = 0$, since it follows from (2.103) that

$$\mathbf{W}(\mathbf{Qn}) \cdot \mathbf{QNQ}^T = -\mathbf{W}(\mathbf{n}) \cdot \mathbf{N}$$

for any given $\mathbf{n} \in \mathbb{S}^2$ and for all $\mathbf{N} \in L(\mathbf{n}, \mathscr{V})$ and $\mathbf{Q} \in O(\mathscr{V}) \backslash SO(\mathscr{V})$, while, by (2.102),

$$\mathbf{P}(\mathbf{Qn}) \cdot \mathbf{QNQ}^T = \mathbf{P}(\mathbf{n}) \cdot \mathbf{N}$$

for all $\mathbf{N} \in L(\mathbf{n}, \mathscr{V})$ and $\mathbf{Q} \in O(\mathscr{V})$. To appreciate this conclusion fully the reader should return to the discussion that follows the Remark of Section 2.4.

Remark 4. It should also be noted that by (2.100) we can write (*cf.* Remark 1 of Section 2.5)

$$\mathbf{P}(\mathbf{n}) \cdot \mathbf{N} = \operatorname{tr} \mathbf{N}. \tag{2.144}$$

Moreover, again by (2.100), we obtain

$$\mathbf{P}(\mathbf{n})\mathbf{N} \cdot \mathbf{N}\mathbf{P}(\mathbf{n}) = \mathbf{N} \cdot (\mathbf{N} - \mathbf{Nn} \otimes \mathbf{n}) = \mathbf{N} \cdot \mathbf{N} - \mathbf{Nn} \cdot \mathbf{Nn} \tag{2.145}$$

for all $\mathbf{N} \in L(\mathbf{n}, \mathscr{V})$. Thus, the last two terms on the right-hand side of (2.139) can also be written as

$$(\beta_4 + \beta_5)\mathbf{N} \cdot \mathbf{N} - \beta_5 |\mathbf{Nn}|^2.$$

2.7 References

Bowen, R.M. and Wang, C.-C. (1976) *Introduction to vectors and tensors*, Volumes 1 and 2, Plenum Press, London.

Cauchy, A.-L. (1850) Mémoire sur les systèmes isotropes de points matériels, *Mém. Acad. Sci. Paris*, **22**, 615–654.

Fraenkel, L.E. (1979) On regularity of the boundary in the theory of Sobolev spaces, *Proc. London Math. Soc.*, **39**, 385–427.

Gurtin, M.E. and Murdoch, A.I. (1974–75) A continuum theory of elastic material surfaces, *Arch. Rational Mech. Anal.*, **57**, 291–323.

Halmos, P.R. (1958) *Finite-dimensional vector spaces*, Van Nostrand, Princeton.

Kellogg, O.D. (1929) *Foundations of potential theory*, Springer, Berlin.
Lang, S. (1985) *Differential manifolds*, Springer, New York.
Noll, W. (1973) Lectures on the foundations of continuum mechanics and thermodynamics, *Arch. Rational Mech. Anal.*, **52**, 62–92.
Noll, W. (1986) Continuum mechanics and geometric integration theory, in: *Categories in continuum physics, Lecture Notes in Mathematics No. 1174*, Springer, New York, 17–29.
Noll, W. (1987) *Finite-dimensional spaces. Algebra, geometry, and analysis*, Vol. I, Martinus Nijhoff Publishers, Dordrecht.
Noll, W. and Virga, E.G. (1988) Fit regions and functions of bounded variation, *Arch. Rational Mech. Anal.*, **102**, 1–21.
Noll, W. and Virga, E.G. (1990) On edge interactions and surface tension, *Arch. Rational Mech. Anal.*, **111**, 1–31.
Podio-Guidugli, P. and Virga, E.G. (1987) Transversely isotropic elasticity tensors, *Proc. R. Soc. London*, **A411**, 85–93.
Truesdell, C.A. and Noll, W. (1965) *The non-linear field theories of mechanics, Encyclopedia of physics*, Vol. III/3, S. Flügge ed., Springer, Berlin, Second edn, 1992.
Truesdell, C.A. (1991) *A first course in rational continuum mechanics*, Vol. 1, second edition, corrected, revised and augmented, Academic Press, New York.
Vol'pert, A.I. and Hudjaev, S.I. (1985) *Analysis in classes of discontinuous functions and equations of mathematical physics*, Martinus Nijhoff Publishers, Dordrecht.
Wang, C.-C. (1970) A new representation theorem for isotropic functions: an answer to Professor G.F. Smith's criticism of my paper on representations for isotropic functions. Part 1. Scalar-valued isotropic functions, *Arch. Rational Mech. Anal.*, **36**, 165–197.
Wang, C.-C. (1971) Corrigendum to my recent papers on 'Representations for isotropic functions', *Arch. Rational Mech. Anal.*, **43**, 392–395.
Ziemer, W.P. (1989) *Weakly differentiable functions*, Springer, New York.

2.8 Further reading

Gurtin, M.E. (1981) *An introduction to continuum mechanics*, Academic Press, New York.

3

The classical theory

The classical mathematical theory of liquid crystals is a continuum theory based on the works of Oseen (1933), Zocher (1933) and Frank (1958).

Oseen attempted to comprehend the molecular forces that give rise to liquid crystals. He described the orientation of each molecule through a vector of unit length which is directed along the axis of symmetry of the molecule. Assuming that the interaction between two molecules decreases rapidly enough as the distance between them increases and that the density of the substance is constant in space, Oseen wrote the functional of the molecular orientation that is to attain an extremum at every equilibrium configuration.

Oseen's work was criticized by Frank, who perceived more clearly the distinction between statistical and continuum models of liquid crystals. The following passage taken from p. 20 of Frank (1958) is rather enlightening:

> [Oseen] based his argument on the postulate that the energy is expressible as a sum of energies between molecules taken in pairs. This is analogous to the way in which Cauchy set up the theory of elasticity for solids, and in that case it is known that the theory predicted fewer independent elastic constants than actually exist, and we may anticipate a similar consequence with Oseen's theory. It is worth remarking that the controversial conflict between the 'swarm theory' and the 'continuum theory' of liquid crystals is illusory. The swarm theory was a particular hypothetical and approximate approach to the statistical mechanical problem of interpreting properties which can be well defined in terms of a continuum theory. This point is seen less clearly from Oseen's point of departure than from that of the present paper.

The **swarm theory** Frank refers to regards a nematic liquid crystal as a collection of small droplets (the swarms) within which the orientation of the optic axis is constant, being generally different in different droplets (*cf.* Section 1.2 of Chapter 1).

The first step towards Frank's point of departure was taken by Zocher (1933). His continuum theory has three elastic constants while Frank's has one more; however, it is well founded when the optic axis is everywhere parallel to a given plane, as Frank himself pointed out (Frank (1958), p. 25).

Frank's theory, unlike Oseen's, is intended only for nematic and cholesteric liquid crystals; smectic liquid crystals are left aside. Actually, Frank observes:

> The Oseen theory embraces smectic mesophases, but it is not really required for this case. The interpretation of the equilibrium structures assumed by smectic substances under a particular system of external influences may be carried out by essentially geometric arguments alone. (*cf.* Frank (1958), p. 19).

Frank was too optimistic in this regard: a comprehensive mathematical theory for all phases of smectic liquid crystals is still lacking. A first step towards such a theory was taken in Section 7.2 of De Gennes (1974), only for small deformations of the smectic layers; Leslie, Stewart and Nakagawa (1991) have recently extended De Gennes's theory to all deformations. This book does not deal with smectic liquid crystals at all.

The reader should not be induced by Frank's criticism to demean Oseen's work, which besides inspiring the modern continuum theory of liquid crystals put forward, *inter alia*, the free-boundary problem that we shall treat in Chapter 5 below (*cf.* Section 5 of Oseen (1933)).

In this chapter we retrace the origins of Frank's theory and we explain its consequences, following mostly Ericksen's work.

3.1 The bulk free energy

The continuum theory of liquid crystals is a variational theory in which a free energy functional is posited. Since the temperature is taken as given throughout our development, such a functional is indeed the Helmholtz free energy.

In the classical theory the degree of orientation is regarded as a prescribed positive constant:

$$s = s_0 > 0. \tag{3.1}$$

The order parameter s first appeared in equation (1.23) of Chapter 1. The optic axis is described by a unit vector $\mathbf{n} \in \mathbb{S}^2$. We take the region $\mathscr{B}$ of $\mathscr{E}$ occupied by a liquid crystal as a regular region (*cf.* Section 2.3 of Chapter 2) and we assume that the orientation of the optic axis is given everywhere in $\mathscr{B}$ by a mapping of class C^1:

$$\mathbf{n}: \mathscr{B} \to \mathbb{S}^2.$$

The free energy per unit volume σ is customarily assumed to depend on $\mathbf{n}$ and its first gradient: by (2.65) of Chapter 2, the domain of σ is the set

$$F(\mathscr{V}) := \{(\mathbf{n}, \mathbf{N}) \mid \mathbf{n} \in \mathbb{S}^2, \quad \mathbf{N} \in L(\mathbf{n}, \mathscr{V})\},$$

where

$$L(\mathbf{n}, \mathscr{V}) := \{\mathbf{L} \in L(\mathscr{V}) \mid \mathbf{L}^T \mathbf{n} = \mathbf{0}\}$$

(*cf.* Sections 2.5 and 2.6). We shall see in Section 3.9 below what difficulties one encounters when, departing from tradition, one assumes that $\mathbf{n}$ is a mapping of class C^2 and σ depends linearly on its second gradient.

The classical theory takes the bulk free energy of liquid crystals as given by the following functional:

$$\mathscr{F}_B[\mathbf{n}] := \int_{\mathscr{B}} \sigma(\mathbf{n}, \nabla \mathbf{n}) \, dv, \tag{3.2}$$

which is also called the **elastic free energy**. The function σ must obey a list of requirements that represent the main physical features of the materials we intend to model: σ must be **frame-indifferent** and **material-symmetric, even** and **positive definite**; we now examine in detail each of these properties.

3.1.1 Frame-indifference and material symmetry

First, σ must be **frame-indifferent**, that is the free energy per unit volume of the body must be the same in any two frames. A rigorous definition of frame is given in Section I.6 of Truesdell (1991) (more precisely, what is here called simply frame is a rigid frame there, to distinguish it from the deformable frame defined later in the same section). It makes precise the intuitive idea that anything can be observed only on a background against which we place it. Here we identify a frame with the list $(o, \mathbf{e}_1, \mathbf{e}_2, \mathbf{e}_3)$ whose members are a point $o \in \mathscr{E}$ and three orthogonal unit vectors $\mathbf{e}_1, \mathbf{e}_2, \mathbf{e}_3 \in \mathscr{V}$ (a basis of $\mathscr{V}$).

Let the frame $(o, \mathbf{e}_1, \mathbf{e}_2, \mathbf{e}_3)$ be given. For all $p \in \mathscr{E}$ we denote by $\mathbf{x}$ the vector $p - o$. Let $\mathbf{Q}$ be any element of $SO(\mathscr{V})$; it changes the vector $\mathbf{x}$ into $\mathbf{x}^*$:

$$\mathbf{x}^* := \mathbf{Q}\mathbf{x}. \tag{3.3}$$

For each point p of $\mathscr{E}$ we define another point of $\mathscr{E}$ by

$$p^* := o + \mathbf{x}^* = o + \mathbf{Q}(p - o). \tag{3.4}$$

In such a way we induce a transformation of $\mathscr{E}$ into itself that changes $\mathscr{B}$ into the region defined by

$$\mathscr{B}^* := \{p^* \in \mathscr{E} \mid p^* = o + \mathbf{Q}(p - o),\ p \in \mathscr{B}\}. \tag{3.5}$$

$\mathscr{B}^*$ is precisely $\mathscr{B}$ as seen against the frame $(o, \mathbf{e}'_1, \mathbf{e}'_2, \mathbf{e}'_3)$, where

$$\mathbf{e}'_i = \mathbf{Q}^T \mathbf{e}_i \quad \text{for } i = 1, 2, 3.$$

Figure 3.1a illustrates a two dimensional example of the transformation that changes $\mathscr{B}$ into $\mathscr{B}^*$. It is worth noting that $\mathbf{Q}$ must be a rotation to preserve the orientation of the frame.

Let a unit vector field $\mathbf{n}$ be defined on $\mathscr{B}$. One sees it against the frame $(o, \mathbf{e}'_1, \mathbf{e}'_2, \mathbf{e}'_3)$ precisely as another unit vector field $\mathbf{n}^*$ is seen against the frame $(o, \mathbf{e}_1, \mathbf{e}_2, \mathbf{e}_3)$ (see Fig. 3.1b); $\mathbf{n}^*$ is defined in $\mathscr{B}^*$ by

$$\mathbf{n}^*(p^*) = \mathbf{Q}\mathbf{n}(p), \tag{3.6}$$

for all $p^* \in \mathscr{B}$ and $p \in \mathscr{B}$ related through (3.4). The function σ is frame-indifferent if

$$\sigma(\mathbf{n}^*(p^*), \nabla\mathbf{n}^*(p^*)) = \sigma(\mathbf{n}(p), \nabla\mathbf{n}(p)), \tag{3.7}$$

for any $\mathbf{Q}$ which relates p to p^* in (3.4). Since, by (3.4),

$$p = o + \mathbf{Q}^T(p^* - o), \tag{3.8}$$

we easily obtain from (3.6) that

$$\nabla\mathbf{n}^*(p^*) = \mathbf{Q}\nabla\mathbf{n}(p)\mathbf{Q}^T. \tag{3.9}$$

Thus, (3.7) takes the following form:

$$\sigma(\mathbf{Q}\mathbf{n}(p), \mathbf{Q}\nabla\mathbf{n}(p)\mathbf{Q}^T) = \sigma(\mathbf{n}(p), \nabla\mathbf{n}(p)), \tag{3.10}$$

for all $p \in \mathscr{B}$ and $\mathbf{Q} \in SO(\mathscr{V})$. Dropping the dummy p, we reduce (3.10) to a purely algebraic requirement:

$$\sigma(\mathbf{Q}\mathbf{n}, \mathbf{Q}\mathbf{N}\mathbf{Q}^T) = \sigma(\mathbf{n}, \mathbf{N}) \tag{3.11}$$

for any given $\mathbf{n} \in \mathbb{S}^2$, and for all $\mathbf{N} \in L(\mathbf{n}, \mathscr{V})$ and $\mathbf{Q} \in SO(\mathscr{V})$.

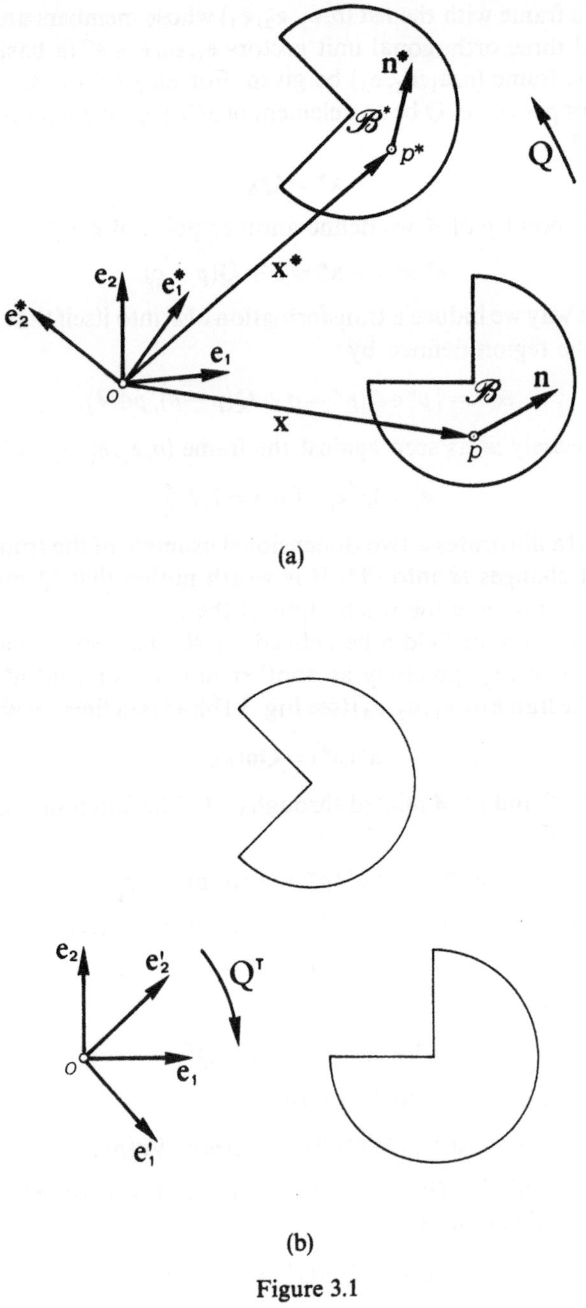

Figure 3.1

Condition (3.11) applies to both nematics and cholesterics: it derives from a change of frame, irrespective of the material that occupies $\mathscr{B}$. There are, however, transformations of $\mathscr{B}$ under which nematics and cholesterics behave differently: they reflect different **material symmetries**. What makes the difference between nematic and cholesteric liquid crystals on a microscopic scale is the different enantiomorphism of their molecules. While nematic molecules, which resemble rods, remain alike after a reflection, cholesteric molecules, which resemble helical springs, suffer a change in chirality under a reflection: right-handed helices are changed into left-handed helices, and vice versa. It is the common opinion that such a microscopic feature has its counterpart also on a macroscopic scale. When $\mathscr{B}$ suffers a reflection, the energy density of nematics must not be affected, whereas that of cholesterics might be.

We now make this idea more precise. Let a point o be given in $\mathscr{E}$ and let $\mathbf{e}$ be an element of $\mathbb{S}^2$. Call $P_o(\mathbf{e})$ the plane through o orthogonal to $\mathbf{e}$. The reflection across $P_o(\mathbf{e})$ is represented by the tensor

$$\mathbf{R_e} := \mathbf{I} - 2\mathbf{e}\otimes\mathbf{e}:$$

the image p_* of a point p in $\mathscr{E}$ is given by

$$p_* = o + \mathbf{R_e}(p - o).$$

The tensor $\mathbf{R_e}$ is both symmetric and orthogonal. A glance at (2.46) of Chapter 2 suffices to show that

$$\mathbf{R_e} = -\mathbf{Q_e}(\pi),$$

where $\mathbf{Q_e}(\pi)$ is the rotation by the angle π about $\mathbf{e}$. Thus, $-\mathbf{R_e}$ belongs to $SO(\mathscr{V})$. Now let the plane $P_o(\mathbf{e})$ be outside $\mathscr{B}$ and call $\mathscr{B}_*$ the image of $\mathscr{B}$ across $P_o(\mathbf{e})$; formally,

$$\mathscr{B}_* := \{p_* \in \mathscr{E} \mid p_* = o + \mathbf{R_e}(p - o), p \in \mathscr{B}\}.$$

An orientation field $\mathbf{n}$ defined on $\mathscr{B}$ is reflected across $P_o(\mathbf{e})$ into a field $\mathbf{n}_*$ defined on $\mathscr{B}_*$ by

$$\mathbf{n}_*(p_*) = \mathbf{R_e}\mathbf{n}(p)$$

for all p_* and p related as above. Furthermore, since $\mathbf{R_e}$ is symmetric, it is easily seen that

$$\nabla\mathbf{n}_*(p_*) = \mathbf{R_e}\mathbf{n}(p)\mathbf{R_e}.$$

Thus, if $\mathscr{B}$ is occupied by a nematic liquid crystal, the free energy density σ must satisfy the following condition:

$$\sigma(\mathbf{n}_*(p_*), \nabla\mathbf{n}(p_*)) = \sigma(\mathbf{n}(p), \nabla\mathbf{n}(p))$$

for all corresponding points $p_* \in \mathscr{B}_*$ and $p \in \mathscr{B}$. This amounts to requiring that

$$\sigma(\mathbf{R}_e\mathbf{n}, \mathbf{R}_e\mathbf{N}\mathbf{R}_e) = \sigma(\mathbf{n}, \mathbf{N})$$

for all $\mathbf{n} \in \mathbb{S}^2$ and $\mathbf{N} \in L(\mathbf{n}, \mathscr{V})$. On the other hand, since $-\mathbf{R}_e \in SO(\mathscr{V}) \cap Sym(\mathscr{V})$, (3.11) implies that

$$\sigma(\mathbf{R}_e\mathbf{n}, \mathbf{R}_e\mathbf{N}\mathbf{R}_e) = \sigma((-\mathbf{R}_e)(-\mathbf{n}), (-\mathbf{R}_e)\mathbf{N}(-\mathbf{R}_e)) = \sigma(-\mathbf{n}, \mathbf{N}),$$

and so material symmetry requires nematics to satisfy

$$\sigma(-\mathbf{n}, \mathbf{N}) = \sigma(\mathbf{n}, \mathbf{N}) \quad \text{for all } \mathbf{n} \in \mathbb{S}^2 \text{ and } \mathbf{N} \in L(\mathbf{n}, \mathscr{V}).$$

It is easily seen that this algebraic requirement, together with (3.11), leads to

$$\sigma(\mathbf{Q}\mathbf{n}, \mathbf{Q}\mathbf{N}\mathbf{Q}^T) = \sigma(\mathbf{n}, \mathbf{N})$$

for any given $\mathbf{n} \in \mathbb{S}^2$ and all $\mathbf{N} \in L(\mathbf{n}, \mathscr{V})$ and $\mathbf{Q} \in O(\mathscr{V})$. Thus, the consequences on σ of both frame-indifference and material symmetry can be stated simultaneously for nematics and cholesterics: *assume that* σ *satisfies* (3.10) *for all* $\mathbf{Q} \in G(\mathscr{V})$, where $G(\mathscr{V})$ is a group that depends on the material:

$$G(\mathscr{V}) := \begin{cases} O(\mathscr{V}) & \text{for nematics,} \\ SO(\mathscr{V}) & \text{for cholesterics.} \end{cases} \tag{3.12}$$

It should be noted here that a nematic phase might also consist of an equal mixture of two species with opposite chirality.

3.1.2 Evenness

As we saw in Section 1.3 of Chapter 1, the direction of $\mathbf{n}$ at a point p of $\mathscr{B}$ can be interpreted, in a sense made precise there, as the average orientation of all molecules that make up the macroscopic particle at p. Generally, in the molecules of both nematic and cholesteric liquid crystals one cannot distinguish the head from the tail, as it were. Thus, $\mathbf{n}$ itself, by its statistical interpretation, cannot be physically different from $-\mathbf{n}$. In particular, this implies that reversing the field $\mathbf{n}$ does not affect the energy density:

$$\sigma(-\mathbf{n}, -\nabla\mathbf{n}) = \sigma(\mathbf{n}, \nabla\mathbf{n}). \tag{3.13}$$

The same conclusion holds also if the molecules are polar, that is with different ends, but at each point $p \in \mathscr{B}$ they are so arranged as to be equally divided into two groups with opposite orientations.

3.1.3 Positive definiteness

The free energy of a system is in general defined to within an additive constant. On the other hand, when neither external actions nor anchoring conditions affect the orientation of a liquid crystal, it naturally relaxes into an undistorted state. To such a state we assign zero free energy, and we call the corresponding orientation **natural**.

Formally, we define a natural orientation as a field $\mathbf{n}$ of the whole space $\mathscr{E}$ into $\mathbb{S}^2$ that makes σ identically zero. Thus, if $\mathbf{n}$ is a natural orientation for a liquid crystal then the field $\mathbf{n}^*: \mathscr{E} \to \mathbb{S}^2$ defined in terms of $\mathbf{n}$ as in (3.6) is also a natural orientation, by (3.10).[1] There is a full class of natural orientations, which we denote by $\mathscr{N}$:

$$\sigma(\mathbf{n}, \nabla\mathbf{n}) = 0 \quad \text{for all } \mathbf{n} \in \mathscr{N}. \tag{3.14}$$

Of course, $\mathscr{N}$ has a constitutive nature: it depends on the specific liquid crystal.

To make precise the intuitive idea that all natural orientations of a liquid crystal represent on the same footing a **reference state** for the energy expended in producing any non-natural orientation, we require σ to attain its minimum on $\mathscr{N}$: that is, in view of (3.14),

$$\sigma(\mathbf{n}, \mathbf{N}) \geqslant 0 \tag{3.15}$$

for any given $\mathbf{n} \in \mathbb{S}^2$ and all $\mathbf{N} \in L(\mathbf{n}, \mathscr{V})$.

We can now interpret in terms of natural orientations the experimental evidence discussed in Section 1.1 of Chapter 1. For nematic liquid crystals $\mathscr{N}$ consists of all constant mappings from $\mathscr{E}$ into $\mathbb{S}^2$. For cholesteric liquid crystals a characteristic member of $\mathscr{N}$ is

$$\mathbf{n}_c := \cos(\tau x_3 + \varphi_0)\mathbf{e}_1 + \sin(\tau x_3 + \varphi_0)\mathbf{e}_2, \tag{3.16}$$

where $\mathbf{e}_1$ and $\mathbf{e}_2$ are unit vectors of a frame, $\mathbf{e}_3 = \mathbf{e}_1 \wedge \mathbf{e}_2$ is the **cholesteric axis** and x_3 is the co-ordinate along it (*cf.* equation (1.1) of Chapter 1). Any other member of $\mathscr{N}$, for a given cholesteric liquid crystal, is still of the form (3.16), with the same scalar τ and possibly a different phase φ_0, but instead of $\mathbf{e}_3$ its cholesteric axis is $\mathbf{Q}\mathbf{e}_3$, for some $\mathbf{Q} \in SO(\mathscr{V})$.

---

[1] It follows from (3.5) that $\mathscr{E}^* = \mathscr{E}$.

Hereafter we take τ as a constitutive constant; it is the **twist** characteristic of each cholesteric liquid crystal.

Remark. If σ satisfies (3.10) for all $\mathbf{Q}\in SO(\mathscr{V})$, then it vanishes on all members of $\mathscr{N}$ whenever it vanishes on $\mathbf{n}_c$.

3.2 Frank's formula

In this section we determine for both nematics and cholesterics the functions $\sigma(\mathbf{n},\mathbf{N})$ that are quadratic polynomials in $\mathbf{N}$ and obey all the requirements laid down in the preceding section except (3.15). The representation formula we shall derive here for such functions is mainly due to Frank (1958), though there are anticipations in the works of Zocher (1925) and Oseen (1933).

To save the reader from having to refer back I recall from Section 2.6 of Chapter 2 (*cf.* Theorem 2.28 and Remark 3) an algebraic result soon to be relevant to our development.

Theorem 3.1. For any given $\mathbf{n}\in\mathbb{S}^2$ let $\varphi(\mathbf{n},\cdot)$ be the scalar-valued function on $L(\mathbf{n},\mathscr{V})$ defined by

$$\varphi(\mathbf{n},\mathbf{N}):=k(\mathbf{n})+\mathbf{K}(\mathbf{n})\cdot\mathbf{N}+\mathbf{N}\cdot\mathbb{K}(\mathbf{n})[\mathbf{N}], \tag{3.17}$$

where $k(\mathbf{n})$ is a scalar, $\mathbf{K}(\mathbf{n})$ is a tensor in $L(\mathbf{n},\mathscr{V})$, and $\mathbb{K}(\mathbf{n})$ is a symmetric tensor in $L^2(\mathbf{n},\mathscr{V})$. The function φ satisfies

$$\varphi(\mathbf{Q}\mathbf{n},\mathbf{Q}\mathbf{N}\mathbf{Q}^T)=\varphi(\mathbf{n},\mathbf{N})\quad\text{for all }\mathbf{Q}\in SO(\mathscr{V}) \tag{3.18}$$

if, and only if,

$$\begin{aligned}\varphi(\mathbf{n},\mathbf{N})=\alpha_0&+\alpha_2\mathbf{P}(\mathbf{n})\cdot\mathbf{N}+\alpha_3\mathbf{W}(\mathbf{n})\cdot\mathbf{N}+\beta_1(\mathbf{W}(\mathbf{n})\cdot\mathbf{N})^2\\&+\beta_2(\mathbf{P}(\mathbf{n})\cdot\mathbf{N})^2+\beta_3(\mathbf{W}(\mathbf{n})\cdot\mathbf{N})(\mathbf{P}(\mathbf{n})\cdot\mathbf{N})\\&+\beta_4\mathbf{N}\cdot\mathbf{N}+\beta_5\mathbf{P}(\mathbf{n})\mathbf{N}\cdot\mathbf{N}(\mathbf{P}(\mathbf{n})),\end{aligned} \tag{3.19}$$

where $\alpha_0,\alpha_2,\alpha_3$ and β_i, $i=1,\dots,5$, are scalars, and $\mathbf{P}(\mathbf{n})$ and $\mathbf{W}(\mathbf{n})$ are defined thus:[2]

$$\mathbf{P}(\mathbf{n})=\mathbf{I}-\mathbf{n}\otimes\mathbf{n}, \tag{3.20}$$

$$\mathbf{W}(\mathbf{n})\mathbf{v}=\mathbf{n}\wedge\mathbf{v}\quad\text{for all }\mathbf{v}\in\mathscr{V}. \tag{3.21}$$

The function φ in (3.17) is defined on the same domain as the function

[2] Equations (3.20) and (3.21) repeat, respectively, (2.100) and (2.101) of Chapter 2.

σ in Section 3.1 above; k is a scalar-valued function on $\mathbb{S}^2$, and $\mathbf{K}$ and $\mathbb{K}$ are mappings from $\mathbb{S}^2$ into $L(\mathscr{V})$ and $L^2(\mathscr{V})$, respectively, such that

$$\mathbf{K}(\mathbf{n})\in L(\mathbf{n},\mathscr{V}),\ \mathbb{K}(\mathbf{n})\in L^2(\mathbf{n},\mathscr{V}),\quad \text{and } \mathbb{K}(\mathbf{n})^T=\mathbb{K}(\mathbf{n}) \text{ for all } \mathbf{n}\in\mathbb{S}^2.$$

Remark 1. If the function φ defined by (3.17) is to satisfy (3.18) for all $\mathbf{Q}\in O(\mathscr{V})$, then both α_3 and β_3 must vanish in (3.19).

As in the preceding section, $\mathscr{B}$ is a given region in space, taken to be a regular region occupied by a liquid crystal.

Theorem 3.2. Let $\mathbf{n}:\mathscr{B}\to\mathbb{S}^2$ be any given mapping of class C^1 and let σ_F be a scalar-valued function of the form

$$\sigma_F(\mathbf{n},\nabla\mathbf{n}):=k(\mathbf{n})+\mathbf{K}(\mathbf{n})\cdot\nabla\mathbf{n}+\nabla\mathbf{n}\cdot\mathbb{K}(\mathbf{n})[\nabla\mathbf{n}], \tag{3.22}$$

where $k(\mathbf{n})$, $\mathbf{K}(\mathbf{n})$, and $\mathbb{K}(\mathbf{n})$ are as in Theorem 3.1. The function σ_F satisfies (3.10) for all $\mathbf{Q}\in G(\mathscr{V})$, (3.13) and (3.14) whenever there are five scalars k_i, $i=0,\dots,4$, such that

$$\begin{aligned}\sigma_F(\mathbf{n},\nabla\mathbf{n})=k_0\tau(\mathbf{n}\cdot\operatorname{curl}\mathbf{n}+\tau)+k_1(\operatorname{div}\mathbf{n})^2+k_2(\mathbf{n}\cdot\operatorname{curl}\mathbf{n}+\tau)^2\\+k_3|\mathbf{n}\wedge\operatorname{curl}\mathbf{n}|^2+(k_2+k_4)(\operatorname{tr}(\nabla\mathbf{n})^2-(\operatorname{div}\mathbf{n})^2),\end{aligned} \tag{3.23}$$

where τ is as in (3.16) for cholesterics, while it is set equal to 0 for nematics.

Equation (3.23) will often be referred to as **Frank's formula**. When in (3.2) σ is given by Frank's formula we call $\mathscr{F}_B$ **Frank's energy functional.**

Before proving Theorem 3.2 we group together in a lemma two identities valid for all unit vector fields of class C^1.

Lemma 3.3. Let $\mathbf{n}$ be a field of class C^1 of any open region of $\mathscr{E}$ into $\mathbb{S}^2$. The following equations hold identically in the domain of $\mathbf{n}$:

$$(\nabla\mathbf{n})\mathbf{n}=-\mathbf{n}\wedge\operatorname{curl}\mathbf{n}, \tag{3.24}$$

$$\nabla\mathbf{n}\cdot\nabla\mathbf{n}=\operatorname{tr}(\nabla\mathbf{n})^2+(\mathbf{n}\cdot\operatorname{curl}\mathbf{n})^2+|\mathbf{n}\wedge\operatorname{curl}\mathbf{n}|^2. \tag{3.25}$$

Proof. We learned in Section 2.2 of Chapter 2 that for every skew tensor $\mathbf{W}\in L(\mathscr{V})$ there is a $\mathbf{w}\in\mathscr{V}$, called the axial vector of $\mathbf{W}$, such that

$$\mathbf{W}\mathbf{v}=\mathbf{w}\wedge\mathbf{v}\quad\text{for all } \mathbf{v}\in\mathscr{V}, \tag{3.26}$$

and

$$\mathbf{W}\cdot\mathbf{W}=2\mathbf{w}\cdot\mathbf{w}. \tag{3.27}$$

When $\mathbf{W}$ is the skew part of $\nabla\mathbf{n}$, its axial vector is

$$\mathbf{w}=\tfrac{1}{2}\operatorname{curl}\mathbf{n} \tag{3.28}$$

(*cf.* equation (2.54) of Chapter 2). Let $\mathbf{S}$ be the symmetric part of $\nabla\mathbf{n}$, so that

$$\nabla\mathbf{n} = \mathbf{S} + \mathbf{W}. \tag{3.29}$$

Since $(\nabla\mathbf{n})^T\mathbf{n} = \mathbf{0}$ (*cf.* equation (2.65) of Chapter 2), we obtain from (3.29) that

$$\mathbf{S}\mathbf{n} = \mathbf{W}\mathbf{n}.$$

Thus, making use of (3.29), (3.28) and (3.26), we get

$$(\nabla\mathbf{n})\mathbf{n} = 2\mathbf{W}\mathbf{n} = \operatorname{curl}\mathbf{n} \wedge \mathbf{n},$$

which is precisely (3.24).

To prove (3.25) we employ (3.29) twice: first to get

$$\nabla\mathbf{n}\cdot\nabla\mathbf{n} = \mathbf{S}\cdot\mathbf{S} + \mathbf{W}\cdot\mathbf{W}, \tag{3.30}$$

and then to compute

$$\operatorname{tr}(\nabla\mathbf{n})^2 = \operatorname{tr}\mathbf{S}^2 + \operatorname{tr}\mathbf{W}^2, \tag{3.31}$$

where use has also been made of the fact that the tensor $\mathbf{L} = \mathbf{S}\mathbf{W} + \mathbf{W}\mathbf{S}$ is skew. Since $\mathbf{S}\cdot\mathbf{S} = \operatorname{tr}\mathbf{S}^2$ and $\mathbf{W}\cdot\mathbf{W} = -\operatorname{tr}\mathbf{W}^2$, we eliminate $\mathbf{S}\cdot\mathbf{S}$ from (3.30) and (3.31), arriving at

$$\nabla\mathbf{n}\cdot\nabla\mathbf{n} = \operatorname{tr}(\nabla\mathbf{n})^2 + 2\mathbf{W}\cdot\mathbf{W}.$$

The desired conclusion then follows from (3.27), (3.28) and the identity

$$|\operatorname{curl}\mathbf{n}|^2 = (\mathbf{n}\cdot\operatorname{curl}\mathbf{n})^2 + |\mathbf{n} \wedge \operatorname{curl}\mathbf{n}|^2, \tag{3.32}$$

whose proof is left as an exercise to the reader.

Proof of Theorem 3.2. In view of what we learned in the preceding section, Theorem 3.1 assures us that requiring σ_F to satisfy (3.10) for all $\mathbf{Q} \in G(\mathscr{V})$ amounts to writing it in the form (3.19), with $\alpha_3 = \beta_3 = 0$ for nematics. Nonetheless, getting from (3.19) to the formula suitable for σ_F, in which $\nabla\mathbf{n}$ is to take the place of $\mathbf{N}$, requires some labour.

First, bearing in mind (2.144) of Chapter 2, we observe that

$$\mathbf{P}(\mathbf{n})\cdot\nabla\mathbf{n} = \operatorname{tr}\nabla\mathbf{n} = \operatorname{div}\mathbf{n}. \tag{3.33}$$

Second, we recall from Section 2.2 of Chapter 2 that if $\mathbf{W}_1$ and $\mathbf{W}_2$ are skew tensors whose axial vectors are $\mathbf{w}_1$ and $\mathbf{w}_2$, respectively, then

$$\mathbf{W}_1\cdot\mathbf{W}_2 = 2\mathbf{w}_1\cdot\mathbf{w}_2,$$

a formula which includes (3.27) as a special case. Thus, also making use of (3.28) and (3.29), we get

$$\mathbf{W}(\mathbf{n})\cdot\nabla\mathbf{n} = \mathbf{W}(\mathbf{n})\cdot\mathbf{W} = \mathbf{n}\cdot\operatorname{curl}\mathbf{n}. \tag{3.34}$$

Third, having in mind (2.145) of Chapter 2, by the preceding lemma we arrive at

$$\mathbf{P}(\mathbf{n})\nabla\mathbf{n}\cdot\nabla\mathbf{n}\mathbf{P}(\mathbf{n}) = \nabla\mathbf{n}\cdot\nabla\mathbf{n} - |(\nabla\mathbf{n})\mathbf{n}|^2 = \operatorname{tr}(\nabla\mathbf{n})^2 + (\mathbf{n}\cdot\operatorname{curl}\mathbf{n})^2. \quad (3.35)$$

Finally, in view of (3.33), (3.34) and (3.35), substituting $\nabla\mathbf{n}$ for $\mathbf{N}$ in (3.19) leads us to

$$\begin{aligned}\sigma_F(\mathbf{n}, \nabla\mathbf{n}) = {} & \alpha_0 + (\alpha_2 + \beta_3\mathbf{n}\cdot\operatorname{curl}\mathbf{n})\operatorname{div}\mathbf{n} + \alpha_3(\mathbf{n}\cdot\operatorname{curl}\mathbf{n}) \\ & + k_1(\operatorname{div}\mathbf{n})^2 + k_2(\mathbf{n}\cdot\operatorname{curl}\mathbf{n})^2 + k_3|\mathbf{n}\wedge\operatorname{curl}\mathbf{n}|^2 \\ & + (k_2 + k_4)(\operatorname{tr}(\nabla\mathbf{n})^2 - (\operatorname{div}\mathbf{n})^2), \end{aligned} \quad (3.36)$$

where we have set

$$\beta_1 =: -k_4, \quad \beta_2 =: k_1 - k_2 - k_4, \quad \beta_4 =: k_3, \quad \beta_5 =: k_2 + k_4 - k_3.$$

Thus, σ_F obeys (3.10) whenever it is of the form (3.36), whence it follows that

$$\sigma_F(-\mathbf{n}, -\nabla\mathbf{n}) - \sigma_F(\mathbf{n}, \nabla\mathbf{n}) = -2(\alpha_2 + \beta_3\mathbf{n}\cdot\operatorname{curl}\mathbf{n})\operatorname{div}\mathbf{n}. \quad (3.37)$$

The right-hand side of (3.37) vanishes identically for all fields $\mathbf{n}$, whenever both α_2 and β_3 vanish, and so σ_F satisfies (3.13) also whenever we set

$$\alpha_2 = \beta_3 = 0$$

in (3.36) for both nematics and cholesterics.

To explore the consequences of (3.14) on (3.36) we now consider nematics and cholesterics separately.

For nematics, besides α_2 and β_3, α_3 must also vanish in (3.36). Furthermore, the class $\mathcal{N}$ introduced in the preceding section consists of all constant fields of $\mathcal{B}$ into $\mathbb{S}^2$, and so σ_F vanishes on $\mathcal{N}$ whenever α_0 vanishes in (3.36).

For cholesterics, σ_F vanishes on all fields of $\mathcal{N}$ whenever it vanishes on the field $\mathbf{n}_c$ given by (3.16) (*cf.* the Remark of Section 3.1). It is easily seen that

$$\operatorname{div}\mathbf{n}_c = 0, \quad \operatorname{curl}\mathbf{n}_c = -\tau\mathbf{n}_c \quad \text{and} \quad (\nabla\mathbf{n}_c)^2 = \mathbf{0}, \quad (3.38)$$

and so it follows from (3.36) that $\sigma_F(\mathbf{n}_c, \nabla\mathbf{n}_c) = 0$ whenever

$$\alpha_3 = k_2\tau + \frac{\alpha_0}{\tau}.$$

If in (3.36) we set

$$\alpha_0 = (k_0 + k_2)\tau^2,$$

we can represent through the same formula the free energy density of both nematics and cholesterics, assigning $\tau = 0$ to the former. This completes the proof of (3.23).

Remark 2. If in (3.23) we set

$$k_0 = 0, \tag{3.39}$$

σ_F still obeys (3.10), (3.13), (3.14) and, furthermore, it acquires a more elegant form. Such an aesthetic reason prompted Frank (1958) to adopt (3.39) in his paper. We shall see in Section 3.4 below that in our setting (3.39) is indeed a consequence of (3.15). It is to be noted, however, that for cholesteric liquid crystals (3.15) will also imply that

$$k_2 + k_4 = 0,$$

an identity that Frank never assumed.

Remark 3. The reader might wonder why in (3.23) the two terms proportional to $(\operatorname{div}\mathbf{n})^2$ are not grouped together. This is because the function defined by

$$\sigma_L(\mathbf{n}, \nabla\mathbf{n}) := \operatorname{tr}(\nabla\mathbf{n})^2 - (\operatorname{div}\mathbf{n})^2 \tag{3.40}$$

will be shown in Section 3.8 to be a **null Lagrangian**, that is a free energy density which does not contribute to the equilibrium equations of the free energy functional $\mathscr{F}_B$.

The constants k_1, k_2, k_3 and k_4, also called **Frank's constants**, are material moduli characteristic of each liquid crystal, just as τ. The next two sections are devoted to illuminating the properties of these constants relevant to our development.

3.3 Frank's constants

Here, following Frank (1958), we aim to interpret the material moduli k_1, k_2, k_3 and k_4, by producing four different orientation fields, each with a distortion energy σ_F which is proportional to only one of the terms present in (3.23) when $\tau = 0$. We shall see that such a task is easily accomplished for k_1, k_2 and k_3, while it is not for k_4; we can exhibit an orientation field whose energy density is just approximately proportional to $(k_2 + k_4)$ in a neighbourhood of a point in space.

Remark 1. Before proceeding farther, I warn the reader that the material moduli in σ_F were differently labelled by Frank: he employed the symbols k_{11}, k_{22}, k_{33} and k_{24} which are related to ours through

$$k_1 = \tfrac{1}{2}k_{11}, \quad k_2 = \tfrac{1}{2}k_{22}, \quad k_3 = \tfrac{1}{2}k_{33}, \quad k_4 = \tfrac{1}{2}k_{24} \tag{3.41}$$

(*cf.* equation (25) of Frank (1958)).

We call k_1, k_3 and k_2 **splay, bend** and **twist moduli**, respectively, while for $(k_2 + k_4)$ we adopt the name of **saddle-splay modulus**, as suggested by Frank (1958). The reasons for giving these names are explained in order below.

3.3.1 Splay modulus

In a system of cylindrical co-ordinates (r, ϑ, z) with origin in $o \in \mathscr{E}$ we denote by $\mathbf{e}_r, \mathbf{e}_\vartheta$ and $\mathbf{e}_z$ the co-ordinate vectors and by e_z the axis through o parallel to $\mathbf{e}_z$, so that

$$e_z := \{p \in \mathscr{E} \,|\, p - o = z\mathbf{e}_z, z \in \mathbb{R}\} \tag{3.42}$$

and

$$p - o = r\mathbf{e}_r + z\mathbf{e}_z \quad \text{for all } p \in \mathscr{E} \backslash e_z \tag{3.43}$$

(*cf.* Section 2.3 of Chapter 2 for further details).

Let $\mathbf{n}_s : \mathscr{E} \backslash e_z \to \mathbb{S}^2$ be the field defined by

$$\mathbf{n}_s := \mathbf{e}_r. \tag{3.44}$$

Recalling equation (2.78) of Chapter 2, we have

$$\nabla \mathbf{n}_s = \frac{1}{r} \mathbf{e}_\vartheta \otimes \mathbf{e}_\vartheta,$$

whence it follows that

$$\operatorname{div} \mathbf{n}_s = \frac{1}{r}, \quad \operatorname{curl} \mathbf{n}_s = \mathbf{0}, \quad \operatorname{tr}(\nabla \mathbf{n}_s)^2 = \frac{1}{r^2}. \tag{3.45}$$

Making use of (3.45) in (3.23) with $\tau = 0$, we arrive at

$$\sigma_F(\mathbf{n}_s, \nabla \mathbf{n}_s) = k_1 \frac{1}{r^2}, \tag{3.46}$$

and so, away from e_z, the energy density associated with $\mathbf{n}_s$ depends only on k_1.

We refer to $\mathbf{n}_s$ as the **splay field**.

Remark 2. The splay field is discontinuous along e_z; we say that e_z is a defect of $\mathbf{n}_s$. Furthermore, we shall see in Section 3.6 that $\mathbf{n}_s$ is a solution of the equilibrium equations for $\mathscr{F}_B$, but, however its defect may resemble a defect actually observed in capillary tubes, it cannot be an energy minimizer, since by (3.46) $\mathscr{F}_B[\mathbf{n}_s]$ is unbounded in any cylinder of axis e_z whenever $k_1 \neq 0$. We shall discuss at length in Chapter 6 this and other issues related to defects.

3.3.2 Bend modulus

In the same system of co-ordinates employed above let $\mathbf{n}_b: \mathscr{E}\backslash e_z \to \mathbb{S}^2$ be the field defined by

$$\mathbf{n}_b := \mathbf{e}_\vartheta. \tag{3.47}$$

By equation (2.79) of Chapter 2, we have

$$\nabla \mathbf{n}_b = -\frac{1}{r}\mathbf{e}_r \otimes \mathbf{e}_\vartheta,$$

whence

$$\operatorname{div} \mathbf{n}_b = 0, \quad \operatorname{curl} \mathbf{n}_b = \frac{1}{r^2}\mathbf{e}_z, \quad (\nabla \mathbf{n}_b)\mathbf{n}_b = -\frac{1}{r}\mathbf{e}_r, \quad (\nabla \mathbf{n}_b)^2 = 0. \tag{3.48}$$

Making use of (3.47), (3.48) and (3.24), from (3.23) with $\tau = 0$ we arrive at

$$\sigma_F(\mathbf{n}_b, \nabla \mathbf{n}_b) = k_3 \frac{1}{r^2}. \tag{3.49}$$

Thus, away from e_z, the energy density of $\mathbf{n}_b$, often called the **bend field**, is proportional to k_3.

Remark 3. Comparing (3.49) and (3.46), we see that both $\mathbf{n}_s$ and $\mathbf{n}_b$ have defects of infinite energy in any cylinder containing e_z. Furthermore, when $k_1 = k_3$ they have just the same energy density.

3.3.3 Twist modulus

We recall here from Section 3.1 the natural orientation characteristic of cholesterics (*cf.* equation (3.16)), but we write t instead of τ because we are dealing with nematics rather than with cholesterics:

$$\mathbf{n}_c(p) = \cos(t x_3 + \varphi_0)\mathbf{e}_1 + \sin(t x_3 + \varphi_0)\mathbf{e}_2. \tag{3.50}$$

In (3.50) a frame $(o, \mathbf{e}_1, \mathbf{e}_2, \mathbf{e}_3)$ has been employed, x_3 denotes the co-ordinate of p along $\mathbf{e}_3$ and φ_0 is an arbitrary angle. By (3.38), we see that when $\tau = 0$ in (3.23) we get

$$\sigma_F(\mathbf{n}_c, \nabla \mathbf{n}_c) = k_2 t^2. \tag{3.51}$$

Thus, for nematics, $\mathbf{n}_c$ represents an orientation whose energy, for any given $t \neq 0$, is proportional to k_2.

The field $\mathbf{n}_c$ is also called the **twist field** and sometimes the scalar t is also referred to as the twist.

Remark 4. Our development applies here only to nematics, but it is enlightening also for cholesterics, which may be regarded as twisted nematics. We should remember this note when in Section 4.4 of Chapter 4 we study a peculiar transition from cholesterics to nematics, which takes place without any change in the temperature.

3.3.4 Saddle-splay modulus

To accomplish the task we have undertaken in this section, we must also provide an interpretation for the last addend in (3.23). We shall do so here in an approximate way that we now make clear.

In a frame like that employed in (3.50) every point $p \in \mathscr{E}$ can be represented by

$$p - o = x_1 \mathbf{e}_1 + x_2 \mathbf{e}_2 + x_3 \mathbf{e}_3,$$

where x_1, x_2 and x_3 are the co-ordinates of p and $o \in \mathscr{E}$ is the origin of the frame. Then the equation

$$\varphi(x_1, x_2, x_3) = 0,$$

where

$$\varphi(x_1, x_2, x_3) := x_3 - x_1 x_2, \tag{3.52}$$

represents a **saddle-surface** of $\mathscr{E}$ containing the origin. The field

$$\mathbf{n}_\varphi := \frac{\nabla \varphi}{|\nabla \varphi|} \tag{3.53}$$

gives a unit normal at all points of this surface. We employ (3.53) to define a unit vector field in the whole of $\mathscr{E}$:

$$\mathbf{n}_\varphi = \frac{1}{\sqrt{1 + x_1^2 + x_2^2}} (-x_2 \mathbf{e}_1 - x_1 \mathbf{e}_2 + \mathbf{e}_3). \tag{3.54}$$

In the small cylinder of axis parallel to $\mathbf{e}_3$ defined by

$$\mathscr{C}_\varepsilon := \{p \in \mathscr{E} \,|\, x_1^2 + x_2^2 < \varepsilon^2\}$$

we have that

$$\mathbf{n}_\varphi = -x_2\mathbf{e}_1 - x_1\mathbf{e}_2 + \mathbf{e}_3 + o(\varepsilon),$$
$$\nabla\mathbf{n}_\varphi = -\mathbf{e}_1 \otimes \mathbf{e}_2 - \mathbf{e}_2 \otimes \mathbf{e}_1 - x_1\mathbf{e}_3 \otimes \mathbf{e}_1 - x_2\mathbf{e}_3 \otimes \mathbf{e}_2 + o(\varepsilon),$$

and so

$$\operatorname{div}\mathbf{n}_\varphi = o(\varepsilon), \quad \operatorname{curl}\mathbf{n}_\varphi \cdot \mathbf{n}_\varphi = o(\varepsilon),$$
$$(\nabla\mathbf{n}_\varphi)\mathbf{n}_\varphi = -x_1\mathbf{e}_1 - x_2\mathbf{e}_2 + o(\varepsilon), \quad \operatorname{tr}(\nabla\mathbf{n}_\varphi)^2 = 2 + o(\varepsilon^2). \tag{3.55}$$

Making use of (3.55) in (3.23), we easily conclude that when $\tau = 0$

$$\sigma_F(\mathbf{n}_\varphi, \nabla\mathbf{n}_\varphi) = 2(k_2 + k_4) + o(\varepsilon). \tag{3.56}$$

Thus, $\mathbf{n}_\varphi$ describes an orientation whose energy density is proportional to $(k_2 + k_4)$, at least in the vicinity of the axis e_z.

We call $\mathbf{n}_\varphi$ a **saddle-splay field.**

Intuitively, for a generic orientation $\mathbf{n}$, the various moduli present in (3.23) measure the energy associated with each of the above special fields, to the extent that they can be adjusted to approximate $\mathbf{n}$. In other words, they express how the energy of a liquid crystal is sensitive, as it were, to four independent components of the orientation field.

If in (3.23) we set

$$\tau = 0, \quad k_1 = k_2 = k_3 =: \kappa \quad \text{and} \quad k_4 = 0, \tag{3.57}$$

then, by (3.25), σ_F reads as

$$\sigma_F(\mathbf{n}, \nabla\mathbf{n}) = \kappa|\nabla\mathbf{n}|^2. \tag{3.58}$$

We call such a simple form of σ_F the **one-constant approximation** to Frank's formula. It is often employed in the early stages of the study of many specific problems, though it might sometimes be misleading. Stephen and Straley (1974) refer to (3.58) as the **theoretician's energy** (*cf.* p. 626), implying that it can be useful only for qualitative considerations.

As is clear from $(3.57)_1$, (3.58) applies only to nematics. An approximation that parallels (3.58) for cholesterics has been proposed by Ericksen (1976) (*cf.* equation (35)); it can be given the form

$$\sigma_F(\mathbf{n}, \nabla\mathbf{n}) = k(|\nabla\mathbf{n}|^2 - \tau^2) + h(\mathbf{n} \cdot \operatorname{curl}\mathbf{n} + \tau)^2, \tag{3.59}$$

which follows from (3.23) when

$$k_0 = -2k, \quad k_1 = k_2 = k, \quad k_3 = k + h, \quad k_4 = -h. \tag{3.60}$$

For thermotropic liquid crystals the material moduli k_1, k_2, k_3 and k_4 depend on the temperature T. For PAA at $T = 125°\text{C}$, for example, k_1, k_2 and k_3 take the values below:

$$k_1 = 2.3 \times 10^{-7}\,\text{dyne}, \quad k_2 = 1.5 \times 10^{-7}\,\text{dyne}, \\ k_3 = 4.8 \times 10^{-7}\,\text{dyne}. \tag{3.61}$$

They have been determined by Zvetkov in 1937 through an experiment whose underlying principles will be described in Chapter 4. (More precisely, in (3.61) use has also been made of additional experimental data obtained by Zvetkov and Sosnovsky in 1943.) The values of k_1, k_2 and k_3 for PAA at temperatures other than 125°C, as well as those for MBBA at various temperatures are listed from different sources in De Gennes (1974) (*cf.* pp. 65, 66) and in Stephen and Straley (1974) (*cf.* p. 700). It is clear from such a collection of data that k_1, k_2 and k_3 all decrease as T increases towards the transition temperature T_{NI}. More recent measurements of these constants for a number of nematic liquid crystals at different temperatures are reported by Bunning, Faber and Sherrell (1981) and Bradshaw, Raynes, Bunning and Faber (1985).

The reader might wonder why the material modulus k_4 is never mentioned among the experimental data available for liquid crystals. The reason is that until very recently no experiment was devised to measure k_4. This is indeed a consequence of Remark 3 of Section 3.2. In the experimental settings that allow measurements of k_1, k_2 and k_3, σ_L does not even contribute to the value of the energy functional $\mathscr{F}_B$ (*cf.* Section 3.8 below). Thus, the value of the saddle-splay modulus cannot affect the observed orientations.

To measure k_4, one has to conceive an experiment where σ_L plays a role. Several papers have recently been published to fill this need. Strigazzi (1987) considers a thin interstice between two coaxial cylinders where a nematic liquid crystal is subject to a magnetic field parallel to the axis of the cylinders. The orientation of the optic axis suffers an instability at a critical strength of the field, which depends on k_4. Allender, Crawford and Doane (1991) study the stable orientation field inside a cylinder and find that it differs according to whether the radius is above or below a critical value; such a transition is related to k_4 in a way which is too involved to be explained briefly here. Lavrentovich (1991) shows that a measurable feature exhibited by some hybridly aligned nematic films, such as those described in Section 1.4 of Chapter 1, is indeed a consequence of σ_L, and so he is able to estimate k_4 in terms of a characteristic value of the other Frank's constants. Finally, Žumer and Kralj (1992) study the influence of k_4 on the equilibrium configurations of the orientation in droplets of nematic liquid crystals dispersed in an isotropic liquid. The phase diagrams that they have computed may lead to measurements of k_4 by observing how the orientation changes in a single drop subject to a magnetic field of increasing strength.

Though most of the experiments performed so far are affected by large errors, the values of k_4 seem to be comparable to those of the other Frank's constants.

This issue is still far from being resolved. Here we only note that Nehring and Saupe (1971) derived from their molecular theory of liquid crystals the following relation:

$$k_4 = \tfrac{1}{2}(k_1 - k_2) \tag{3.62}$$

(*cf.* their equation (13)). If (3.62) were valid, Frank's constants would reduce to three. This is not surprising since a molecular theory usually leads to less macroscopic moduli than a genuine macroscopic theory, as is the case, for example, in the classical theory of elasticity where Cauchy's relations are analogous to (3.62) (*cf. e.g.* Love (1944), p. 100).

At present it seems that there is no good reason in favour of (3.62). In this book we shall continue to regard k_4 as an independent modulus.

3.4 Ericksen's inequalities

In Section 3.1 we have listed all the requirements that the energy density σ of both nematics and cholesterics is expected to satisfy. In Section 3.2 we have explored for σ_F the consequences of all these requirements, except the last one, arriving at formula (3.23), where Frank's constants made their first appearance. It is time now to take into account the requirement left aside so far. We derive in this section the conditions that Frank's constants must satisfy to render σ_F positive definite. As usual, we shall consider below nematics and cholesterics separately.

3.4.1 Nematics

For nematic liquid crystals $\tau = 0$, and so (3.23) reads as follows

$$\sigma_F(\mathbf{n}, \mathbf{N}) = k_1(\operatorname{tr}\mathbf{N})^2 + k_2(\mathbf{W}(\mathbf{n})\cdot\mathbf{N})^2 + k_3|\mathbf{N}\mathbf{n}|^2 + (k_2 + k_4)(\operatorname{tr}\mathbf{N}^2 - (\operatorname{tr}\mathbf{N})^2), \tag{3.63}$$

where we have set $\mathbf{N} = \nabla\mathbf{n}$ and we have employed (3.24), (3.33) and (3.34).

Theorem 3.4. The function σ_F in (3.63) satisfies

$$\sigma_F(\mathbf{n}, \mathbf{N}) \geqslant 0 \tag{3.64}$$

for any given $\mathbf{n}\in\mathbb{S}^2$ and all $\mathbf{N}\in L(\mathbf{n}, \mathscr{V})$ if, and only if,

$$2k_1 \geqslant k_2 + k_4, \quad k_2 \geqslant |k_4|, \quad k_3 \geqslant 0. \tag{3.65}$$

Proof. We first observe that if (3.64) holds for a given $\mathbf{n}\in\mathbb{S}^2$ and all $\mathbf{N}\in L(\mathbf{n}, \mathscr{V})$, then it holds also for any other $\mathbf{n}^*\in\mathbb{S}^2$ and all $\mathbf{N}\in L(\mathbf{n}^*, \mathscr{V})$. To see this, we recall that σ_F obeys

$$\sigma_F(\mathbf{Q}\mathbf{n}, \mathbf{Q}\mathbf{N}\mathbf{Q}^T) = \sigma_F(\mathbf{n}, \mathbf{N}) \tag{3.66}$$

for any given $\mathbf{n}\in\mathbb{S}^2$ and for all $\mathbf{N}\in L(\mathbf{n}, \mathscr{V})$ and $\mathbf{Q}\in O(\mathscr{V})$. If $\mathbf{n}$ and $\mathbf{n}^*$ are given vectors of $\mathbb{S}^2$ there is a tensor $\mathbf{Q}\in SO(\mathscr{V})$ such that $\mathbf{n}^* = \mathbf{Q}\mathbf{n}$, and it follows from (3.66) that

$$\sigma_F(\mathbf{n}^*, \mathbf{N}^*) = \sigma_F(\mathbf{n}, \mathbf{Q}^T\mathbf{N}^*\mathbf{Q}) \tag{3.67}$$

for all $\mathbf{N}^*\in L(\mathbf{n}^*, \mathscr{V})$. Since $\mathbf{Q}^T\mathbf{N}^*\mathbf{Q}$ belongs to $L(\mathbf{n}, \mathscr{V})$ whenever $\mathbf{N}^*$ belongs to $L(\mathbf{n}^*, \mathscr{V})$, (3.67) implies that $\sigma_F(\mathbf{n}^*, \cdot)$ is positive definite on $L(\mathbf{n}^*, \mathscr{V})$ if $\sigma_F(\mathbf{n}, \cdot)$ is positive definite on $L(\mathbf{n}, \mathscr{V})$.

Let $\mathbf{n}\in\mathbb{S}^2$ be given and let $(\mathbf{e}_1, \mathbf{e}_2, \mathbf{e}_3)$ be a basis of $\mathscr{V}$ such that

$$\mathbf{e}_3 = \mathbf{n} \quad \text{and} \quad \mathbf{e}_1 \wedge \mathbf{e}_2 = \mathbf{e}_3. \tag{3.68}$$

Thus,

$$\mathbf{W}(\mathbf{n}) = \mathbf{e}_2 \otimes \mathbf{e}_1 - \mathbf{e}_1 \otimes \mathbf{e}_2$$

and every member of $L(\mathbf{n}, \mathscr{V})$ has the form

$$\mathbf{N} = \sum_{i,j=1}^{3} \alpha_{ij}\mathbf{e}_i \otimes \mathbf{e}_j, \tag{3.69}$$

where

$$\alpha_{3j} = 0 \quad \text{for } j = 1, 2, 3. \tag{3.70}$$

It follows from (3.68), (3.69) and (3.70) that

$$\operatorname{tr}\mathbf{N}=\alpha_{11}+\alpha_{22},\quad \mathbf{W}(\mathbf{n})\cdot\mathbf{N}=\alpha_{21}-\alpha_{12},$$
$$\mathbf{N}\mathbf{n}=\alpha_{13}\mathbf{e}_1+\alpha_{23}\mathbf{e}_2,\quad \operatorname{tr}\mathbf{N}^2=\alpha_{11}^2+\alpha_{22}^2+2\alpha_{12}\alpha_{21}. \tag{3.71}$$

Inserting (3.71) into (3.63) we get

$$\sigma_F(\mathbf{n},\mathbf{N})=k_1(\alpha_{11}^2+\alpha_{22}^2)+2(k_1-k_2-k_4)\alpha_{11}\alpha_{22}$$
$$+k_2(\alpha_{12}^2+\alpha_{21}^2)+2k_4\alpha_{12}\alpha_{21}+k_3(\alpha_{13}^2+\alpha_{23}^2). \tag{3.72}$$

The right-hand side of (3.72) is the sum of three independent quadratic functions, and so (3.64) is satisfied whenever all these functions are not negative. Clearly, the last one is so if, and only if, $(3.65)_3$ is satisfied. The first two both have the following form:

$$f(x,y)=k(x^2+y^2)+2hxy. \tag{3.73}$$

It is an easy exercise to check that the function defined in $\mathbb{R}^2$ by (3.73) attains its minimum at the origin if, and only if, the parameters k and h satisfy

$$k\geqslant|h|. \tag{3.74}$$

Thus, coming back to (3.72), we conclude that (3.64) holds if, and only if, Frank's constants satisfy $(3.65)_3$ and

$$k_1\geqslant|k_1-k_2-k_4|,\quad k_2\geqslant|k_4|. \tag{3.75}$$

The second inequality in (3.75) is just the same as $(3.65)_2$, while the first and the following inequalities are equivalent:

$$2k_1\geqslant k_2+k_4,\quad k_2+k_4\geqslant 0. \tag{3.76}$$

Now, $(3.76)_1$ is precisely $(3.65)_1$, and $(3.76)_2$ is a consequence of $(3.65)_2$.

Inequalities (3.65) were first derived by Ericksen (1966): they will be referred to as **Ericksen's inequalities**.

Remark 1. If k_4 were to be related to k_1 and k_2 through (3.62), $(3.65)_{1,2}$ would become

$$3k_1\geqslant k_2,\quad k_2\geqslant 0. \tag{3.77}$$

The values of Frank's constants listed in (3.61) satisfy (3.77) and so do other experimental data (*cf.* Ericksen (1976), p. 239), but, of course, this does not prove (3.62).

It is worth noting that inequalities (3.65) are also compatible with the one-constant approximation introduced in the preceding section.

3.4.2 Cholesterics

For cholesteric liquid crystals $\tau \neq 0$ and formula (3.23) can be rewritten as

$$\sigma_F(\mathbf{n}, \mathbf{N}) = k_0\tau(\mathbf{W}(\mathbf{n})\cdot\mathbf{N} + \tau) + k_1(\operatorname{tr}\mathbf{N})^2 + k_2(\mathbf{W}(\mathbf{n})\cdot\mathbf{N} + \tau)^2 + k_3|\mathbf{N}\mathbf{n}|^2 + (k_2 + k_4)(\operatorname{tr}\mathbf{N}^2 - (\operatorname{tr}\mathbf{N})^2), \tag{3.78}$$

where $\nabla\mathbf{n}$ has been replaced by $\mathbf{N}$ and use has been made of (3.24), (3.33) and (3.34).

Theorem 3.5. The function σ_F in (3.78) is such that

$$\sigma_F(\mathbf{n}, \mathbf{N}) \geqslant 0$$

for any given $\mathbf{n} \in \mathbb{S}^2$ and all $\mathbf{N} \in L(\mathbf{n}, \mathscr{V})$ if, and only if,

$$k_0 = 0, \quad k_2 + k_4 = 0, \tag{3.79}$$

and

$$k_1 \geqslant 0, \quad k_2 \geqslant 0, \quad k_3 \geqslant 0. \tag{3.80}$$

Proof. We learned in proving Theorem 3.4 that σ_F is not negative for any given $\mathbf{n} \in \mathbb{S}^2$ and all $\mathbf{N} \in L(\mathbf{n}, \mathscr{V})$ whenever it is not negative for just one $\mathbf{n} \in \mathbb{S}^2$ and all $\mathbf{N} \in L(\mathbf{n}, \mathscr{V})$. Thus, we select the same basis of $\mathscr{V}$ as in (3.68) and we write $\mathbf{N}$ as in (3.69), taking into account (3.70). A computation which parallels that performed in the proof of Theorem 3.4 leads us to

$$\sigma_F(\mathbf{n}, \mathbf{N}) = k_1(\alpha_{11}^2 + \alpha_{22}^2) + 2(k_1 - k_2 - k_4)\alpha_{11}\alpha_{22} + k_2(\alpha_{21}^2 + \alpha_{12}^2) + 2k_4\alpha_{12}\alpha_{21} + \tau(2k_2 + k_0)(\alpha_{21} - \alpha_{12}) + \tau^2(k_0 + k_2) + k_3(\alpha_{13}^2 + \alpha_{23}^2). \tag{3.81}$$

We write $\mathbf{n}_c$, the prototype of all natural orientations of cholesterics, taking $\mathbf{e}_2$ as cholesteric axis (*cf.* (3.16)):

$$\mathbf{n}_c = \cos\tau x_2\mathbf{e}_3 + \sin\tau x_2\mathbf{e}_1; \tag{3.82}$$

here $\varphi_0 = 0$ and x_2 denotes the co-ordinate along $\mathbf{e}_2$. It follows from (3.82) that

$$\nabla\mathbf{n}_c = -\tau\sin\tau x_2\mathbf{e}_3 \otimes \mathbf{e}_2 + \tau\cos\tau x_2\mathbf{e}_1 \otimes \mathbf{e}_2. \tag{3.83}$$

For $x_2 = 0$, $\mathbf{n}_c = \mathbf{e}_3$ and $\nabla\mathbf{n}_c = \tau\mathbf{e}_1 \otimes \mathbf{e}_2$: the latter is a member of

$L(\mathbf{e}_3, \mathscr{V})$ expressed by (3.69) with

$$\alpha_{12}=\tau, \quad \text{and} \quad \alpha_{ij}=0 \quad \text{for all } i\neq 1 \text{ and } j\neq 2. \tag{3.84}$$

It is an easy consequence of (3.81) that σ_F vanishes when (3.84) hold; thus, σ_F is not negative if, and only if, it attains its minimum when (3.84) hold. The right-hand side of (3.81) is the sum of three independent quadratics. If we apply to the first and the last of these functions the same argument as was employed in the proof of Theorem 3.4, we arrive at the following inequalities

$$k_1 \geqslant |k_1-k_2-k_4|, \quad k_3 \geqslant 0. \tag{3.85}$$

On the other hand, if we require the second quadratic on the right-hand side of (3.81) to be stationary for $\alpha_{21}=0$ and $\alpha_{12}=\tau$, we get

$$-\tau k_0=0 \quad \text{and} \quad \tau(k_0+2k_2+2k_4)=0,$$

whence we have (3.79), since $\tau\neq 0$. By $(3.79)_2$, $(3.85)_1$ becomes $(3.80)_1$. We can now complete the proof of the theorem, arriving at $(3.80)_2$ by the same argument that led us to $(3.65)_2$ in Theorem 3.4.

The conclusion reached in Theorem 3.5 has been criticized by Jenkins (1971), who objected that condition $(3.79)_2$ is too restrictive for cholesterics. In Jenkins's view any free energy quadratic in the gradient of the orientation field, such as Frank's, should more properly be regarded as an approximation which is as appropriate as the orientation field is close, in a sense to be made precise, to a natural orientation of the liquid crystal. Thus, if Frank's formula is adequate to represent a quadratic approximation to the free energy of nematics, whose natural orientations are constant fields, this may not be the case for cholesterics, whose natural orientations possess a uniform twist.

Let $\mathbf{n}_c$ be a natural orientation of a cholesteric liquid crystal as above. Jenkins (1971) writes $\mathbf{n}$ as

$$\mathbf{n}=\mathbf{n}_c+\mathbf{u}, \tag{3.86}$$

where, since $\mathbf{n}$ is a unit vector, $\mathbf{u}$ is subject to

$$2\mathbf{n}_c\cdot\mathbf{u}+\mathbf{u}\cdot\mathbf{u}=0; \tag{3.87}$$

he proposes to write the free energy density as a function of $\mathbf{u}$ and $\nabla\mathbf{u}$, in which only terms linear and quadratic in both variables are retained:

$$\sigma_J^*(\mathbf{n}_c, \nabla\mathbf{n}_c; \mathbf{u}, \nabla\mathbf{u}):=\mathbf{b}\cdot\mathbf{u}+\mathbf{B}\cdot\nabla\mathbf{u}+\mathbf{u}\cdot\mathbf{C}\mathbf{u}+\mathbf{u}\cdot\mathfrak{a}[\nabla\mathbf{u}]+\nabla\mathbf{u}\cdot\mathbb{A}[\nabla\mathbf{u}], \tag{3.88}$$

where $\mathbf{b}$ is a vector of $\mathscr{V}$, $\mathbf{B}$ and $\mathbf{C}$ are tensors of $L(\mathscr{V})$, $\mathfrak{a}$ is a linear mapping which takes $L(\mathscr{V})$ into $\mathscr{V}$ (that is, a third-order tensor), and $\mathbb{A}$ is a tensor of $L^2(\mathscr{V})$. All these tensors depend on $\mathbf{n}_c$ and $\nabla\mathbf{n}_c$, as Jenkins regards σ_J^* as a series expansion about the natural orientation of a function which depends on $\mathbf{n}$ and $\nabla\mathbf{n}$. We have not made such a dependence explicit to avoid clutter.

If we apply to σ_J^* the requirements laid down for σ in Section 3.1, with the aid of (3.86) we easily see that σ_J^* must be such that

$$\sigma_J^*(\mathbf{Q}\mathbf{n}_c, \mathbf{Q}\nabla\mathbf{n}_c\mathbf{Q}^T; \mathbf{Q}\mathbf{u}, \mathbf{Q}\nabla\mathbf{u}\mathbf{Q}^T) = \sigma_J^*(\mathbf{n}_c, \nabla\mathbf{n}_c; \mathbf{u}, \nabla\mathbf{u}), \tag{3.89}$$

$$\sigma_J^*(-\mathbf{n}_c, -\nabla\mathbf{n}_c; -\mathbf{u}, -\nabla\mathbf{u}) = \sigma_J^*(\mathbf{n}_c, \nabla\mathbf{n}_c; \mathbf{u}, \nabla\mathbf{u}), \tag{3.90}$$

$$\sigma_J^*(\mathbf{n}_c, \nabla\mathbf{n}_c; \mathbf{u}, \nabla\mathbf{u}) \geqslant 0, \tag{3.91}$$

for all $\mathbf{Q}\in SO(\mathscr{V})$ and all fields $\mathbf{u}$ that satisfy (3.87). The reader should compare these equations and (3.10), (3.13) and (3.15). Equation (3.14) does not impose any further constraint on σ_J^*, since it follows from (3.88) that σ_J^* vanishes when $\mathbf{u} \equiv 0$.

Jenkins has determined the most general function of the form (3.88) that obeys (3.89), (3.90) and (3.91); with a slight change in notation the formula he found can be written as follows (*cf.* (3.91) and (5.20)–(5.22) of Jenkins (1971)):

$$\begin{aligned}
&\sigma_J^*(\mathbf{n}_c, \nabla\mathbf{n}_c; \mathbf{u}, \nabla\mathbf{u}) \\
&\quad = \beta_1|(\nabla\mathbf{n}_c)\mathbf{u}|^2 + 2\beta_1(\nabla\mathbf{n}_c)\mathbf{u}\cdot(\nabla\mathbf{u})\mathbf{n}_c \\
&\qquad + \beta_1\tau^{-2}|(\nabla\mathbf{n}_c)^T(\nabla\mathbf{u})\mathbf{n}_c|^2 + \beta_2\tau^{-2}|(\nabla\mathbf{n}_c)(\nabla\mathbf{u})\mathbf{n}_c|^2 \\
&\qquad + \beta_3((\nabla\mathbf{n}_c)\cdot(\nabla\mathbf{u}))^2 + \beta_4((\nabla\mathbf{n}_c)^T\cdot(\nabla\mathbf{u}))^2 \\
&\qquad + \beta_5(|(\nabla\mathbf{n}_c)(\nabla\mathbf{u})(\nabla\mathbf{n}_c)^T|^2 + |(\nabla\mathbf{n}_c)^T(\nabla\mathbf{u})(\nabla\mathbf{n}_c)|^2) \\
&\qquad + \beta_5\tau^2((\nabla\mathbf{n}_c)(\nabla\mathbf{u})\cdot(\nabla\mathbf{u})(\nabla\mathbf{n}_c) + (\nabla\mathbf{n}_c)^T(\nabla\mathbf{u})\cdot(\nabla\mathbf{u})(\nabla\mathbf{n}_c)^T) \\
&\qquad + \beta_6((\nabla\mathbf{n}_c)(\nabla\mathbf{u})\cdot(\nabla\mathbf{u})(\nabla\mathbf{n}_c)^T + (\nabla\mathbf{n}_c)^T(\nabla\mathbf{u})\cdot(\nabla\mathbf{u})(\nabla\mathbf{n}_c)),
\end{aligned} \tag{3.92}$$

where the scalars β_i, $i = 1, \ldots, 6$, may depend on τ and must satisfy the inequalities

$$\beta_i \geqslant 0 \quad \text{for } i = 1, \ldots, 5 \quad \text{and} \quad \beta_6^2 - \beta_3\beta_4 \geqslant 0. \tag{3.93}$$

Formula (3.92) is very complicated. Building upon it, Jenkins proposed a simpler formula to replace Frank's, that is

$$\begin{aligned}
\sigma_J(\mathbf{n}, \nabla\mathbf{n}) := {}& \alpha_1(\mathbf{n}\cdot\operatorname{curl}\mathbf{n} + \tau)^2 + \alpha_2|(\nabla(\mathbf{n})\mathbf{n})|^2 + \alpha_3(\operatorname{div}\mathbf{n})^2 \\
&+ 2\alpha_4(\mathbf{A}\cdot\mathbf{A} - \tfrac{1}{2}\tau^2)(\mathbf{n}\cdot\operatorname{curl}\mathbf{n} + \tau) + \alpha_5|(\nabla\mathbf{n})(\operatorname{curl}\mathbf{n})|^2 \\
&+ \alpha_6(\mathbf{A}\cdot\mathbf{A} - \tfrac{1}{2}\tau^2)^2,
\end{aligned} \tag{3.94}$$

where

$$\mathbf{A} := \tfrac{1}{2}(\nabla\mathbf{n} + (\nabla\mathbf{n})^T).$$

What links σ_J and σ_J^* is the formula

$$\sigma_J(\mathbf{n}_c + \mathbf{u}, \nabla\mathbf{n}_c + \nabla\mathbf{u}) = \sigma_J^*(\mathbf{n}_c, \nabla\mathbf{n}_c; \mathbf{u}, \nabla\mathbf{u}) + \sigma_2(\mathbf{n}_c, \nabla\mathbf{n}_c; \mathbf{u}, \nabla\mathbf{u}), \quad (3.95)$$

where σ_2 is infinitesimal of order higher than 2 with respect to both $\mathbf{u}$ and $\nabla\mathbf{u}$. In other words, σ_J^* is precisely the quadratic approximation to σ_J in the vicinity of the field $\mathbf{n}_c$.

Remark 2. Similarly, if in Frank's formula for nematics we write $\mathbf{e} + \mathbf{u}$ for $\mathbf{n}$, with $\mathbf{e}$ a constant unit vector field, we have

$$\sigma_F(\mathbf{n}, \nabla\mathbf{n}) = \sigma_F^*(\mathbf{e}; \mathbf{u}, \nabla\mathbf{u}) + \sigma_2(\mathbf{e}; \mathbf{u}, \nabla\mathbf{u}),$$

where

$$\begin{aligned}\sigma_F^*(\mathbf{e}; \mathbf{u}, \nabla\mathbf{u}) := {} & k_1(\operatorname{div}\mathbf{u})^2 + k_2(\mathbf{e}\cdot\operatorname{curl}\mathbf{u})^2 + k_3|(\nabla\mathbf{u})\mathbf{e}|^2 \\ & + (k_2 + k_4)(\operatorname{tr}(\nabla\mathbf{u})^2 - (\operatorname{div}\mathbf{u})^2)\end{aligned}$$

and σ_2 is infinitesimal of order higher than 2 with respect to both $\mathbf{u}$ and $\nabla\mathbf{u}$. We reach just the same conclusion as in Theorem 3.4 above, if we require σ_F^* to be positive definite for all fields $\mathbf{u}$ such that

$$2\mathbf{e}\cdot\mathbf{u} + \mathbf{u}\cdot\mathbf{u} = 0.$$

In proving (3.95), Jenkins (1971) established the equalities

$$\begin{aligned}&2\beta_1 = \alpha_2, \quad 2\beta_2 = \alpha_2 + \tau\alpha_5, \quad 2\beta_3 = \tau^{-2}\alpha_1 - 2\tau^{-1}\alpha_4 + 4\alpha_6, \\ &2\beta_4 = \tau^{-2}\alpha_1, \quad 2\beta_5 = \alpha_3, \quad 2\beta_6 = -\tau^{-2}\alpha_1 + 2\tau^{-1}\alpha_4. \end{aligned} \quad (3.96)$$

If in (3.94) we treat $(\mathbf{n}\cdot\operatorname{curl}\mathbf{n} + \tau)$, $|(\nabla\mathbf{n})\mathbf{n}|$, $\operatorname{div}\mathbf{n}$, and $(\mathbf{A}\cdot\mathbf{A} - \frac{1}{2}\tau^2)$ as if they were independent variables, we easily see that the inequalities

$$\alpha_i \geqslant 0 \quad \text{for } i = 2, 3, 5 \quad \text{and} \quad \alpha_1 + \alpha_6 \geqslant 0, \quad \alpha_4^2 - \alpha_1\alpha_6 \leqslant 0 \quad (3.97)$$

ensure that σ_J attains its minimum for $\mathbf{n} \equiv \mathbf{n}_c$. With the aid of (3.96), inequalities (3.97) also imply (3.93), as it should be.

Remark 3. If in (3.94) we set equal to 0 all coefficients but α_3, since $\operatorname{div}\mathbf{n}_c = 0$ we find out that the sum of all terms that in (3.92) are multiplied by β_5 is simply proportional to $(\operatorname{div}\mathbf{u})^2$. To prove this directly, one might compute all the aforementioned terms of (3.92) at a point where $x_2 = 0$. There, by (3.82) and (3.83), we have

$$\mathbf{n}_c = \mathbf{e}_3, \quad \nabla\mathbf{n}_c = \tau\mathbf{e}_1 \otimes \mathbf{e}_2. \quad (3.98)$$

Taking **u** in the form

$$\mathbf{u} = u_1\mathbf{e}_1 + u_2\mathbf{e}_2 + u_3\mathbf{e}_3,$$

we get from (3.87) that

$$u_3 = o(|\mathbf{u}|) \quad \text{and} \quad (\nabla\mathbf{u})^T\mathbf{e}_3 + \tau(\mathbf{e}_2 \otimes \mathbf{e}_1)\mathbf{u} = o(|\mathbf{u}|).$$

Thus, to within terms infinitesimal of order higher than 1 with respect to **u**, $\nabla\mathbf{u}$ can be written as

$$\nabla\mathbf{u} = \sum_{i,j=1}^{3} u_{ij}\mathbf{e}_i \otimes \mathbf{e}_j \quad \text{with} \quad u_{33} = u_{31} = 0 \quad \text{and} \quad u_{32} = -\tau u_1. \tag{3.99}$$

Putting together (3.98) and (3.99), we readily arrive at the desired conclusion.

Formula (3.94) provides just one function of **n** and $\nabla\mathbf{n}$ which agrees with σ_J^* to within terms more than quadratic in the field that expresses the discrepancy between **n** and the natural orientation $\mathbf{n}_c$. Though it appears as the simplest function with such a property that one would imagine, it clearly fails to be quadratic in $\nabla\mathbf{n}$. Maybe this is what has prevented researchers from making extensive use of it. Furthermore, Nehring and Saupe (1971) argued that the distorsion in the natural orientations of cholesterics should also call for terms involving second-order derivatives of **n** in the free energy density. They proposed another formula which is occasionally employed; we shall comment briefly on it in Section 3.9 below.

Nevertheless, Frank's formula is commonly adopted. De Gennes (1974) justifies such a choice by estimating the order of magnitude of the corrections to Frank's formula proposed by Jenkins: he finds them to be negligible in all practical situations (*cf.* p. 239). Besides, Frank's formula for cholesterics was successfully employed by De Gennes (1968) to predict an instability which shortly afterwards was confirmed experimentally by Durand, Leger, Rondalez and Veyssie (1969) (*cf.* Section 4.4 of Chapter 4).

In this book we keep Frank's formula and hold valid the conclusions of Theorem 3.5 above, though the chances are that Jenkins's formula might be needed to explain new phenomena.

3.5 Equilibrium equations

In this section we shall arrive at the equilibrium equations of the functional $\mathscr{F}_B$ defined in (3.2) above. We take $\mathscr{B}$ as a given regular region in space occupied by a liquid crystal and σ, the free energy density per unit volume, as a function of class C^2 which satisfies the requirements posited in Section 3.1. Clearly, the conclusions that we shall reach here apply, in particular, when $\sigma = \sigma_F$.

Let $\mathscr{S}$ be a closed subset of the closure of $\mathscr{B}$. We assume that the optic axis is described by a field of class C^2 away from $\mathscr{S}$. The set $\mathscr{S}$, which may well be empty, is called the **singular set** of $\mathbf{n}$: it is to be interpreted as a **defect** of $\mathbf{n}$. In Chapter 6 we shall give a fairly exhaustive account of defects in liquid crystals. Here we focus attention on the equations that govern the optic axis at regular points of $\mathscr{B}$. Our development is merely local: it does not either imply or resort to global properties of $\mathscr{B}$ or $\mathscr{S}$.

We say that $\mathbf{n}: \mathscr{B} \cup \partial^* \mathscr{B} \setminus \mathscr{S} \to \mathbb{S}^2$ represents an **equilibrium configuration** for the optic axis if $\mathscr{F}_B$ is stationary at $\mathbf{n}$ among all unit vector fields of class C^2 on $\mathscr{B} \cup \partial^* \mathscr{B} \setminus \mathscr{S}$ that satisfy appropriate boundary conditions. Before proceeding further we briefly discuss these latter.

Here $\mathbf{n}$ will be prescribed on the reduced boundary $\partial^* \mathscr{B}$ because I think that a physically preferred direction such as $\mathbf{n}$ might be assigned only where the most elementary geometrically preferred direction, *i.e.* the normal, is defined. Besides the case when $\mathbf{n}$ is completely free on $\partial^* \mathscr{B}$, two other cases are often encountered: in the one $\mathbf{n}$ is set equal to a given field $\mathbf{n}_0$ defined on $\partial^* \mathscr{B}$, in the other only the inner product of $\mathbf{n}$ with the normal $\mathbf{v}$ to $\mathscr{B}$ is prescribed on $\partial^* \mathscr{B}$. When $\mathbf{n}$ is free at the boundary we say that there is **no anchoring**; when $\mathbf{n}$ is set equal to $\mathbf{n}_0$ on $\partial^* \mathscr{B}$ we say that it is subject to **strong anchoring**; finally, when only $\mathbf{n} \cdot \mathbf{v}$ is prescribed on $\partial^* \mathscr{B}$ we say that $\mathbf{n}$ is subject to **conical anchoring**. There are two extreme instances of conical anchoring worth noting: they occur, respectively, when $\mathbf{n}$ is parallel to $\mathbf{v}$ and when $\mathbf{n}$ is orthogonal to $\mathbf{v}$. If $\mathbf{n}$ is parallel to $\mathbf{v}$ the conical anchoring becomes indeed a strong anchoring; this boundary condition is often called **homeotropic**. If, on the other hand, $\mathbf{n}$ is orthogonal to $\mathbf{v}$, the boundary condition is called **tangential**.

All these boundary conditions for $\mathbf{n}$ will be considered in more detail below. We should first see when $\mathscr{F}_B$ is stationary in general, regardless of any specific boundary condition, and so we need to compute variations of $\mathscr{F}_B$. It should be noted in passing that boundary conditions are often responsible for the appearance of defects in liquid crystals. For example, when a genuine conical anchoring is imposed on the boundary of a convex region $\mathscr{B}$, the orientation field must have a defect at least on $\partial \mathscr{B}$.

Let $\varepsilon_0 > 0$ be given and let $\mathbf{u}: \mathscr{B} \cup \partial^*\mathscr{B} \setminus \mathscr{S} \to \mathscr{V}$ be a field of class C^2. For every $\varepsilon \in [-\varepsilon_0, \varepsilon_0]$ we define the field $\mathbf{n}_\varepsilon$ of $\mathscr{B} \cup \partial^*\mathscr{B} \setminus \mathscr{S}$ into $\mathbb{S}^2$ by

$$\mathbf{n}_\varepsilon(p) := \frac{\mathbf{n}(p) + \varepsilon\mathbf{u}(p)}{|\mathbf{n}(p) + \varepsilon\mathbf{u}(p)|} \quad \text{for all } p \in \mathscr{B} \cup \partial^*\mathscr{B} \setminus \mathscr{S}. \tag{3.100}$$

The field $\mathbf{u}$ is to be defined on $\partial^*\mathscr{B}$ so as to ensure that $\mathbf{n}_\varepsilon$ satisfies for all $\varepsilon \in [-\varepsilon_0, \varepsilon_0]$ the same boundary conditions to which $\mathbf{n}$ is subject. We shall see below how we need to restrict the values of $\mathbf{u}$ on $\partial^*\mathscr{B}$ so as to make $\mathbf{n}_\varepsilon$ satisfy the anchoring conditions for $\mathbf{n}$ mentioned above.

The field $\mathbf{n}_\varepsilon$ is of class C^2 for all $\varepsilon \in [-\varepsilon_0, \varepsilon_0]$. Furthermore, the mapping $\varepsilon \mapsto \mathbf{n}_\varepsilon(p)$ defines for each $p \in \mathscr{B} \cup \partial^*\mathscr{B} \setminus \mathscr{S}$ a curve in $\mathbb{S}^2$, which for $\varepsilon = 0$ crosses $\mathbf{n}(p)$. It follows from (3.100) that

$$\mathbf{n}_\varepsilon(p) = \mathbf{n}(p) + \varepsilon\mathbf{P}(\mathbf{n}(p))\mathbf{u}(p) + o(\varepsilon) \quad \text{for all } p \in \mathscr{B} \cup \partial^*\mathscr{B} \setminus \mathscr{S}. \tag{3.101}$$

As ε spans the interval $[-\varepsilon_0, \varepsilon_0]$, $\mathbf{n}_\varepsilon$ describes a **path** of configurations about $\mathbf{n}$, which depends on $\mathbf{u}$. We apply the term **first variation** of $\mathscr{F}_B$ at $\mathbf{n}$ to the functional $\delta\mathscr{F}_B(\mathbf{n})$ defined by

$$\delta\mathscr{F}_B(\mathbf{n})[\mathbf{u}] := \frac{d}{d\varepsilon}\mathscr{F}_B[\mathbf{n}_\varepsilon]\Big|_{\varepsilon=0}. \tag{3.102}$$

As is clear from (3.101), $\delta\mathscr{F}_B(\mathbf{n})[\mathbf{u}]$ depends linearly on $\mathbf{u}$.

Definition 3.1. We say that $\mathscr{F}_B$ is **stationary** at $\mathbf{n}$ if $\delta\mathscr{F}_B(\mathbf{n})$ vanishes identically.

Thus, $\mathbf{n}$ is an equilibrium configuration for the optic axis whenever

$$\delta\mathscr{F}_B(\mathbf{n}) \equiv 0. \tag{3.103}$$

The equilibrium equations for $\mathscr{F}_B$ will assure us that (3.103) is satisfied. Before deriving them we need to prove a preliminary result.

Lemma 3.6. Let $(\mathbf{n}, \mathbf{N}) \mapsto \sigma(\mathbf{n}, \mathbf{N})$ be a mapping of class C^1 which for any given $\mathbf{n} \in \mathbb{S}^2$ assigns a real to all $\mathbf{N} \in L(\mathbf{n}, \mathscr{V})$. The derivatives of σ then obey

$$\mathbf{P}(\mathbf{n})\frac{\partial\sigma}{\partial\mathbf{n}} = \frac{\partial\sigma}{\partial\mathbf{n}}, \quad \mathbf{P}(\mathbf{n})\frac{\partial\sigma}{\partial\mathbf{N}} = \frac{\partial\sigma}{\partial\mathbf{N}} \tag{3.104}$$

where $\mathbf{P}(\mathbf{n})$ is defined as in (3.20).

Proof. Proving (3.104) is just a matter of properly understanding how the derivatives of σ are defined. To this end we follow here almost the same line of thought as outlined in Section 2.3 of Chapter 2 in regard to the differentiation of curves, fields and deformations.

For a given $\mathbf{n}$ in $\mathbb{S}^2$ let $\mathbf{N}$ be chosen in $L(\mathbf{n},\mathscr{V})$. Let $\mathbf{v}:[-1,1]\to\mathscr{V}$ be a curve of class C^1 which takes values in $\mathbb{S}^2$ and $\mathbf{V}:[-1,1]\to L(\mathscr{V})$ a curve of class C^1 such that

$$\mathbf{v}(0)=\mathbf{n},\quad \mathbf{V}(0)=\mathbf{N}, \tag{3.105}$$

and

$$\mathbf{V}(t)\in L(\mathbf{v}(t),\mathscr{V})\quad \text{for all } t\in[-1,1]. \tag{3.106}$$

Thus, σ may be evaluated on the pair $(\mathbf{v}(t),\mathbf{V}(t))$ for each $t\in[-1,1]$. Furthermore, it easily follows from (3.105) and (3.106) that

$$\dot{\mathbf{v}}(0)=\mathbf{P}(\mathbf{n})\dot{\mathbf{v}}(0), \tag{3.107}$$

$$\dot{\mathbf{V}}(0)=\mathbf{P}(\mathbf{n})\dot{\mathbf{V}}(0)-(\mathbf{n}\otimes\dot{\mathbf{v}}(0))\mathbf{N}. \tag{3.108}$$

Saying that σ is differentiable at $(\mathbf{n},\mathbf{N})$ means that there are precisely one vector $\partial\sigma/\partial\mathbf{n}$ and one tensor $\partial\sigma/\partial\mathbf{N}$ such that

$$\frac{d}{dt}\sigma(\mathbf{v}(t),\mathbf{V}(t))|_{t=0}=\frac{\partial\sigma}{\partial\mathbf{n}}\cdot\dot{\mathbf{v}}(0)+\frac{\partial\sigma}{\partial\mathbf{N}}\cdot\dot{\mathbf{V}}(0) \tag{3.109}$$

for all curves $\mathbf{v}$ and $\mathbf{V}$ as above. If we now insert both (3.107) and (3.108) in (3.109), we arrive at

$$\frac{d}{dt}\sigma(\mathbf{v}(t),\mathbf{V}(t))|_{t=0}=\left(\mathbf{P}(\mathbf{n})\frac{\partial\sigma}{\partial\mathbf{n}}-\mathbf{N}\left(\frac{\partial\sigma}{\partial\mathbf{N}}\right)^T\mathbf{n}\right)\cdot\dot{\mathbf{v}}(0)+\mathbf{P}(\mathbf{n})\frac{\partial\sigma}{\partial\mathbf{N}}\cdot\dot{\mathbf{V}}(0).$$

Since both $\partial\sigma/\partial\mathbf{n}$ and $\partial\sigma/\partial\mathbf{N}$ are uniquely determined by (3.109), they must obey

$$\frac{\partial\sigma}{\partial\mathbf{n}}=\mathbf{P}(\mathbf{n})\frac{\partial\sigma}{\partial\mathbf{n}}-\mathbf{N}\left(\frac{\partial\sigma}{\partial\mathbf{N}}\right)^T\mathbf{n},$$

$$\frac{\partial\sigma}{\partial\mathbf{N}}=\mathbf{P}(\mathbf{n})\frac{\partial\sigma}{\partial\mathbf{N}},$$

whence the desired conclusion follows.

We are now in a position to compute the first variation of the functional $\mathscr{F}_B$.

Theorem 3.7. Let $\mathbf{n}$ and $\mathbf{u}$ be mappings of class C^2 of $\mathscr{B}\cup\partial^*\mathscr{B}\setminus\mathscr{S}$

into $\mathbb{S}^2$ and $\mathscr{V}$, respectively. Then

$$\delta\mathscr{F}_B(\mathbf{n})[\mathbf{u}] = \int_{\mathscr{B}\setminus\mathscr{S}} \mathbf{P}(\mathbf{n})\left(\frac{\partial\sigma}{\partial\mathbf{n}} - \operatorname{div}\left(\frac{\partial\sigma}{\partial\nabla\mathbf{n}}\right)\right)\cdot\mathbf{u}\,dv + \int_{\partial^*\mathscr{B}\setminus\mathscr{S}} \left(\frac{\partial\sigma}{\partial\nabla\mathbf{n}}\right)\mathbf{v}\cdot\mathbf{u}\,da, \tag{3.110}$$

where σ is the free energy density per unit volume and $\mathbf{v}$ is the outer normal to $\mathscr{B}$.

Proof. It follows from (3.101) that

$$\mathscr{F}_B[\mathbf{n}_\varepsilon] = \mathscr{F}_B[\mathbf{n}] + \varepsilon\int_{\mathscr{B}\setminus\mathscr{S}}\left(\frac{\partial\sigma}{\partial\mathbf{n}}\cdot\mathbf{P}(\mathbf{n})\mathbf{u} + \frac{\partial\sigma}{\partial\nabla\mathbf{n}}\cdot\nabla(\mathbf{P}(\mathbf{n})\mathbf{u})\right)dv + o(\varepsilon).$$

Since

$$\frac{\partial\sigma}{\partial\nabla\mathbf{n}}\cdot\nabla(\mathbf{P}(\mathbf{n})\mathbf{u}) = \operatorname{div}\left(\left(\frac{\partial\sigma}{\partial\nabla\mathbf{n}}\right)^T\mathbf{P}(\mathbf{n})\mathbf{u}\right) - \operatorname{div}\left(\frac{\partial\sigma}{\partial\nabla\mathbf{n}}\right)\cdot\mathbf{P}(\mathbf{n})\mathbf{u},$$

making use of the Divergence Theorem and Lemma 3.6 above, the reader will easily arrive at the desired result.

The equation obeyed in $\mathscr{B}\setminus\mathscr{S}$ by an equilibrium configuration readily follows from (3.110). There the two integrals are indeed independent of each other, and so for an equilibrium configuration $\mathbf{n}$ they must both vanish for all fields $\mathbf{u}$. Thus, as the reader would learn from any classical book on calculus of variations (see, for example, Chapter IV of Courant and Hilbert (1989), Vol. 1), the integrands of both integrals in (3.110) must vanish identically. Since the values of $\mathbf{u}$ can be chosen arbitrarily in $\mathscr{B}\setminus\mathscr{S}$, $\mathbf{n}$ must then satisfy the following equation

$$\frac{\partial\sigma}{\partial\mathbf{n}} - \operatorname{div}\left(\frac{\partial\sigma}{\partial\nabla\mathbf{n}}\right) = \lambda\mathbf{n} \quad \text{in } \mathscr{B}\setminus\mathscr{S}, \tag{3.111}$$

where λ is an arbitrary scalar field defined in $\mathscr{B}\setminus\mathscr{S}$, which can be interpreted as the **Lagrange multiplier** associated with the constraint on the length of $\mathbf{n}$; (3.111) is the **Euler–Lagrange equation** of $\mathscr{F}_B$.

Another formula for λ, which is worth knowing, can now be derived. If we apply $(3.104)_2$ to the free energy density σ, we see that

$$\left(\frac{\partial\sigma}{\partial\nabla\mathbf{n}}\right)^T\mathbf{n} = \mathbf{0} \quad \text{in } \mathscr{B}\setminus\mathscr{S}. \tag{3.112}$$

Similarly, it follows from $(3.104)_1$ that

$$\frac{\partial\sigma}{\partial\mathbf{n}}\cdot\mathbf{n}=0 \quad \text{in } \mathscr{B}\backslash\mathscr{S}. \tag{3.113}$$

Equation (3.112) then implies that

$$\operatorname{div}\left(\frac{\partial\sigma}{\partial\mathbf{n}}\right)\cdot\mathbf{n}=-\frac{\partial\sigma}{\partial\nabla\mathbf{n}}\cdot\nabla\mathbf{n}. \tag{3.114}$$

If we take the inner product of both sides of (3.111) with **n**, making use of both (3.113) and (3.114), we see that

$$\lambda=\frac{\partial\sigma}{\partial\nabla\mathbf{n}}\cdot\nabla\mathbf{n} \quad \text{in } \mathscr{B}\backslash\mathscr{S}, \tag{3.115}$$

for all equilibrium configurations.

The equations satisfied on $\partial^*\mathscr{B}\backslash\mathscr{S}$ by an equilibrium configuration are different for different anchoring conditions imposed on **n**.

3.5.1 No anchoring

If there is no anchoring condition for **n** on $\partial^*\mathscr{B}\backslash\mathscr{S}$ then the second integral in (3.110) can be treated in precisely the same way as the first. Since the values of **u** are completely arbitrary also on $\partial^*\mathscr{B}\backslash\mathscr{S}$, we conclude that

$$\left(\frac{\partial\sigma}{\partial\nabla\mathbf{n}}\right)\mathbf{v}=\mathbf{0} \quad \text{on } \partial^*\mathscr{B}\backslash\mathscr{S}, \tag{3.116}$$

which is the **natural boundary condition** for **n**.

3.5.2 Strong anchoring

When a strong anchoring condition applies for **n** on $\partial^*\mathscr{B}\backslash\mathscr{S}$, **u** must vanish identically there, so that the field $\mathbf{n}_\varepsilon$ defined by (3.100) obeys the same condition as **n** for all $\varepsilon\in[-\varepsilon_0,\varepsilon_0]$. Thus, the surface integral in (3.110) vanishes for all admissible **u**, and so no other condition must be obeyed by **n** on $\partial^*\mathscr{B}\backslash\mathscr{S}$ besides that required by the anchoring.

3.5.3 Conical anchoring

Suppose now that **n** is subject to the condition

$$\mathbf{n}\cdot\mathbf{v}=\alpha_0 \quad \text{on } \partial^*\mathscr{B}\backslash\mathscr{S}, \tag{3.117}$$

where $\alpha_0 : \partial^*\mathscr{B}\setminus\mathscr{S} \to [-1, 1]$ is a given mapping of class C^1. It follows from (3.101) that $\mathbf{n}_\varepsilon$ also obeys (3.117), at least up to the first order in ε, whenever $\mathbf{u}$ is such that

$$\mathbf{P}(\mathbf{n})\mathbf{v}\cdot\mathbf{u} = 0 \quad \text{on } \partial^*\mathscr{B}\setminus\mathscr{S}. \tag{3.118}$$

Thus, we see from (3.110) that an equilibrium configuration $\mathbf{n}$ subject to (3.117) must satisfy the following condition at every point of $\partial^*\mathscr{B}\setminus\mathscr{S}$:

$$\left(\frac{\partial \sigma}{\partial \nabla \mathbf{n}}\right)\mathbf{v}\cdot\mathbf{u} = 0 \tag{3.119}$$

for all $\mathbf{u}$ that obey (3.118). Since (3.119) must hold everywhere on $\partial^*\mathscr{B}\setminus\mathscr{S}$, it is ultimately an algebraic requirement which falls within the category envisaged in the lemma below.

Lemma 3.8. Let $\mathbf{v}$ be given in the linear space $\mathscr{V}$ and let $\mathscr{U}$ be the linear subspace of $\mathscr{V}$ defined by

$$\mathscr{U} := \{\mathbf{u} \in \mathscr{V} \mid \mathbf{u}\cdot\mathbf{v} = 0\}.$$

Then the only vectors $\mathbf{x}$ of $\mathscr{V}$ that satisfy

$$\mathbf{x}\cdot\mathbf{u} = 0 \quad \text{for all } \mathbf{u} \in \mathscr{U}$$

have the form

$$\mathbf{x} = \lambda\mathbf{v}$$

for some scalar λ.

Proof. $\mathscr{U}$ is the orthogonal complement of span $\{\mathbf{v}\}$. Thus, $\mathscr{U}$ and span $\{\mathbf{v}\}$ decompose $\mathscr{V}$. Clearly, then, if a vector $\mathbf{x}$ belongs to $\mathscr{U}^\perp$ it must be parallel to $\mathbf{v}$ (*cf.* Section 2.2).

Applying Lemma 3.8 to (3.119) and (3.118), we arrive at the following boundary condition for $\mathbf{n}$ which supplements (3.117):

$$\left(\frac{\partial \sigma}{\partial \nabla \mathbf{n}}\right)\mathbf{v} = \lambda\mathbf{P}(\mathbf{n})\mathbf{v} \quad \text{on } \partial^*\mathscr{B}\setminus\mathscr{S}, \tag{3.120}$$

where λ is now to be regarded as the extension up to $\partial^*\mathscr{B}\setminus\mathscr{S}$ of the scalar field that appears in (3.111).

Here we have just considered the case that the same anchoring condition holds on the whole of $\partial^*\mathscr{B}$, with the possible exception of the singular set $\mathscr{S}$. It may well happen that different anchoring conditions are to be imposed on different parts of $\partial^*\mathscr{B}$. If this is the case, the appropriate boundary conditions for $\mathbf{n}$ would follow from a judicious application of the discussion above.

I finally remark that throughout this section the region $\mathscr{B}$ has been taken as given, and so it has been held fixed in deriving the equilibrium equations for $\mathscr{F}_B$. This applies to most situations that we encounter in this book. There are, however, interesting problems, which we shall treat in Chapter 5 below, where the boundary of the region occupied by a liquid crystal is free to adjust to the surrounding environment. When this is the case, a surface energy functional is to be added to $\mathscr{F}_B$ and accordingly the analysis leading to the equations obeyed by the equilibrium configurations is to be largely recast. Section 5.2 of Chapter 5 is entirely devoted to this issue.

3.6 General solutions

As we saw in Section 3.1 above, the free energy density σ of both nematic and cholesteric liquid crystals must be such that

$$\sigma(\mathbf{Qn}, \mathbf{Q}\nabla\mathbf{n}\mathbf{Q}^T) = \sigma(\mathbf{n}, \nabla\mathbf{n}), \tag{3.121}$$

for all $\mathbf{Q} \in SO(\mathscr{V})$ (*cf.* equations (3.10) and (3.12)). Following Ericksen (1967a and b), we now ask the question:

Are there orientation fields $\mathbf{n}$ of class C^2 defined in the whole space $\mathscr{E}$, possibly away from a singular set $\mathscr{S}$, which solve the equilibrium equation (3.111) for any function σ that satisfies (3.121) and for some choice of λ, which may be different for different energies?

Every such solution will be called **general**, if there are any. Here we seek the general solutions of (3.111) and we shall find all of them. We shall see that all natural orientations defined in Section 3.1 are indeed general solutions, and that there are also general solutions whose singular set is not empty.

Theorem 3.9. Let $\mathbf{n}$ be a mapping of class C^2 into $\mathbb{S}^2$ defined in the whole space $\mathscr{E}$, possibly away from a singular set $\mathscr{S}$. It solves (3.111) for all σ that satisfy (3.121) and for some scalar field λ on $\mathscr{E}\setminus\mathscr{S}$ if, and only if, it can be given one of the following forms:

$$\mathbf{n}_0 \equiv \mathbf{e}_1, \tag{3.122}$$

$$\mathbf{n}_1(p)=\frac{p-o}{|p-o|}\quad \text{for all } p\neq o, \tag{3.123}$$

$$\mathbf{n}_2(p)=\frac{\mathbf{P}(\mathbf{e}_1)(p-o)}{|\mathbf{P}(\mathbf{e}_1)(p-o)|}\quad \text{for all } p\in\mathscr{E}$$

$$\text{such that } \mathbf{P}(\mathbf{e}_1)(p-o)\neq\mathbf{0}, \tag{3.124}$$

$$\mathbf{n}_3(p)=\cos(tx_1+\alpha)\mathbf{e}_2+\sin(tx_1+\alpha)\mathbf{e}_3, \tag{3.125}$$

where $(o,\mathbf{e}_1,\mathbf{e}_2,\mathbf{e}_3)$ is a frame of $\mathscr{E}$, x_1 denotes the co-ordinate of p along $\mathbf{e}_1$, and both t and α are constants.

The proof of this theorem will rest on two preliminary results, which we prove first.

Lemma 3.10. Let σ be a mapping of class C^1 which satisfies (3.121). Then for any given field $\mathbf{n}$ of class C^2 the mappings defined by

$$\mathbf{h}(\mathbf{n},\nabla\mathbf{n}):=\frac{\partial\sigma}{\partial\mathbf{n}},\quad \mathbf{H}(\mathbf{n},\nabla\mathbf{n}):=\frac{\partial\sigma}{\partial\nabla\mathbf{n}} \tag{3.126}$$

satisfy

$$\mathbf{h}(\mathbf{Q}\mathbf{n},\mathbf{Q}\nabla\mathbf{n}\mathbf{Q}^T)=\mathbf{Q}\mathbf{h}(\mathbf{n},\nabla\mathbf{n}), \tag{3.127}$$

$$\operatorname{div}_{p^*}\mathbf{H}(\mathbf{Q}\mathbf{n},\mathbf{Q}\nabla\mathbf{n}\mathbf{Q}^T)=\mathbf{Q}\operatorname{div}_p\mathbf{H}(\mathbf{n},\nabla\mathbf{n}),\quad \text{for all } \mathbf{Q}\in SO(\mathscr{V}), \tag{3.128}$$

where p^* and p, which are related through (3.4), indicate the variable of differentiation.

Proof. Differentiating both sides of (3.121) with respect to $\mathbf{n}$, we arrive at

$$\mathbf{h}(\mathbf{Q}\mathbf{n},\mathbf{Q}\nabla\mathbf{n}\mathbf{Q}^T)\mathbf{Q}^T=\mathbf{h}(\mathbf{n},\nabla\mathbf{n}),$$

whence (3.127) follows. Similarly, we prove that

$$\mathbf{Q}^T\mathbf{H}(\mathbf{Q}\mathbf{n},\mathbf{Q}\nabla\mathbf{n}\mathbf{Q}^T)\mathbf{Q}=\mathbf{H}(\mathbf{n},\nabla\mathbf{n})\quad \text{for all } \mathbf{Q}\in SO(\mathscr{V}). \tag{3.129}$$

Computing the divergence of both sides of (3.129) with respect to p and making use of (3.4), we arrive at

$$\mathbf{Q}^T\operatorname{div}_{p^*}\mathbf{H}(\mathbf{Q}\mathbf{n},\mathbf{Q}\nabla\mathbf{n}\mathbf{Q}^T)=\operatorname{div}_p\mathbf{H}(\mathbf{n},\nabla\mathbf{n}),$$

whence (3.128) follows.

Lemma 3.11. If a field $\mathbf{n}$ of class C^1 into $\mathbb{S}^2$ is subject to

$$\mathbf{n}\cdot\mathbf{e} = 0,$$

where $\mathbf{e}$ is a given unit vector, then the mappings $\mathbf{h}$ and $\mathbf{H}$ defined by (3.126) satisfy

$$\mathbf{h}\cdot\mathbf{e} = 0 \quad \text{and} \quad \mathbf{H}^T\mathbf{e} = \mathbf{0}.$$

Proof. The proof of this lemma closely parallels that of Lemma 3.10 in the preceding section. The reader will easily arrive at the equations

$$\mathbf{P}(\mathbf{e})\mathbf{h} = \mathbf{h} \quad \text{and} \quad \mathbf{P}(\mathbf{e})\mathbf{H} = \mathbf{H},$$

moving from the following premisses:

$$\mathbf{P}(\mathbf{e})\mathbf{n} = \mathbf{n} \quad \text{and} \quad \mathbf{P}(\mathbf{e})\nabla\mathbf{n} = \nabla\mathbf{n}.$$

Proof of Theorem 3.9. The following proof is divided into two parts. First, we find conditions that must be obeyed by all general solutions of (3.111). Then, applying the lemmas above, we show that all orientation fields which satisfy these conditions are indeed solutions of (3.111) for some appropriate choice of λ.

3.6.1 Necessary conditions

If $\mathbf{n}$ is a general solution of (3.111) then it must solve the equations that we obtain from (3.111) on substituting for σ each of the following functions

$$\sigma_1 := \mathbf{n}\cdot\operatorname{curl}\mathbf{n}, \quad \sigma_2 := (\operatorname{div}\mathbf{n})^2, \quad \sigma_3 := |\nabla\mathbf{n}|^2, \tag{3.130}$$

all of which obey (3.121).

Let $\sigma = \sigma_1$; since $\mathbf{n}\cdot\operatorname{curl}\mathbf{n} = \mathbf{W}(\mathbf{n})\cdot\nabla\mathbf{n}$ the fields defined in (3.126) read

$$\mathbf{h} = \mathbf{P}(\mathbf{n})\operatorname{curl}\mathbf{n} \quad \text{and} \quad \mathbf{H} = \mathbf{W}(\mathbf{n}),$$

and so equation (3.111) becomes

$$\mathbf{P}(\mathbf{n})\operatorname{curl}\mathbf{n} + \operatorname{curl}\mathbf{n} = \lambda\mathbf{n},$$

which implies that

$$\operatorname{curl}\mathbf{n} = \lambda_1\mathbf{n}, \tag{3.131}$$

where λ_1 is a scalar field to be determined. Since $\mathbf{n}$ is taken to be of class C^2, (3.131) implies that λ_1 must be at least of class C^1. Similarly, with a little more labour we see that when $\sigma = \sigma_2$

$$\mathbf{h} = \mathbf{0} \quad \text{and} \quad \mathbf{H} = 2(\operatorname{div}\mathbf{n})\mathbf{P}(\mathbf{n}),$$

and so (3.111) becomes

$$\mathbf{P}(\mathbf{n})\nabla \operatorname{div}\mathbf{n} - (\nabla\mathbf{n})\mathbf{n} = (\lambda + (\operatorname{div}\mathbf{n})^2)\mathbf{n}. \qquad (3.132)$$

Because $(\nabla\mathbf{n})\mathbf{n}$ is everywhere orthogonal to $\mathbf{n}$ (*cf.* (3.24)), (3.132) and the following equations are equivalent:

$$\lambda = -(\operatorname{div}\mathbf{n})^2, \qquad (3.133)$$

$$\mathbf{P}(\mathbf{n})\nabla \operatorname{div}\mathbf{n} = (\nabla\mathbf{n})\mathbf{n}. \qquad (3.134)$$

Finally, if $\sigma = \sigma_3$ then

$$\mathbf{h} = \mathbf{0} \quad \text{and} \quad \mathbf{H} = 2\nabla\mathbf{n},$$

and (3.111) can be written as follows:

$$\Delta\mathbf{n} = \lambda_3\mathbf{n}, \qquad (3.135)$$

where $\Delta\mathbf{n}$ is the Laplacian of $\mathbf{n}$.

Thus, a general solution of (3.111) must also solve (3.131), (3.134) and (3.135). By (3.24), (3.131) implies that

$$(\nabla\mathbf{n})\mathbf{n} = \mathbf{0}, \qquad (3.136)$$

and so (3.134) can be replaced by

$$\nabla \operatorname{div}\mathbf{n} = \lambda_2\mathbf{n}, \qquad (3.137)$$

where λ_2 is an unknown scalar field.

We first assume that $\lambda_2 \neq 0$ at some point p_* of $\mathscr{E}$, and so, since λ_2 is continuous, there is a neighbourhood $\mathscr{U}_*$ of p_* where it does not vanish. There, by (3.137), $\mathbf{n}$ is proportional to the gradient of a scalar, and so its curl must vanish; this conclusion agrees with (3.131) only if

$$\lambda_1 = 0. \qquad (3.138)$$

If (3.138) holds, then, at least in $\mathscr{U}_*$, $\mathbf{n}$ can be given the form

$$\mathbf{n} = \nabla\varphi, \qquad (3.139)$$

where φ is a scalar field of class C^1. Making use of (3.139) in (3.135), we arrive at the equation

$$\nabla(\Delta\varphi) = \lambda_3 \nabla\varphi,$$

which can be solved only if $\Delta\varphi$ can be expressed as a differentiable function of φ, that is only if

$$\Delta\varphi = f(\varphi). \tag{3.140}$$

If this is the case, then

$$\lambda_3 = f'(\varphi), \tag{3.141}$$

where f' is the derivative of f.

Now, equation (3.136) tells us that the vector lines of $\mathbf{n}$ are straight lines, which by (3.139) must be orthogonal to the surfaces where φ is constant. Let $\mathscr{S}_\varphi$ be one of these surfaces; since $\mathbf{n}$ and one normal $\mathbf{v}$ of $\mathscr{S}_\varphi$ coincide (*cf.* (3.139)), we have that

$$\nabla_s \mathbf{v} = (\nabla^2\varphi)(\mathbf{I} - \nabla\varphi \otimes \nabla\varphi),$$

and so by (3.136) the mean curvature of $\mathscr{S}_\varphi$ is given by

$$h_{\mathscr{S}_\varphi} = \tfrac{1}{2}\Delta\varphi = \tfrac{1}{2} f(\varphi).$$

Thus, $\mathscr{S}_\varphi$ is a surface with constant mean curvature. Likewise, we arrive at the following formula for the Gaussian curvature of $\mathscr{S}_\varphi$ (*cf. e.g.* p. 204 of Spivak (1975) Vol. 3)

$$k_{\mathscr{S}_\varphi} = \tfrac{1}{2}((\Delta\varphi)^2 - \operatorname{tr}(\nabla^2\varphi)^2).$$

Whenever (3.139) holds, $\nabla\mathbf{n}$ is a symmetric tensor field and it is easily seen that

$$\operatorname{tr}(\nabla^2\varphi)^2 = -\Delta\mathbf{n}\cdot\mathbf{n}.$$

Making use of (3.135) and (3.141) in this latter equation, we conclude that the Gaussian curvature of $\mathscr{S}_\varphi$ is also constant.

A theorem of differential geometry (see *e.g.* Ericksen (1954), pp. 473–74) tells us that families of surfaces such that both mean and Gaussian curvature are constant and with straight lines as normal trajectories may only be parallel planes, concentric spheres, or coaxial circular cyclinders. Thus, we conclude that if $\mathbf{n}$ solves equation (3.111) when σ is set equal to each of the three functions in (3.130), then it is of the form (3.139) with φ equal to one of the functions defined below:

$$\varphi_0(p) := \mathbf{e}\cdot(p-o), \quad \varphi_1(p) := |p-o|, \quad \varphi_2(p) := |\mathbf{P}(\mathbf{e})(p-o)|, \tag{3.142}$$

where o is any point of $\mathscr{E}$ and $\mathbf{e}$ is an arbitrary vector of $\mathbb{S}^2$.

The orientation field corresponding to φ_0 is $\mathbf{n}_0 = \mathbf{e}$. Though the surfaces orthogonal to the vector lines of $\mathbf{n}_0$ are clearly parallel planes, and so $\mathbf{n}_0$ complies with the geometrical requirement derived above, strictly speaking, we should have left φ_0 out of (3.142) since it follows from (3.137) that the scalar field λ_2 associated with it vanishes identically, while we have so far assumed that $\lambda_2 \neq 0$. On the other hand, it is clear that $\mathbf{n}_0$ solves equations (3.131), (3.135) and (3.137) with $\lambda_1 = \lambda_2 = \lambda_3 = 0$. This makes it legitimate to list φ_0 in (3.142).

The above analysis was originally confined to $\mathscr{U}_*$, the neighbourhood of p_* where λ_2 does not vanish. The conclusions that we have reached hold, however, in the whole space, because joining together surfaces on which different functions in (3.142) take constant values would give rise to a mapping $\mathbf{n}$ of class C^1 which fails to be of class C^2, contrary to the definition of a general solution.

We now consider the case when $\lambda_2 = 0$ in (3.137); this amounts to taking $\operatorname{div}\mathbf{n}$ as a constant. We compute $\operatorname{curl}(\operatorname{curl}\mathbf{n})$ in two different ways: making use first of (3.135) and (3.137), and then of (3.131):

$$\operatorname{curl}(\operatorname{curl}\mathbf{n}) = \nabla(\operatorname{div}\mathbf{n}) - \Delta\mathbf{n} = -\lambda_3\mathbf{n},$$

$$\operatorname{curl}(\operatorname{curl}\mathbf{n}) = \lambda_1 \operatorname{curl}\mathbf{n} + \nabla\lambda_1 \wedge \mathbf{n} = \lambda_1^2\mathbf{n} + \nabla\lambda_1 \wedge \mathbf{n}.$$

It follows from these equations that

$$\lambda_1^2\mathbf{n} + \nabla\lambda_1 \wedge \mathbf{n} = -\lambda_3\mathbf{n},$$

whence

$$\lambda_1^2 = -\lambda_3, \tag{3.143}$$

$$\nabla\lambda_1 \wedge \mathbf{n} = \mathbf{0}. \tag{3.144}$$

Suppose that $\nabla\lambda_1 \neq 0$. Then it follows from (3.144) that

$$\mathbf{n} = \gamma\nabla\lambda_1$$

for some scalar field γ of class C^1 and so, by (3.131), we have

$$\operatorname{curl}\mathbf{n} = \nabla\gamma \wedge \nabla\lambda_1 = \lambda_1\gamma\nabla\lambda_1.$$

The latter equation implies that

$$\nabla\gamma = 0 \quad \text{and} \quad \lambda_1\gamma = 0;$$

since γ cannot vanish, we obtain that $\lambda_1 = 0$ and hence $\nabla\lambda_1 = 0$, which is a contradiction.

This shows that λ_1 must be constant, and by (3.143) so also is λ_3. If they both vanish, it follows from the identity

$$|\nabla\mathbf{n}|^2 + \Delta\mathbf{n}\cdot\mathbf{n} = \operatorname{div}((\nabla\mathbf{n})^T\mathbf{n}) = 0$$

that $\mathbf{n}$ must be constant, which is just the solution $\mathbf{n}_0$ of (3.131), (3.135) and (3.137) that we have already encountered.

Accordingly, we assume that neither λ_1 nor λ_3 vanishes. Thus, taking the divergence of the mappings on both sides of (3.131) we learn that $\operatorname{div}\mathbf{n} = 0$. Furthermore, since $(\nabla\mathbf{n})\mathbf{n} = \mathbf{0}$ by (3.131) and (3.24), it follows from the identity

$$\operatorname{div}((\nabla\mathbf{n})\mathbf{n}) = \operatorname{tr}(\nabla\mathbf{n})^2 + \nabla\operatorname{div}\mathbf{n}\cdot\mathbf{n}$$

and (3.137) that $\operatorname{tr}(\nabla\mathbf{n})^2 = 0$.

Collecting all the information we have about $\mathbf{n}$, we see that it must satisfy

$$\operatorname{tr}(\nabla\mathbf{n}) = 0,\ \operatorname{tr}(\nabla\mathbf{n})^2 = 0,\ (\nabla\mathbf{n})\mathbf{n} = \mathbf{0},\ (\nabla\mathbf{n})^T\mathbf{n} = \mathbf{0}. \tag{3.145}$$

Let the frame $(p, \mathbf{e}_1, \mathbf{e}_2, \mathbf{e}_3)$ of $\mathscr{E}$ be chosen in such a way that $\mathbf{n}(p) = \mathbf{e}_3$. Then $\nabla\mathbf{n}(p)$ obeys all the requirements in (3.145) whenever there are three scalars, γ_1, γ_2 and γ_3, such that

$$\gamma_1^2 + \gamma_2\gamma_3 = 0 \tag{3.146}$$

and

$$\nabla\mathbf{n}(p) = \gamma_1(\mathbf{e}_1\otimes\mathbf{e}_1 - \mathbf{e}_2\otimes\mathbf{e}_2) + \gamma_2\mathbf{e}_1\otimes\mathbf{e}_2 + \gamma_3\mathbf{e}_2\otimes\mathbf{e}_1. \tag{3.147}$$

It is easily seen that for all values of γ_1, γ_2, γ_3 that satisfy (3.146) there is precisely one unit vector $\mathbf{v}$ orthogonal to $\mathbf{n}(p)$ and such that $\nabla\mathbf{n}(p)\mathbf{v} = \mathbf{0}$. Actually, if we choose $\mathbf{e}_2 = \mathbf{v}$ then $\gamma_1 = \gamma_2 = 0$, and

$$(\operatorname{curl}\mathbf{n})(p) = \gamma_3\mathbf{n}(p),$$

which is consistent with (3.131) only if $\gamma_3 = \lambda_1$.

Since the point p was arbitrary, we conclude that there are two fields $\mathbf{v}$ and $\mathbf{w}$ of class C^1 into $\mathbb{S}^2$ such that

$$\mathbf{v}\cdot\mathbf{w} = 0, \quad \mathbf{v}\cdot\mathbf{n} = 0, \quad \mathbf{w}\cdot\mathbf{n} = 0, \tag{3.148}$$

and

$$\nabla\mathbf{n} = \lambda_1\mathbf{v}\otimes\mathbf{w}. \tag{3.149}$$

Computing $\Delta\mathbf{n}$ with the aid of (3.149) and employing (3.135), we arrive at

$$\lambda_3 \mathbf{n} = \lambda_1((\operatorname{div} \mathbf{w})\mathbf{v} + (\nabla \mathbf{v})\mathbf{w}),$$

whence, since $\mathbf{n}$ and $\mathbf{v}$ are orthogonal and $\lambda_1 \neq 0$, it follows that

$$\operatorname{div} \mathbf{w} + \mathbf{v} \cdot (\nabla \mathbf{v})\mathbf{w} = 0.$$

Because $\mathbf{v}$ is a unit vector field, $(\nabla \mathbf{v})^T \mathbf{v} = \mathbf{0}$ and so $\operatorname{div} \mathbf{w} = 0$. We now prove that $\nabla \mathbf{w}$ is a symmetric tensor field. In any basis $e = (\mathbf{e}_1, \mathbf{e}_2, \mathbf{e}_3)$ of $\mathscr{V}$ the second gradient of $\mathbf{n}$ is given by

$$\nabla^2 \mathbf{n} = \sum_{i,j,k=1}^{3} (v_{i,k} w_j + v_i w_{j,k}) \mathbf{e}_i \otimes \mathbf{e}_j \otimes \mathbf{e}_k,$$

where v_i and w_i, $i = 1, 2, 3$, are the components in e of $\mathbf{v}$ and $\mathbf{w}$, respectively, and the comma denotes the partial derivative with respect to a space variable. Since the components of $\nabla^2 \mathbf{n}$ are symmetric in the last two indices, we easily see that

$$v_{i,k} w_j - v_{i,j} w_k = v_i (w_{k,j} - w_{j,k}) \quad \text{for all } i, j, k \in \{1, 2, 3\}, \tag{3.150}$$

whence, since $\mathbf{v} \cdot \mathbf{v} = 1$, we derive an equation which in absolute notation reads as follows:

$$(\nabla \mathbf{v})^T \mathbf{v} \otimes \mathbf{w} - \mathbf{w} \otimes (\nabla \mathbf{v})^T \mathbf{v} = \nabla \mathbf{w} - (\nabla \mathbf{w})^T.$$

Since $(\nabla \mathbf{v})^T \mathbf{v} = \mathbf{0}$ the left-hand side of this equation vanishes and we conclude that $\nabla \mathbf{w}$ is symmetric.

Moreover, from the equations $\mathbf{w} \cdot \mathbf{w} = 1$ and $\mathbf{w} \cdot \mathbf{n} = 0$ the reader will arrive at

$$(\nabla \mathbf{w})\mathbf{w} = \mathbf{0} \quad \text{and} \quad (\nabla \mathbf{w})\mathbf{n} = \mathbf{0}.$$

This means that both $\mathbf{w}$ and $\mathbf{n}$ are eigenvectors of $\nabla \mathbf{w}$ with null eigenvalue. Thus, since $\operatorname{tr} \nabla \mathbf{w} = 0$, we deduce from the Spectral Theorem that $\nabla \mathbf{w} = \mathbf{0}$, that is

$$\mathbf{w} \equiv \mathbf{e}_1. \tag{3.151}$$

Likewise, making use of (3.151) in (3.150), we see that

$$(\nabla \mathbf{v})\mathbf{e}_2 = (\nabla \mathbf{v})\mathbf{e}_3 = \mathbf{0}.$$

Thus, since by (3.151) $\mathbf{v}$ must be orthogonal to $\mathbf{e}_1$ and by (3.148) $\mathbf{n}$ must be orthogonal to both $\mathbf{v}$ and $\mathbf{e}_1$, we arrive at the following formula for $\mathbf{n}$:

$$\mathbf{n}(p) = \cos \gamma(x_1)\mathbf{e}_2 + \sin \gamma(x_1)\mathbf{e}_3, \tag{3.152}$$

where γ is an arbitrary function of class C^2 and x_1 is the co-ordinate of p along $\mathbf{e}_1$. By use of (3.152) we get

$$\operatorname{curl}\mathbf{n}\cdot\mathbf{n} = -\gamma',$$

which is consistent with (3.131) only if

$$\gamma(x_1) = -\lambda_1 x_1 + \alpha,$$

where α is a constant.

We have thus proved that $\mathbf{n}$ solves the equilibrium equations associated with all the special energies defined in (3.130) only if it can be given one of the forms (3.122), (3.123), (3.124), or (3.125). It remains to be shown that these mappings are indeed solutions of the equilibrium equations associated with all energies σ that satisfy (3.121).

3.6.2 Sufficient conditions

We first prove that $\mathbf{n}_1$ is a general solution of (3.111). Let us set $r(p) := |p - o|$ for all $p \neq o$. An easy computation yields

$$\nabla\mathbf{n}_1 = \frac{1}{r}\mathbf{P}(\mathbf{n}_1),$$

and so the mappings $\mathbf{h}$ and $\mathbf{H}$ defined by (3.126) above can be given the form

$$\mathbf{h}(\mathbf{n}_1, \nabla\mathbf{n}_1) = \mathbf{h}_1(r, \mathbf{n}_1), \tag{3.153}$$

$$\mathbf{H}(\mathbf{n}_1, \nabla\mathbf{n}_1) = \mathbf{H}_1(r, \mathbf{n}_1), \tag{3.154}$$

where both $\mathbf{h}_1$ and $\mathbf{H}_1$ are defined on $\mathbb{R}^+ \times \mathbb{S}^2$. Likewise, it can be shown from (3.154) that there is a mapping $\mathbf{d}_1 : \mathbb{R}^+ \times \mathbb{S}^2 \rightarrow \mathscr{V}$ such that

$$\operatorname{div}\mathbf{H} = \mathbf{d}_1(r, \mathbf{n}_1). \tag{3.155}$$

Applying (3.127) and (3.128) to (3.153) and (3.155), respectively, we arrive at the following requirements for $\mathbf{h}_1$ and $\mathbf{d}_1$:

$$\mathbf{h}_1(r, \mathbf{Q}\mathbf{n}_1) = \mathbf{Q}\mathbf{h}_1(r, \mathbf{n}_1) \quad \text{and} \quad \mathbf{d}_1(r, \mathbf{Q}\mathbf{n}_1) = \mathbf{Q}\mathbf{d}_1(r, \mathbf{n}_1)$$

for all $\mathbf{Q} \in SO(\mathscr{V})$. By Theorem 2.27 in Section 2.6 of Chapter 2 there are two scalar functions φ and ψ such that

$$\mathbf{h}_1(r, \mathbf{n}_1) = \varphi(r)\mathbf{n}_1 \quad \text{and} \quad \mathbf{d}_1(r, \mathbf{n}_1) = \psi(r)\mathbf{n}_1.$$

Thus equation (3.111) is solved by $\mathbf{n}_1$, provided we set $\lambda = \varphi - \psi$.

The proof that $\mathbf{n}_2$ is also a general solution of (3.111) proceeds essentially along the same lines and is left here to the reader as an exercise.

We now consider the field $\mathbf{n}_3$ defined by (3.125). A direct computation easily leads us to

$$\nabla \mathbf{n}_3 = \mathbf{n}_3' \otimes \mathbf{e}_1, \tag{3.156}$$

where $\mathbf{n}_3'$ is defined by

$$\mathbf{n}_3' := -t(\sin(tx_1 + \alpha)\mathbf{e}_2 - \cos(tx_1 + \alpha)\mathbf{e}_3).$$

The mapping $x_1 \mapsto \mathbf{n}_3$ can be regarded as a curve on $\mathbb{S}^2$ whose derivative is precisely $\mathbf{n}_3'$; clearly,

$$\mathbf{n}_3' \cdot \mathbf{n}_3 = 0, \quad |\mathbf{n}_3'| = |t|, \quad \mathbf{n}_3' \wedge \mathbf{n}_3 = -t\mathbf{e}_1. \tag{3.157}$$

Evaluating σ on $\mathbf{n}_3$ we see that there is a scalar mapping $\hat{\sigma}$ such that

$$\sigma(\mathbf{n}_3, \nabla \mathbf{n}_3) = \hat{\sigma}(\mathbf{n}_3, \mathbf{n}_3'),$$

whence, by Theorem 2.26 in Section 2.6 of Chapter 2 and $(3.157)_{1,2}$, we conclude that σ obeys (3.121) whenever $\hat{\sigma}$ is given the form

$$\hat{\sigma}(\mathbf{n}_3, \mathbf{n}_3') = \varphi(|t|),$$

where φ is a scalar function of class C^1. In precisely the same way we prove that

$$\nabla \mathbf{n}_3 \cdot \mathbf{H}(\mathbf{n}_3, \nabla \mathbf{n}_3) = \psi(|t|),$$

where ψ is defined as φ. Thus we see that the difference $\sigma(\mathbf{n}_3, \nabla \mathbf{n}_3) - \nabla \mathbf{n}_3 \cdot \mathbf{H}(\mathbf{n}_3, \nabla \mathbf{n}_3)$, being a function of $|t|$ alone, must be constant. This implies that

$$(\sigma(\mathbf{n}_3, \nabla \mathbf{n}_3) - \nabla \mathbf{n}_3 \cdot \mathbf{H}(\mathbf{n}_3, \nabla \mathbf{n}_3))' = 0,$$

where the prime denotes differentiation with respect to x_1. By use of the chain rule this latter equation reduces to

$$\mathbf{h} \cdot \mathbf{n}_3' - \mathbf{H}' \cdot \nabla \mathbf{n}_3 = 0,$$

which by (3.156) leads us to

$$(\mathbf{h} - \mathbf{H}'\mathbf{e}_1) \cdot \mathbf{n}_3' = 0. \tag{3.158}$$

Since both $\mathbf{n}_3$ and $\mathbf{n}_3'$ depend only on x_1, by the use of (3.156) it is easily seen that $\mathbf{H}'\mathbf{e}_1 = \operatorname{div} \mathbf{H}$, and so equation (3.158) can also be written as follows:

$$(\mathbf{h} - \operatorname{div}\mathbf{H})\cdot\mathbf{n}_3' = 0. \tag{3.159}$$

Finally, since $\mathbf{n}_3\cdot\mathbf{e}_1 = 0$, by Lemma 3.11 above, we conclude that both $\mathbf{h}$ and $\operatorname{div}\mathbf{H}$ are orthogonal to $\mathbf{e}_1$, and so

$$(\mathbf{h} - \operatorname{div}\mathbf{H})\cdot\mathbf{e}_1 = 0. \tag{3.160}$$

Equations (3.157) ensure that for all $t \neq 0$ the vectors $\mathbf{n}_3$, $\mathbf{n}_3'$ and $\mathbf{e}_1$ are mutually orthogonal at all points of $\mathscr{E}$. Equations (3.159) and (3.160) then imply that

$$\mathbf{h} - \operatorname{div}\mathbf{H} = \lambda_3\mathbf{n}_3,$$

where the scalar field λ_3, being equal to $\psi(|t|)$ by (3.115), is constant in space.

To complete the proof of Theorem 3.9 it remains to be seen that $\mathbf{n}_0$ is also a general solution of (3.111). This will be an easy exercise for the reader who has followed thus far the unfolding of the proof.

It is clear from the proof of Theorem 3.9 that each of the fields $\mathbf{n}_0$, $\mathbf{n}_1$, $\mathbf{n}_2$ and $\mathbf{n}_3$ defined above represents indeed a class of general solutions: a change of frame would not affect a general solution of (3.111). Thus, $\mathbf{n}_0$ and $\mathbf{n}_3$ represent all natural orientations of nematics and cholesterics. Besides, they are the only general solutions without defects. The singular set of $\mathbf{n}_1$ is the singleton $\{o\}$, while that of $\mathbf{n}_2$ is the line through o parallel to $\mathbf{e}_1$.

Remark 1. We have already encountered in Section 3.3 the fields $\mathbf{n}_2$ and $\mathbf{n}_3$. Indeed they differ only by a change of frame from the splay field $\mathbf{n}_s$ and the twist field $\mathbf{n}_c$.

In Theorem 3.9 we have listed all general solutions of (3.111). They apply to both nematics and cholesterics since (3.121) reflects a requirement which holds for all these materials. Moreover, the orientation fields we have found also solve the equilibrium equations of the energy functional when the energy density is different from that given by Frank's formula in Section 3.2 (*cf.* equation (3.23)).

We now restrict attention to nematics and to Frank's free energy σ_F. That is, we seek all orientation fields $\mathbf{n}$ of class C^2 into $\mathbb{S}^2$ defined in the whole space, possibly away from a singular set $\mathscr{S}$, that solve equation (3.111) with a scalar field λ which may be different for different orientation fields, whenever σ is given by Frank's formula (3.23) with $\tau = 0$. We give the name **Frank's general solution** to any such orientation field. In other words, a Frank's general solution is a mapping which solves the equilibrium equations associated with Frank's energy for all values of Frank's constants. Clearly, each general solution is also Frank's.

Marris (1978) has determined all Frank's general solutions. His main theorem is reproduced below; the proof he gives is very involved and mainly based on arguments for a contradiction, which I have not been able to follow completely.

Theorem 3.12. Let $\mathbf{n}$ be a Frank's general solution other than any general solution listed in Theorem 3.9. Then the vector lines of $\mathbf{n}$ are either concentric circles or a family of circles intersected orthogonally by all members of another family of circles. Conversely, every field $\mathbf{n}$ whose vector lines have these features is a Frank's general solution.

Thus, for example, the bend field $\mathbf{n}_b$ defined in Section 3.3 is a Frank's general solution because its vector lines are concentric circles; its singular set is a straight line like that of the splay field $\mathbf{n}_s$.

Remark 2. To put Marris's theorem in the right perspective it should be noted that among the special energy densities listed in (3.130) σ_2 and σ_3 can be obtained from Frank's formula (3.111) for special choices of Frank's constants, while σ_1 cannot.

Furthermore, in a subsequent paper Marris (1979) proved that the only family of circles which are intersected orthogonally by all members of another family of circles can be represented in Cartesian co-ordinates (x, y) by the equation

$$x^2 + y^2 + 2\eta x + c = 0, \tag{3.161}$$

where c is a given real constant and η is a parameter satisfying $\eta^2 - c \geqslant 0$, which labels the circles in this family. It is easily checked that the circles which satisfy the following equation

$$x^2 + y^2 + 2\mu x - c = 0, \tag{3.162}$$

with the same c as in (3.161) and for μ such that $\mu^2 + c \geqslant 0$, intersect orthogonally all circles represented by (3.161). When $c > 0$, (3.161) represents the circles passing through the points of co-ordinate $(\sqrt{c}, 0)$ and $(-\sqrt{c}, 0)$. We shall encounter in the following section one of Frank's general solutions whose vector lines are members of one of these families of circles with $c = 0$.

3.7 Special solutions

In this section we confine attention to nematic liquid crystals and we take Frank's formula for the free energy desity. We seek solutions of the equilibrium equations for the free energy functional within a special class of orientation fields. We assume that $\mathbf{n}$ is a planar field, that is it delivers vectors everywhere parallel to a given plane. We make use of a frame $(o, \mathbf{e}_1, \mathbf{e}_2, \mathbf{e}_3)$ such that the preferred plane is orthogonal to $\mathbf{e}_3$. Moreover, we take $\mathscr{B}$, the region in space occupied by the liquid crystal, as a hollow cylinder with axis parallel to $\mathbf{e}_3$. Thus, in a system of cylindrical co-ordinates (r, ϑ, z) with $\mathbf{e}_z = \mathbf{e}_3$, $\mathscr{B}$ is represented in the form

$$\mathscr{B} = \{p = o + r\mathbf{e}_r + z\mathbf{e}_z | r \in]r_0, R[,\ z \in]0, L[\}, \tag{3.163}$$

where $o \in \mathscr{E}$ and $\mathbf{e}_r$ is the radial unit vector of the movable frame $(p, \mathbf{e}_r, \mathbf{e}_\vartheta, \mathbf{e}_z)$. The unit vectors $\mathbf{e}_r$ and $\mathbf{e}_\vartheta$ are related to $\mathbf{e}_1$ and $\mathbf{e}_2$ through the formulae

$$\mathbf{e}_r = \cos\vartheta\mathbf{e}_1 + \sin\vartheta\mathbf{e}_2, \quad \mathbf{e}_\vartheta = -\sin\vartheta\mathbf{e}_1 + \cos\vartheta\mathbf{e}_2.$$

In $\mathscr{B}$ we write $\mathbf{n}$ in the following form:

$$\mathbf{n} = \cos\chi(\vartheta)\mathbf{e}_r + \sin\chi(\vartheta)\mathbf{e}_\vartheta. \tag{3.164}$$

The function χ is assumed to be of class C^2 in $[0, 2\pi]$ and such that

$$\chi(2\pi) - \chi(0) = m\pi \quad \text{for some } m \in \mathbb{Z}; \tag{3.165}$$

it gives the angle between $\mathbf{n}$ and the radial unit vector of the movable frame.

The reason why the right-hand side of (3.165) need not be a multiple of 2π, as the reader might perhaps expect, will become clear shortly below, when we define Frank's index.

Clearly, the function ψ defined on $[0, 2\pi]$ by

$$\psi(\vartheta) := \chi(\vartheta) + \vartheta \tag{3.166}$$

obeys (3.165) and represents the angle between $\mathbf{n}$ and the unit vector $\mathbf{e}_1$, which is fixed in space.

A computation which employs the method illustrated in Section 2.3 of Chapter 2 (*cf.*, in particular, formulae (2.78) and (2.79)) easily leads to

$$\nabla\mathbf{n} = \frac{1}{r}(\chi' + 1)(\cos\chi\mathbf{e}_\vartheta \otimes \mathbf{e}_\vartheta - \sin\chi\mathbf{e}_r \otimes \mathbf{e}_\vartheta), \tag{3.167}$$

where a prime denotes differentiation with respect to ϑ. Making use of (3.167) we readily arrive at

$$\operatorname{div}\mathbf{n} = \frac{1}{r}(\chi' + 1)\cos\chi,$$

$$\operatorname{tr}(\nabla\mathbf{n})^2 = \frac{1}{r^2}(\chi' + 1)^2\cos^2\chi,$$

$$\operatorname{curl}\mathbf{n} = \frac{1}{r}(\chi' + 1)\sin\chi\mathbf{e}_z.$$

Inserting these formulae in (3.23), when $\tau = 0$ we obtain

$$\sigma_F = \frac{1}{r^2}(\chi' + 1)^2(k_1\cos^2\chi + k_3\sin^2\chi). \tag{3.168}$$

It is to be noted that the orientation fields in the class described by (3.164) do not contribute to the energy density either through the twist modulus or through the saddle-splay modulus, even if they do not vanish.

Integrating over $\mathscr{B}$ the free energy density given by (3.168), we easily arrive at

$$\mathscr{F}_B[\mathbf{n}] = L\log\left(\frac{R}{r_0}\right)F[\chi], \tag{3.169}$$

where F is the functional defined by

$$F[\chi] = \int_0^{2\pi}(\chi' + 1)^2(k_1\cos^2\chi + k_3\sin^2\chi)d\vartheta. \tag{3.170}$$

We know from (3.65) that both k_1 and k_3 are not negative. Here we assume that they do not both vanish, to exclude the trivial case that $F \equiv 0$.

Thus, seeking the orientation fields $\mathbf{n}$ which make $\mathscr{F}_B$ stationary within the class described by (3.164) amounts to seeking the functions χ of class C^2 in $[0, 2\pi]$ which satisfy (3.165) and make F stationary.

The functional defined by (3.170) can be regarded as an instance of the functionals represented by the general formula

$$I[u] := \int_\alpha^\beta f(t, u, u')dt, \tag{3.171}$$

where f is a real function of class C^2 and the functions u in the domain of I are of class C^2 on $[\alpha, \beta]$ and satisfy

$$u(\beta) - u(\alpha) = a, \tag{3.172}$$

where a is a given constant. We say that the functions **admissible** for I are members of the class

$$\mathscr{A}_I := \{u \in C^2(\alpha, \beta) | (3.172) \text{ holds}\}.$$

We now deduce the equilibrium equations for I in $\mathscr{A}_I$ following the same lines of thought outlined in Section 3.5 for the functional $\mathscr{F}_B$ and then we apply to F the outcome of this analysis. The reader already familiar with elementary calculus of variations may jump to equations (3.175) and (3.176).

Let $\varepsilon_0 > 0$ be given and let v be a scalar function of class C^2 on $[\alpha, \beta]$. If u is a function of $\mathscr{A}_I$ then the function u_ε defined by

$$u_\varepsilon := u + \varepsilon v \tag{3.173}$$

also belongs to $\mathscr{A}_I$ for all $\varepsilon \in [-\varepsilon_0, \varepsilon_0]$ if, and only if,

$$v(\beta) - v(\alpha) = 0. \tag{3.174}$$

We say that I is stationary at $u \in \mathscr{A}_I$ if its first variation

$$\delta I(u)[v] := \frac{d}{d\varepsilon} I[u_\varepsilon]|_{\varepsilon=0}$$

vanishes for all $v \in C^2(\alpha, \beta)$ that satisfy (3.174).

It easily follows from (3.173) that

$$I[u_\varepsilon] = I[u] + \varepsilon \int_\alpha^\beta \left(\frac{\partial f}{\partial u} v + \frac{\partial f}{\partial u'} v' \right) dt + o(\varepsilon),$$

where a prime now denotes differentiation with respect to t. Hence, by use of the identity

$$\frac{\partial f}{\partial u'} v' = \left(\frac{\partial f}{\partial u'} v \right)' - \left(\frac{\partial f}{\partial u'} \right)' v,$$

we arrive at

$$\delta I(u)[v] = \int_\alpha^\beta \left(\frac{\partial f}{\partial u} - \left(\frac{\partial f}{\partial u'} \right)' \right) v\, dt + \left(\frac{\partial f}{\partial u'} v \right)_{t=\beta} - \left(\frac{\partial f}{\partial u'} v \right)_{t=\alpha}.$$

Thus, by (3.174) and a classical result of calculus of variations (see *e.g.* p. 185 of Courant and Hilbert (1989), Vol. 1), I is stationary at u if, and only if,

$$\left(\frac{\partial f}{\partial u'}\right)' - \frac{\partial f}{\partial u} = 0 \quad \text{in}]\alpha, \beta[, \tag{3.175}$$

$$\frac{\partial f}{\partial u'}(\beta, u(\beta), u'(\beta)) = \frac{\partial f}{\partial u'}(\alpha, u(\alpha), u'(\alpha)). \tag{3.176}$$

For the functional F defined by (3.170) and subject to (3.165) equations (3.175) and (3.176) become, respectively,

$$(k_1 \cos^2 \chi + k_3 \sin^2 \chi)\chi'' + (k_3 - k_1)(\chi'^2 - 1)\cos \chi \sin \chi = 0, \tag{3.177}$$

$$(\chi'(2\pi) - \chi'(0))(k_1 \cos^2 \chi(2\pi) + k_3 \sin^2 \chi(2\pi)) = 0. \tag{3.178}$$

When both k_1 and k_3 are positive the latter equation reduces to

$$\chi'(2\pi) = \chi'(0). \tag{3.179}$$

We now consider some special solutions of (3.177). Suppose first that

$$k_1 = k_3 > 0.$$

Then (3.177) becomes

$$\chi'' = 0,$$

whose solutions are

$$\chi(\vartheta) = \gamma\vartheta + \psi_0 \tag{3.180}$$

with γ and ψ_0 constants. All functions in (3.180) satisfy (3.179); they satisfy (3.165) also whenever $2\gamma \in \mathbb{Z}$. These special solutions were found by Frank (1958); by (3.166) they imply that

$$\psi(\vartheta) = \tfrac{1}{2}n\vartheta + \psi_0, \tag{3.181}$$

where $n := 2(\gamma + 1)$, which belongs to $\mathbb{Z}$, is called **Frank's index**. When $|n|$ is even it equals the number of times that $\mathbf{n}(p)$ winds on the equator of $\mathbb{S}^2$ while p describes a trajectory which encloses the axis of $\mathcal{B}$. On the other hand, when $|n|$ is odd $\mathbf{n}(p)$ turns into its opposite after p has described a complete loop around the axis of $\mathcal{B}$.

Strictly speaking, when $|n|$ is odd $\mathbf{n}$ is discontinuous on a plane section which cuts both bases of $\mathcal{B}$ along a radius. Such a defect of $\mathbf{n}$ is indeed fictitious because it just reflects the fact that $\mathbb{S}^2$ is not the most appropriate manifold to describe the orientation of liquid crystals (*cf.* Section 1.3 of Chapter 1). Here the planar symmetry imposed on $\mathbf{n}$ allows us to remedy this failure of the mathematical model that we have adopted, by simply identifying a member of $\mathbb{S}^2$ and its opposite. In general, however, detecting fictitious defects of $\mathbf{n}$ would not be so easy.

When n is positive both $\mathbf{n}$ and $\mathbf{e}_r$ wind in the same direction around the axis of $\mathscr{B}$; when n is negative they wind in opposite directions.

Figure 3.2 represents the vector lines of the orientation field $\mathbf{n}$ for different choices of Frank's index. It could be shown that choosing an angle ψ_0 different from zero in (3.181) would simply affect these pictures by a rotation about the origin, for all values of n except $n = 2$.

Both the splay and bend fields are among the special solutions (3.181) with Frank's index equal to 2: they occur for $\psi_0 = 0$ and $\psi_0 = \pi/2$, respectively.

By the use of (3.180), (3.170) and (3.169), recalling that we have set $k_1 = k_3$, we easily express in terms of Frank's index the energy of the fields $\mathbf{n}$ corresponding to (3.181):

$$\mathscr{F}_B[\mathbf{n}] = \frac{\pi}{2} n^2 L k_1 \ln\left(\frac{R}{r_0}\right). \tag{3.182}$$

Thus, all solutions with equal or opposite Frank's index share the same energy.

Clearly, when the inner lateral boundary of $\mathscr{B}$ shrinks to its axis, that is when $r_0 \to 0$, the free energy of these solutions is unbounded. However, some of these fields $\mathbf{n}$ solve the general equilibrium equations for $\mathscr{F}_B$ even in this limiting case, as is the case for the splay and bend fields which are both Frank's general solutions.

We now let k_1 and k_3 differ from each other; equation (3.177) then possesses two special classes of solutions which are easily found: they occur when

$$(\chi')^2 = 1,$$

and when

$$\chi' = 0 \quad \text{and} \quad \cos\chi \sin\chi = 0,$$

respectively. In the one case

$$\psi(\vartheta) = 2\vartheta + \psi_0 \quad \text{or} \quad \psi(\vartheta) = \psi_0, \tag{3.183}$$

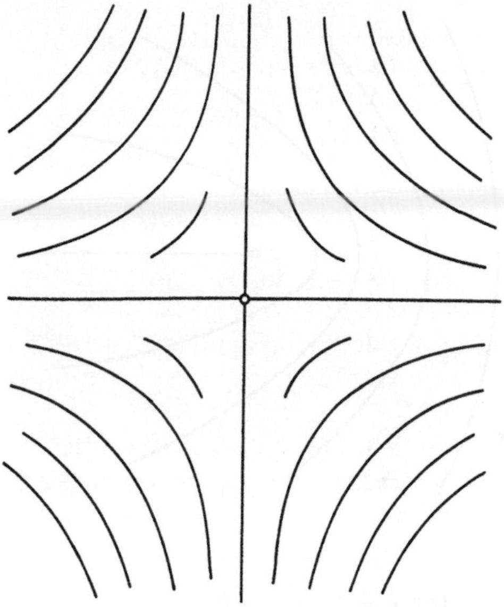

(a) $n = 2 \quad \psi_0 = 0$

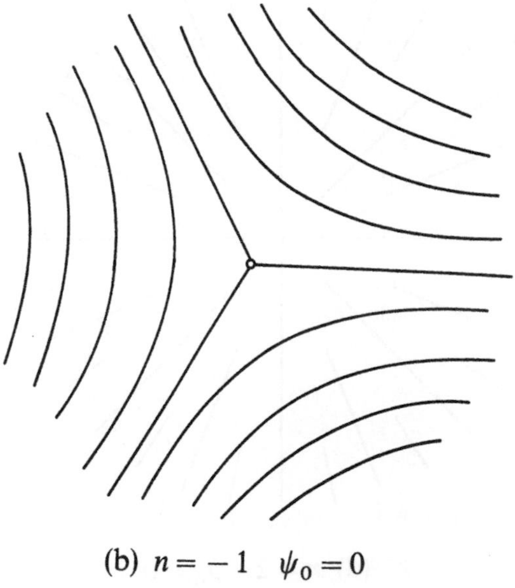

(b) $n = -1 \quad \psi_0 = 0$

Figure 3.2 (*a to f*)

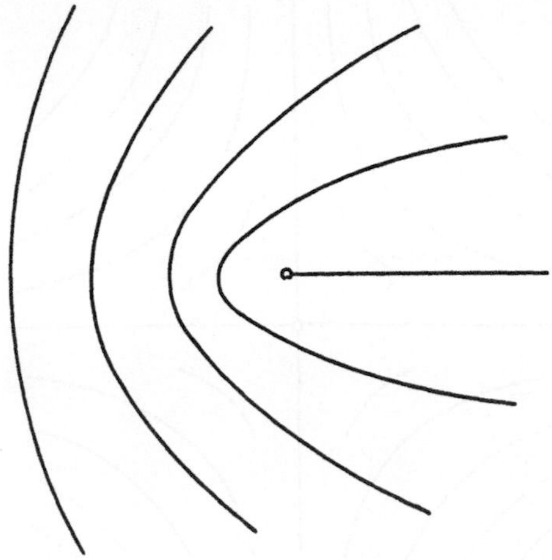

(c) $n=1 \quad \psi_0=0$

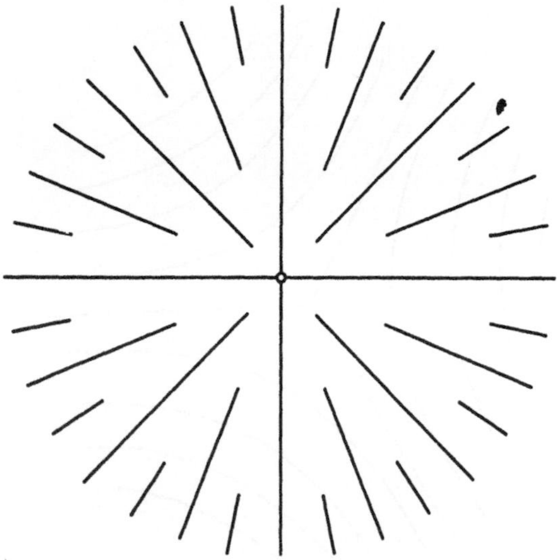

(d) $n=2 \quad \psi_0=0$

Figure 3.2 (*cont'd*)

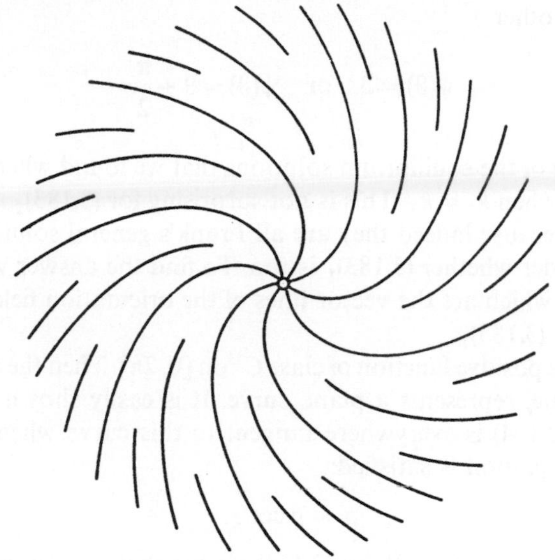

(e) $n=2 \quad \psi_0=\dfrac{\pi}{4}$

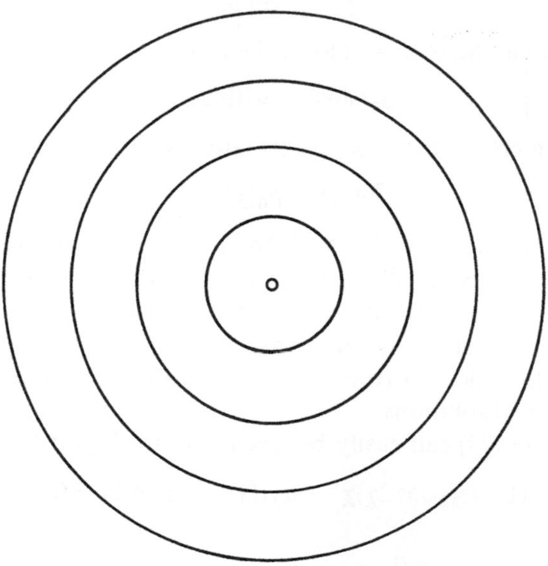

(f) $n=2 \quad \psi_0=\dfrac{\pi}{2}$

Figure 3.2 (*cont'd*)

and in the other

$$\psi(\vartheta) = \vartheta \quad \text{or} \quad \psi(\vartheta) = \vartheta + \frac{\pi}{2}. \tag{3.184}$$

Thus, some of the equilibrium solutions that we found when $k_1 = k_3$ are still so when $k_1 \neq k_3$. This is not surprising for $(3.183)_2$ and both of (3.184) because indeed they are all Frank's general solutions. We might wonder whether $(3.183)_1$ is too. To find the answer we ought to find out which are the vector lines of the orientation field corresponding to $(3.183)_1$.

Let ρ be a positive function of class C^1 on $[0, 2\pi]$. Then the mapping $\vartheta \mapsto o + \rho(\vartheta)\mathbf{e}_r$ represents a plane curve. It is easily shown that the field $\mathbf{n}$ in (3.164) is everywhere tangent to this curve whenever the following equation is satisfied:

$$\rho' = \rho \cot \chi. \tag{3.185}$$

If we set $\psi_0 = 0$ in (3.183), by (3.166) we see that the desired vector lines solve the equation

$$\rho' = \rho \cot \vartheta,$$

and so they can be represented in the form

$$\rho = \rho_0 |\sin \vartheta| \quad \text{with } \rho_0 > 0. \tag{3.186}$$

In Cartesian co-ordinates (x, y) this latter equation becomes

$$x^2 + y^2 - \rho_0 |y| = 0,$$

which reads like (3.162) with $c = 0$ and $\mu = -\rho_0$ or $\mu = \rho_0$, for $y > 0$ or $y < 0$, respectively. Thus, these curves, which are sketched in Fig. 3.3, belong to a family of circles intersected orthogonally by all circles of another family, and so by Theorem 3.12 of the last section the orientation field corresponding to $(3.183)_1$ is indeed one of Frank's general solutions.

Equation (3.177) can easily be given a more elegant form:

$$(1 - \alpha_{31} \cos 2\chi)\chi'' + \alpha_{31}(\chi'^2 - 1)\sin 2\chi = 0, \tag{3.187}$$

where

$$\alpha_{31} := \frac{k_3 - k_1}{k_3 + k_1}. \tag{3.188}$$

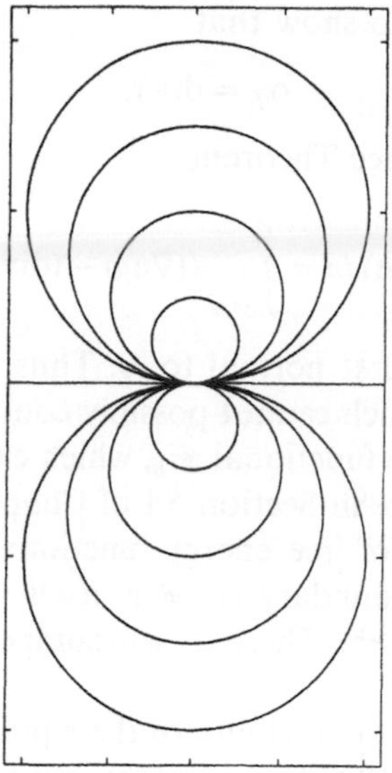

Figure 3.3

Dzyaloshinskiĭ (1970) has carried out a fairly exhaustive analysis of equation (3.187). Readers wishing to learn more about the special equilibrium solutions considered here are referred to his paper.

3.8 Null Lagrangians

Null Lagrangian is the name given to any energy density that does not contribute to the equilibrium equations for the energy functional $\mathscr{F}_B$. In Section 3.2 (*cf.* Remark 3) I claimed that the function defined by

$$\sigma_L(\mathbf{n}, \nabla\mathbf{n}) := \operatorname{tr}(\nabla\mathbf{n})^2 - (\operatorname{div}\mathbf{n})^2 \tag{3.189}$$

is a null Lagrangian. We see here why it is so and in what sense it represents a prototype of all null Lagrangians for liquid crystals.

Let $\mathbf{n}$ be a field of class C^2 of $\bar{\mathscr{B}}$, the closure of a regular region $\mathscr{B}$, into $\mathbb{S}^2$, and let $\mathbf{v}$ be the vector field defined on $\mathscr{B}$ by

$$\mathbf{v} := (\nabla\mathbf{n})\mathbf{n} - (\operatorname{div}\mathbf{n})\mathbf{n}. \tag{3.190}$$

It is an easy exercise to show that

$$\sigma_L = \operatorname{div} \mathbf{v}.$$

Then, by the Divergence Theorem,

$$\int_{\mathscr{B}} \sigma_L(\mathbf{n}, \nabla\mathbf{n})dv = \int_{\partial^*\mathscr{B}} ((\nabla\mathbf{n})\mathbf{n} - (\operatorname{div}\mathbf{n})\mathbf{n})\cdot\boldsymbol{\nu}\, da, \tag{3.191}$$

where $\boldsymbol{\nu}$ is the outer unit normal to $\mathscr{B}$. Thus, σ_L indeed gives rise to a surface energy which cannot possibly contribute to the equilibrium equations for the functional $\mathscr{F}_B$, which expresses the bulk free energy. We shall derive in Section 5.1 of Chapter 5 the equilibrium equations for a class of free energy functionals which seem to be adequate when the boundary of $\mathscr{B}$ is itself free to adjust to the surrounding environment. There σ_L will compete with these genuine surface energies.

Though σ_L does not contribute to the equilibrium equations for $\mathscr{F}_B$ in the bulk, it generally enters the boundary conditions and affects the value of the free energy. One case is, however, exceptional and deserves notice: it occurs when the orientation field is subject to strong anchoring conditions on the whole of $\partial^*\mathscr{B}$.

We now show that if

$$\mathbf{n}|_{\partial^*\mathscr{B}} = \mathbf{n}_0, \tag{3.192}$$

where $\mathbf{n}_0$ is a given field of class C^1 of $\partial^*\mathscr{B}$ into $\mathbb{S}^2$, then the integral of σ_L over $\mathscr{B}$ depends only on $\mathbf{n}_0$, and so it takes the same value for all fields $\mathbf{n}$ which obey (3.192).

Let $\nabla_s\mathbf{n}$ be the surface gradient of $\mathbf{n}$ on $\partial^*\mathscr{B}$ as defined in Section 2.3 of Chapter 2:

$$\nabla_s\mathbf{n} := \nabla\mathbf{n}(\mathbf{I} - \boldsymbol{\nu}\otimes\boldsymbol{\nu}). \tag{3.193}$$

It follows from (3.193) that

$$(\nabla_s\mathbf{n})\mathbf{n}\cdot\boldsymbol{\nu} = (\nabla\mathbf{n})\mathbf{n}\cdot\boldsymbol{\nu} - (\mathbf{n}\cdot\boldsymbol{\nu})(\nabla\mathbf{n})\boldsymbol{\nu}\cdot\boldsymbol{\nu},$$
$$(\operatorname{tr}\nabla_s\mathbf{n})(\mathbf{n}\cdot\boldsymbol{\nu}) = (\operatorname{tr}\nabla\mathbf{n})(\mathbf{n}\cdot\boldsymbol{\nu}) - (\mathbf{n}\cdot\boldsymbol{\nu})(\nabla\mathbf{n})\boldsymbol{\nu}\cdot\boldsymbol{\nu},$$

and so

$$((\nabla_s\mathbf{n})\mathbf{n} - (\operatorname{div}_s\mathbf{n})\mathbf{n})\cdot\boldsymbol{\nu} = ((\nabla\mathbf{n})\mathbf{n} - (\operatorname{div}\mathbf{n})\mathbf{n})\cdot\boldsymbol{\nu}. \tag{3.194}$$

When (3.192) applies the function on the left-hand side of (3.194) depends only on $\mathbf{n}_0$, since it involves only the surface derivatives of

$\mathbf{n}$. Thus, by (3.191), we conclude that when $\mathbf{n}$ is subject to strong anchoring conditions on the whole of $\partial^*\mathscr{B}$, the null Lagrangian σ_L contributes the same value to the total free energy functional for all admissible orientation fields.

Ericksen (1962) has considered null Lagrangians within a more general theory of liquid crystals, which allows the field $\mathbf{n}$ to take values in the whole of $\mathscr{V}$, not only on $\mathbb{S}^2$. Ericksen addressed the following question: *Which are the smooth scalar-valued functions* $\chi:(\mathbf{n},\mathbf{N})\mapsto\chi(\mathbf{n},\mathbf{N})$ *on* $\mathscr{V}\times L(\mathscr{V})$ *such that the equation*

$$\operatorname{div}\left(\frac{\partial\chi}{\partial\nabla\mathbf{n}}\right)-\left(\frac{\partial\chi}{\partial\mathbf{n}}\right)=\mathbf{0} \tag{3.195}$$

is satisfied by every vector field $\mathbf{n}$ *of class* C^2 *on* $\mathscr{B}$?

Each such function would indeed be a null Lagrangian, for adding it to any scalar-valued function ψ of class C^2 on $\mathscr{V}\times L(\mathscr{V})$ does not affect the equilibrium equation

$$\operatorname{div}\left(\frac{\partial\psi}{\partial\nabla\mathbf{n}}\right)-\frac{\partial\psi}{\partial\mathbf{n}}+\mathbf{f}=\mathbf{0},$$

where $\mathbf{f}$: $(p,\mathbf{n})\mapsto\mathbf{f}(p,\mathbf{n})$ is any continuous vector-valued mapping defined on $\mathscr{B}\times\mathscr{V}$. Ericksen (1962) proved the following theorem:

Theorem 3.13. Let $\chi:(\mathbf{n},\mathbf{N})\mapsto\chi(\mathbf{n},\mathbf{N})$ be a scalar-valued function of class C^4 on $\mathscr{V}\times L(\mathscr{V})$ such that

$$\chi(-\mathbf{n},-\mathbf{N})=\chi(\mathbf{n},\mathbf{N})$$

and

$$\chi(\mathbf{Q}\mathbf{n},\mathbf{Q}\mathbf{N}\mathbf{Q}^T)=\chi(\mathbf{n},\mathbf{N})$$

for all $\mathbf{n}\in\mathscr{V}$, $\mathbf{N}\in L(\mathscr{V})$, and $\mathbf{Q}\in SO(\mathscr{V})$. Then χ satisfies (3.195) for all vector fields $\mathbf{n}$ of class C^2 on $\mathscr{B}$ if, and only if, there is a scalar-valued function H on $\mathbb{R}^+$ such that

$$\begin{aligned}\chi(\mathbf{n},\mathbf{N})=&\,H(\mathbf{n}\cdot\mathbf{n})((\operatorname{tr}\mathbf{N})^2-\operatorname{tr}\mathbf{N}^2)\\&+2H'(\mathbf{n}\cdot\mathbf{n})((\operatorname{tr}\mathbf{N})\mathbf{n}\cdot\mathbf{N}\mathbf{n}-\mathbf{n}\cdot\mathbf{N}^2\mathbf{n}),\end{aligned} \tag{3.196}$$

where H' is the first derivative of H.

The proof of this theorem is omitted here because it is not central to our development, but it should be noted that whenever (3.196) is evaluated on a pair $(\mathbf{n},\mathbf{N})$ such that $\mathbf{n}\in\mathbb{S}^2$ and $\mathbf{N}\in L(\mathbf{n},\mathscr{V})$, then it reads

$$\chi(\mathbf{n}, \mathbf{N}) = H\sigma_L(\mathbf{n}, \mathbf{N}),$$

where H is now a constant. In this sense (3.196) generalizes the formula for σ_L above.

There is another result which I wish to prove here, though it is not strictly germane to null Lagrangians. It is an identity, due to Ericksen (1961), which applies to all energy densities σ that are frame-indifferent.

Theorem 3.14. Let σ be a mapping of class C^1 which for every $\mathbf{n} \in \mathbb{S}^2$ assigns a scalar to all $\mathbf{N} \in L(\mathbf{n}, \mathscr{V})$. If it satisfies the equation

$$\sigma(\mathbf{Qn}, \mathbf{QNQ}^T) = \sigma(\mathbf{n}, \mathbf{N}) \quad \text{for all } \mathbf{Q} \in SO(\mathscr{V}), \tag{3.197}$$

then the tensor

$$\frac{\partial \sigma}{\partial \mathbf{n}} \otimes \mathbf{n} + \left(\frac{\partial \sigma}{\partial \mathbf{N}}\right)\mathbf{N}^T + \left(\frac{\partial \sigma}{\partial \mathbf{N}}\right)^T \mathbf{N} \tag{3.198}$$

must be symmetric for any given $\mathbf{n} \in \mathbb{S}^2$ and all $\mathbf{N} \in L(\mathbf{n}, \mathscr{V})$.

Proof. By Theorems 2.8 and 2.9 in Section 2.2 of Chapter 2, the tensor defined by

$$\mathbf{Q}(\vartheta) := \mathbf{I} + \sin\vartheta\,\mathbf{W} + (1 - \cos\vartheta)\mathbf{W}^2 \tag{3.199}$$

belongs to $SO(\mathscr{V})$ for any skew tensor $\mathbf{W}$ and for all $\vartheta \in [0, 2\pi[$. If (3.197) holds, the function defined by

$$\vartheta \mapsto \sigma(\mathbf{Q}(\vartheta)\mathbf{n}, \mathbf{Q}(\vartheta)\mathbf{N}\mathbf{Q}(\vartheta)^T)$$

must be constant. Requiring its first derivative to vanish at $\vartheta = 0$, with the aid of (3.199) we arrive at the equation

$$\frac{\partial \sigma}{\partial \mathbf{n}} \cdot \mathbf{Wn} + \frac{\partial \sigma}{\partial \mathbf{N}} \cdot (\mathbf{WN} - \mathbf{NW}) = 0,$$

whence it follows that

$$\mathbf{W} \cdot \left(\frac{\partial \sigma}{\partial \mathbf{n}} \otimes \mathbf{n} + \left(\frac{\partial \sigma}{\partial \mathbf{N}}\right)\mathbf{N}^T + \left(\frac{\partial \sigma}{\partial \mathbf{N}}\right)^T \mathbf{N}\right) = 0. \tag{3.200}$$

Since the skew tensor $\mathbf{W}$ can be chosen arbitrarily in (3.200), by the Decomposition Theorem of Second-Order Tensors (*cf.* Section 2.2 of Chapter 2), the tensor (3.198) must be symmetric, which is the desired conclusion.

Comparing (3.197) and (3.12), we see that the conclusion reached in Theorem 3.14 applies to both nematic and cholesteric liquid crystals. It is an instructive exercise to show that the function σ_F given by Frank's formula (3.23) obeys (3.200).

3.9 Second-order energies

Nehring and Saupe (1971), motivated by their molecular theory of liquid crystals, allowed the free energy density to depend also on the second gradient of the orientation field. They considered a general formula for σ which differs from (3.22), the prototype of Frank's free energy density, by a function linear in $\nabla^2\mathbf{n}$. Requiring such an energy to be frame-indifferent and invariant under reflection of $\mathbf{n}$, they proved that it must differ from σ_F, as given by Frank's formula (3.23), only by

$$\sigma_{13}(\mathbf{n}, \nabla\mathbf{n}, \nabla^2\mathbf{n}) := k_{13}\,\mathrm{div}((\mathrm{div}\,\mathbf{n})\mathbf{n}), \tag{3.201}$$

where k_{13} is a material modulus.

Clearly, σ_{13} does not contribute to the equilibrium equations for the energy functional $\mathscr{F}_B$ because, by the Divergence Theorem,

$$\int_{\mathscr{B}} \sigma_{13}(\mathbf{n}, \nabla\mathbf{n}, \nabla^2\mathbf{n})\,dv = \int_{\partial^*\mathscr{B}} (\mathrm{div}\,\mathbf{n})\mathbf{n}\cdot\mathbf{v}\,da. \tag{3.202}$$

Thus σ_{13} is a null Lagrangian. Nonetheless, it is quite different from σ_L, the null Lagrangian that we have studied in the preceding section. While the contribution of σ_L to the total free energy depends only on the trace of $\mathbf{n}$ on $\partial^*\mathscr{B}$, that of σ_{13} inevitably depends also on the trace of $\nabla\mathbf{n}$, as is clear from (3.202). Thus, when $\mathbf{n}$ is subject to strong anchoring conditions on $\partial^*\mathscr{B}$, but $\nabla\mathbf{n}$ is not prescribed there, apart from its surface part $\nabla_s\mathbf{n}$, σ_{13} contributes to the total free energy through a surface integral over $\partial^*\mathscr{B}$ which depends on the part of $\nabla\mathbf{n}$ that remains free. Such a surface free energy has received much attention in the past decades from both experimentalists and theoreticians.

I show here that variational problems for the total free energy functional $\mathscr{F}_B$ may in general appear to be ill-posed when k_{13} does not vanish. The example that I illustrate below has been largely inspired by a paper of Barbero and Oldano (1989).

Let $\mathscr{B}$ be the region in space between two parallel square plates, $2d$ apart: it is often called a **nematic cell**. In a Cartesian frame

$(o, \mathbf{e}_1, \mathbf{e}_2, \mathbf{e}_3)$ $\mathscr{B}$ can be represented as follows:

$$\mathscr{B} = \{p = o + x\mathbf{e}_1 + y\mathbf{e}_2 + z\mathbf{e}_3 | x, y \in]0, \ell[, z \in]-d, d[\},$$

where o is a given point in the space $\mathscr{E}$. Thus, $\mathbf{e}_3$ is the unit normal to $\mathscr{B}$ at the upper plate (where $z = d$) and $-\mathbf{e}_3$ is that at the lower plate. We denote these plates by $\mathscr{S}^+$ and $\mathscr{S}^-$, respectively, and we prescribe $\mathbf{n}$ to be constant on both of them:

$$\mathbf{n}|_{\mathscr{S}^+} = \mathbf{n}|_{\mathscr{S}^-} = \cos\vartheta_0\mathbf{e}_3 + \sin\vartheta_0\mathbf{e}_1, \tag{3.203}$$

where ϑ_0 is given in $[0, 2\pi]$.

We seek the minimizers of the functional

$$\mathscr{F}_B[\mathbf{n}] = \int_{\mathscr{B}} (\sigma_F + \sigma_{13})dv \tag{3.204}$$

within the class of all orientation fields $\mathbf{n}$ such that

$$\mathbf{n}(p) = \cos\vartheta(z)\mathbf{e}_3 + \sin\vartheta(z)\mathbf{e}_1, \tag{3.205}$$

where ϑ: $[-d, d] \to [0, 2\pi]$ is a function of class C^1 subject to

$$\vartheta(d) = \vartheta(-d) = \vartheta_0. \tag{3.206}$$

It follows from (3.205) that

$$\nabla\mathbf{n} = -\vartheta'(\sin\vartheta\mathbf{e}_3 \otimes \mathbf{e}_3 - \cos\vartheta\mathbf{e}_1 \otimes \mathbf{e}_3), \tag{3.207}$$

whence

$$\operatorname{div}\mathbf{n} = -\vartheta'\sin\vartheta,$$
$$\operatorname{curl}\mathbf{n} = \vartheta'\cos\vartheta\mathbf{e}_2,$$
$$(\nabla\mathbf{n})\mathbf{n} = -\vartheta'\cos\vartheta(\sin\vartheta\mathbf{e}_3 - \cos\vartheta\mathbf{e}_1).$$

From these equations, making use of (3.23) and (3.24), we see that when $\mathscr{B}$ is occupied by a nematic liquid crystal

$$\sigma_F = (k_1\sin^2\vartheta + k_3\cos^2\vartheta)\vartheta'^2. \tag{3.208}$$

By (3.202), in the special class of orientation fields described by (3.205) and (3.206) the functional $\mathscr{F}_B$ then reduces to

$$\mathscr{F}_B[\mathbf{n}] = \ell^2 F[\vartheta],$$

where

$$F[\vartheta] := \int_{-d}^{+d} \{k_1\sin^2\vartheta + k_3\cos^2\vartheta\}\vartheta'^2 dz + \tfrac{1}{2}k_{13}\sin 2\vartheta_0(\vartheta'(d) - \vartheta'(-d)). \tag{3.209}$$

If $\sin 2\vartheta_0 = 0$ the functional defined by (3.209) would not possess any term due to a surface energy. Thus, in the following we assume that $\sin 2\vartheta_0 \neq 0$.

We now show that the functional F does not attain a minimum in the following class of admissible functions:

$$\mathscr{A}_F := \{\vartheta\colon [-d, d] \to [0, 2\pi] \,|\, \vartheta \text{ is of class } C^1 \text{ and (3.206) holds}\}.$$

More precisely, we construct a sequence of functions in $\mathscr{A}_F$ along which the limiting value of F diverges to $-\infty$.

For simplicity, in (3.209) we take

$$k_1 = k_3 = k > 0; \tag{3.210}$$

the conclusion which follows would still be valid if $k_1 \neq k_3$, only its derivation becomes less direct (*cf.* the Remark below).

Let $\varepsilon_0 > 0$ be given; for each ε in $[-\varepsilon_0, \varepsilon_0]$ and each r in $]0, d[$ we define a function of $\mathscr{A}_F$ as follows:

$$\vartheta_r(-z) = \vartheta_r(z) \quad \text{for all } z \in [-d, d], \tag{3.211}$$

$$\vartheta_r(z) := \begin{cases} \vartheta_0 + \varepsilon & \text{if } z \in [0, d - r], \\ \vartheta_0 + \varepsilon - \dfrac{\varepsilon}{r^2}(z - d + r)^2 & \text{if } z \in [d - r, d]. \end{cases} \tag{3.212}$$

Figure 3.4 illustrates a member of the family of functions which is generated by letting r vary in $]0, d[$: in this example ε is given a negative value.

When (3.210) applies, a simple computation leads us to

$$F[\vartheta_r] = \frac{4\varepsilon^2}{r}\left(\frac{2k}{3} + \frac{k_{13}}{2\varepsilon}\sin 2\vartheta_0\right). \tag{3.213}$$

Setting $r = \varepsilon^2$ in (3.213) we obtain that either

$$\lim_{\varepsilon \to 0^+} F[\vartheta_{\varepsilon^2}] = -\infty \quad \text{or} \quad \lim_{\varepsilon \to 0^-} F[\vartheta_{\varepsilon^2}] = -\infty, \tag{3.214}$$

depending on the sign of k_{13}. This shows that for all values of k_{13} different from zero the functional F is not bounded from below in $\mathscr{A}_F$, and so it does not attain its minimum in this class.

Remark. Hinov (1990) has proposed to relate k_{13} to k_1 and k_3 through the formula

$$k_{13} = k_1 - k_3. \tag{3.215}$$

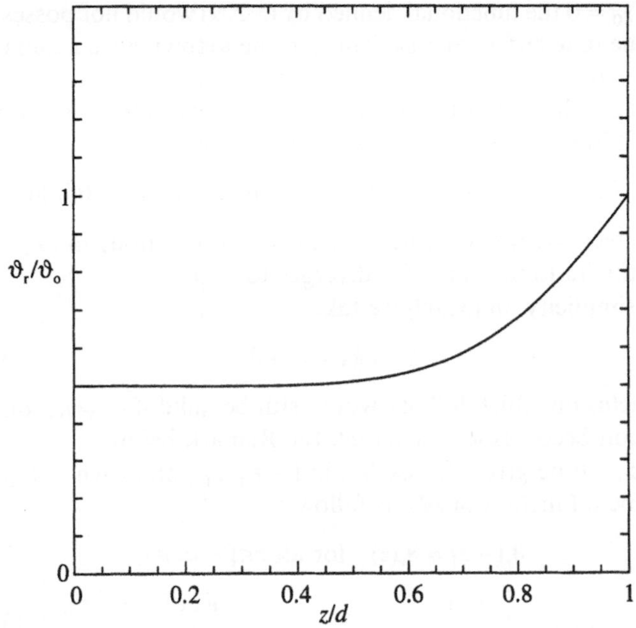

Figure 3.4

Of course, when (3.210) applies (3.215) requires k_{13} to vanish, and so the argument illustrated above is no longer conclusive. Nevertheless, with little more labour, when $k_1 \neq k_3$ we arrive at

$$F[\vartheta_r] = \frac{4\varepsilon^2}{r}\left(\int_0^1 \{k_1 \sin 2(\vartheta_0 + \varepsilon - \varepsilon\xi^2) + k_3 \cos^2(\vartheta_0 + \varepsilon - \varepsilon\xi^2)\}\xi^2 d\xi + \frac{k_{13}}{2\varepsilon}\sin 2\vartheta_0\right),$$

whence (3.214) follows again. Thus, (3.215), like any other relation between k_{13} and Frank's constants, cannot eliminate the difficulty that we have encountered here.

We have learned that adding σ_{13} to σ_F leads to an energy functional which in general does not attain its minimum within a class of admissible orientation fields where the energy minimizer would indeed be the uniform field, were it only the case that $k_{13} = 0$. Since only terms linear in $\nabla^2\mathbf{n}$ were considered in Nehring and

Saupe's energy density, we may wonder whether the difficulty above can be overcome by allowing terms quadratic in $\nabla^2\mathbf{n}$ also to enter the energy density. It seems that this way was first taken by Brand and Pleiner (1984), though in a quite different context. Here we simply remark that the functional defined by

$$F^*[\vartheta]:=F[\vartheta]+\int_{-d}^{+d} k^*\vartheta''^2\,dz, \tag{3.216}$$

where F is as in (3.209), does attain its minimum in $\mathscr{A}_F$ for all $k^*>0$. Furthermore, the minimizers of F^* are functions of class C^4 in $[-d,d]$.

It is an easy exercise to derive the Euler–Lagrange equations of F^* in $]-d,d[$ and the equilibrium conditions at the end-points of this interval (*cf. e.g.* p. 190 of Courant and Hilbert (1989), Vol. 1):

$$k^*\vartheta^{(iv)}-k\vartheta''=0 \quad \text{in }]-d,\, d[, \tag{3.217}$$

$$\vartheta''(z)=\frac{k_{13}}{4k^*}\sin 2\vartheta(z) \quad \text{for } z\in\{-d,\, d\}, \tag{3.218}$$

where, for simplicity, (3.210) has again been employed. There is only one solution of (3.217) and (3.218) subject to (3.206); it is the minimizer of F^* in $\mathscr{A}_F$:

$$\vartheta^*(z):=\frac{k_{13}}{4k}\sin 2\vartheta_0\left(\frac{\cosh\sqrt{\dfrac{k}{k^*}}\,z}{\cosh\sqrt{\dfrac{k}{k^*}}\,d}-1\right)+\vartheta_0. \tag{3.219}$$

Figure 3.5 illustrates the graph of ϑ^* for special values of k_{13}/k, k/k^*, and ϑ_0.

When in (3.219), for any given $k>0$, we take the limit as $k^*\to 0^+$, ϑ^* converges to the function

$$\vartheta_0^*(z):=\begin{cases}\vartheta_0 & \text{for } z\in\{-d,d\},\\[2ex] \vartheta_0-\dfrac{k_{13}}{4k}\sin 2\vartheta_0 & \text{for } z\in]-d,d[.\end{cases}$$

Clearly, ϑ_0^* is not a member of $\mathscr{A}_F$, unless $\vartheta_0=0$.

Barbero and Strigazzi (1989) have indeed employed the functional F^* to study the same instability that we shall consider within the

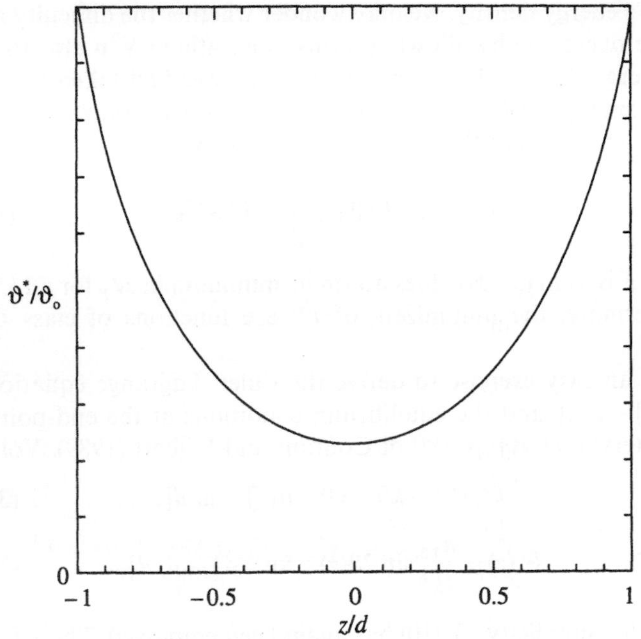

Figure 3.5

classical theory in Section 4.2 of the next chapter. Here we do not pursue these lines of thought: throughout this book we assume that

$$k_{13} = 0,$$

and no energy density depending on the second gradient of the orientation field will ever be taken into account.

3.10 References

Allender, D.W., Crawford, G.P. and Doane, J.W. (1991) Determination of the liquid-crystal surface elastic constant K_{24}, *Phys. Rev. Lett.*, **67**, 1442–1445.

Barbero, G. and Oldano, C. (1989) On the k_{13}-dependent energy term in nematic liquid crystals, *Mol. Cryst. Liq. Cryst.*, **168**, 1–5.

Barbero, G. and Strigazzi, A. (1989) Second order elasticity in nematics: a new anchoring source, *Liq. Crystals*, **5**, 693–696.

Bradshaw, M.J., Raynes, E.P. Bunning, J.D. and Faber, T.E. (1985) The Frank constants of some nematic liquid crystals, *J. Phys. (Paris)*, **46**, 1513–1520.

Brand, H.R. and Pleiner, H. (1984) Macroscopic dynamics of chiral smectic C, *J. Phys. (Paris)*, **45**, 563–573.

Bunning, J.D., Faber, T.E. and Sherrell, P.L. (1981) The Frank constants of nematic 5CB at atmospheric pressure, *J. Phys. (Paris)*, **42**, 1175–1182.

Courant, R. and Hilbert, D. (1989) *Methods of mathematical physics*, Volumes 1 & 2, J. Wiley & Sons, New York. Republication in the Wiley Classics Library Series of the first English edition (1953), Interscience Publishers.

De Gennes, P.G. (1968) Calcul de la distorsion d'une structure cholesterique par un champ magnêtique, *Solid State Commun.*, **6**, 163–165.

De Gennes, P.G. (1974) *The physics of liquid crystals*, Clarendon Press, Oxford.

Durand, G., Leger, L., Rondalez, F. and Veyssie, M. (1969) Magnetically induced cholesteric-to-nematic phase transition in liquid crystals, *Phys. Rev. Lett.*, **22**, 227–228.

Dzyaloshinskiĭ, I.E. (1970) Theory of disclinations in liquid crystals, *Sov. Phys. JETP*, **31**, 773–777.

Ericksen, J.L. (1954) Deformations possible in every isotropic, incompressible, perfectly elastic body, *ZAMP*, **5**, 466–489.

Ericksen, J.L. (1961) Conservation laws for liquid crystals, *Trans. Soc. Rheol.*, **5**, 23–34.

Ericksen, J.L. (1962) Nilpotent energies in liquid crystal theory, *Arch. Rational Mech. Anal.*, **10**, 189–196.

Ericksen, J.L. (1966) Inequalities in liquid crystals theory, *Phys. Fluids*, **9**, 1205–1207.

Ericksen, J.L. (1967a) General solutions in the hydrostatic theory of liquid crystals, *Trans. Soc. Rheol.*, **11**, 5–14.

Ericksen, J.L. (1967b) Twisting of liquid crystals, *J. Fluid Mech.*, **27**, 59–64.

Ericksen, J.L. (1976) Equilibrium theory of liquid crystals, in: *Advances in liquid crystals*, Vol. 2, G.H. Brown ed., Academic Press, New York.

Frank, F.C. (1958) On the theory of liquid crystals, *Discuss. Faraday Soc.*, **25**, 19–28.

Hinov, H.P. (1990) Further theoretical proofs for the existence of a relation between the elastic constants $k'_{11}(k_{11})$, $k'_{33}(k_{33})$ and k_{13} in nematics: Three- and one-dimensional cases, *Mol. Cryst. Liq. Cryst.*, **178**, 53–64.

Jenkins, J.T. (1971) Cholesteric energies, *J. Fluid. Mech.*, **45**, 465–475.

Lavrentovich, O.D. (1991) Hybrid aligned nematic films with horizontally degenerate boundary conditions and saddle-splay elastic term, *Physica Scripta*, **T39**, 394–395.

Leslie, F.M., Stewart, I.W. and Nakagawa, M. (1991) A continuum theory for smectic *C* liquid crystals, *Mol. Cryst. Liq. Cryst.*, **198**, 443–454.

Love, A.E.H. (1944) *A treatise on the mathematical theory of elasticity*, Dover, New York. Republication for the fourth edition (1927), Cambridge University Press.

Marris, A.W. (1978) Universal solutions in the hydrostatics of nematic liquid crystals, *Arch. Rational Mech. Anal.*, **67**, 251–303.

Marris, A.W. (1979) Addition to 'Universal solutions in the hydrostatics of nematic liquid crystals', *Arch. Rational Mech. Anal.*, **69**, 323–333.

Nehring, J. and Saupe, A. (1971) On the elastic theory of uniaxial liquid crystals, *J. Chem. Phys.*, **54**, 337–343.

Oseen, C.W. (1933) The theory of liquid crystals, *Trans. Faraday Soc.*, **29**, 883–899.

Spivak, M. (1975) *A comprehensive introduction to differential geometry*, Volume 3, Publish or Perish, Boston.

Strigazzi, A. (1987) On the surface-type elastic constant K_{24} in nematics, *Mol. Cryst. Liq. Cryst.*, **152**, 435–445.

Truesdell, C.A. (1991) *A first course in rational continuum mechanics*, Vol. 1, second edition, corrected, revised, and augmented, Academic Press, New York.

Zocher, H. (1925) Über freiwillige Strukturbildung in Solen, *Z. Anorg. Chem.*, **147**, 91–110.

Zocher, H. (1933) The effect of a magnetic field on the nematic state, *Trans. Faraday Soc.*, **29**, 945–957.

Žumer, S. and Kralj, S. (1992) Influence of K_{24} on the structure of nematic liquid crystal droplets, *Liq. Crystals*, **12**, 613–624.

3.11 Further reading

The bulk free energy

Barbero, G. (1991) Some considerations on the elastic theory for nematic liquid crystals, *Mol. Cryst. Liq. Cryst.*, **195**, 199–220.

Ericksen, J.L. (1967) Continuum theory of liquid crystals, *Appl. Mech. Rev.*, **20**, 1029–1032.

Ericksen, J.L. (1969) Continuum theory of liquid crystals of nematic type, *Mol. Cryst. Liq. Cryst.*, **7**, 153–164.

Ericksen, J.L. (1977) The mechanics of nematic liquid crystals, in: *The mechanics of viscoelastic fluids*, R.S. Rivlin ed., The American Society of Mechanical Engineers, New York.

Goossens, W.J.A. (1985) Bulk, interfacial and anchoring energies of liquid crystals, *Mol. Cryst. Liq. Cryst.*, **124**, 305–331.

Leslie, F.M. (1966) Some constitutive equations for anisotropic fluids, *Quart. J. Mech. Appl. Math.*, **19**, 357–370.

Leslie, F.M. (1968) Some constitutive equations for liquid crystals, *Arch Rational Mech. Anal.*, **28**, 265–283.

Leslie, F.M. (1987) Some topics in equilibrium theory of liquid crystals, in: *Theory and applications of liquid crystals*, J.L. Ericksen and D. Kinderlehrer eds., IMA Volumes in Mathematics and its Applications, **5**, 211–234.

Longa, L. and Trebin, H.-R. (1989) Integrity basis approach to the elastic free energy functional of liquid crystals. I. Classification of basic elastic modes, *Liq. Crystals*, **5**, 617–622.

Frank's constants

Di Lisi, G., Rosenblatt, C., Griffin, A.C. and Hari, U. (1990) Splay elasticity in an oligomeric liquid crystal, *Liq. Crystals*, **8**, 437–443.

Faetti, S. and Fronzoni, L. (1978) Molecular orientation in nematic liquid crystal films with two free surfaces, *Solid State Commun.*, **25**, 1087–1090.

Faetti, S., Gatti, M. and Palleschi, V. (1986) Measurements of surface elastic torques in liquid crystals: a method to measure elastic constants and anchoring energies, *Revue Phys. Appl.*, **21**, 451–461.

Faetti, S. and Palleschi, V. (1987) The twist elastic constant and anchoring energy of the nematic liquid crystal 4-*n*-octyl-4′-cyanobiphenyl, *Liq. Crystals*, **2**, 261–268.

Govers, E. and Vertogen, G. (1989) The elastic constants of nematics, *Liq. Crystals*, **5**, 323–326.

Haller, I. (1972) Elastic constants of the nematic liquid crystalline phase of *p*-methoxybenzylidene-*p*-*n*-butylaniline (MBBA), *J. Chem. Phys.*, **57**, 1400–1405.

Equilibrium equations

Ericksen, J.L. (1960a) Anisotropic fluids, *Arch. Rational Mech. Anal.*, **4**, 231–237.

Ericksen, J.L. (1960b) Theory of anisotropic fluids, *Trans. Soc. Rheol.*, **4**, 29–39.

Ericksen, J.L. (1960c) Transversely isotropic fluids, *Kolloid Z.*, **173**, 117–122.

Ericksen, J.L. (1962) Hydrostatic theory of liquid crystals, *Arch. Rational Mech. Anal.*, **9**, 371–378.

Leslie, F.M. (1992) Continuum theory for nematic liquid crystals, *Continuum Mech. Thermodyn.*, **4**, 167–175.

Special solutions

Anisimov, S.I. and Dzyaloshinskiĭ, I.E. (1973) A new type of disclination in liquid crystals and the stability of disclinations of various types, *Sov. Phys. JETP*, **36**, 774–779.

Coleman, B.D. and Jenkins, J.T. (1991) On a class of solutions in the theory of nematic phases, in: *Nematics. Mathematical and physical aspects*, J.-M. Coron, J.-M. Ghidaglia and F. Hélein eds., Kluwer Academic Publishers, Dordrecht, 93–105.

Second-order energies

Barbero, G. (1990) Non-local terms in the elastic free energy of distorted nematics, *Mod. Phys. Lett. B*, **4**, 983–988.

Barbero, G., Gabbasova, Z. and Kosevich, Yu.A. (1991) On the generalization of the Rapini–Papoular expression of the surface energy for nematic liquid crystals, *J. Phys. II France*, **1**, 1505–1513.

Barbero, G. and Oldano, C. (1989) Distortion in the surface layers of nematic liquid crystals: an elastic constants approach, *Mol. Cryst. Liq. Cryst.*, **170**, 99–106.

Barbero, G., Sparavigna, A. and Strigazzi, A. (1990) The structure of the distortion free-energy density in nematics: second-order elasticity and surface terms, *Nuovo Cimento*, **12D**, 1259–1272.

Yokoyama, H. (1988) Surface anchoring of nematic liquid crystals, *Mol. Cryst. Liq. Cryst.*, **165**, 265–316.

4

Instabilities

When a liquid crystal is subject to an electric or magnetic field the bulk free energy functional acquires a term which expresses the energy of interaction with the applied field. In the absence of other causes which may affect the orientation of the optic axis, these fields can induce two opposite effects on it, depending on the sign of a material constant; they either orient it in the same direction along which they are applied or in any direction orthogonal to that.

If the optic axis is somehow prescribed on the boundary of the region occupied by the liquid crystal, then the elastic free energy competes against the electric or the magnetic energy and the orientation field that minimizes the total free energy functional may be strongly affected by the applied field. Generally, the configuration of minimal energy remains unchanged as long as the intensity of the external field is sufficiently small; however, when it exceeds a critical value the configuration of minimal energy changes: that which once was stable becomes unstable and the system settles in a new stable configuration.

In this chapter we treat within the theory outlined in Chapter 3 the most classical instabilities driven by electric or magnetic fields.

4.1 Electric and magnetic energies

In Section 2.5 of Chapter 2, to illustrate an algebraic theorem we have already written the equations which express the electric displacement $\mathbf{d}$ and the magnetization $\mathbf{m}$ in a material with a preferred direction represented by a vector $\mathbf{n}$ of $\mathbb{S}^2$:

$$\mathbf{d} = \varepsilon_\perp \mathbf{e} + \varepsilon_a(\mathbf{e}\cdot\mathbf{n})\mathbf{n}, \tag{4.1}$$

$$\mathbf{m} = \chi_\perp \mathbf{h} + \chi_a(\mathbf{h}\cdot\mathbf{n})\mathbf{n}, \tag{4.2}$$

where **e** and **h** are the electric and magnetic fields which give rise to **d** and **m**, respectively.

In (4.1) $\varepsilon_\perp$ represents the dielectric constant exhibited by the material when **e** is applied in any direction orthogonal to **n**, while ε_a is the **dielectric anisotropy**, that is the difference between $\varepsilon_\parallel$ and $\varepsilon_\perp$, the former being the dielectric constant relative to electric fields parallel to **n**. Also the material constants $\chi_\perp$ and χ_a, which appear in (4.2), can be interpreted in a similar way: χ_a is the **diamagnetic anisotropy**; it is defined as $\chi_a := \chi_\perp - \chi_\parallel$, where the susceptibilities $\chi_\parallel$ and $\chi_\perp$ play precisely the same role as the dielectric constants $\varepsilon_\parallel$ and $\varepsilon_\perp$.

Liquid crystals are materials with a preferred direction which varies in space: for them **n** is the orientation field. Equations (4.1) and (4.2) still apply, but they are to be interpreted as relations which, at every point of the region occupied by the material, deliver the electric displacement and the magnetization which correspond to the orientation **n** and the fields **e** and **h** at that point. Thus, even if **e** and **h** are uniform fields, **d** and **m** vary in space, whenever **n** does.

A consequence of both (4.1) and (4.2) is to be noted: the field which represents the material response of a liquid crystal to either an electric or magnetic field is parallel to the external action which has stimulated it only if the action is itself a vector parallel or orthogonal to the optic axis at that point. Under all other circumstances, the external action and the material response are not parallel vectors, unless the corresponding anisotropy vanishes. As pointed out in Section 2.5 of Chapter 2, this is indeed a characteristic feature of all transversely isotropic materials.

Liquid crystals are **diamagnetic** materials, that is both $\chi_\parallel$ and $\chi_\perp$ in (4.2) are negative moduli; their typical magnitudes range from 10^{-7} to 10^{-6} in c.g.s. electromagnetic units. Usually, the diamagnetic anisotropy χ_a of nematics is positive and its order of magnitude is 10^{-7}. Its value depends on the temperature: for example, $\chi_a = 1.21 \times 10^{-7}$ for PAA at 122°C and $\chi_a = 1.23 \times 10^{-7}$ for MBBA at 19°C, as we learn on p. 81 of De Gennes (1974). For cholesterics χ_a is usually negative and its absolute value tends to be much smaller than for nematics, being of the order of 10^{-9} in c.g.s. electromagnetic units (*cf.* p. 248 of Ericksen (1976)).

In (4.1) **e** represents the electric field suffered by the liquid crystal; it is taken to be equal to the field created from outside the material by external electrodes. In other words, we assume that **e** does not get distorted in the liquid crystal. In practice, such an assumption is not always justified, but it does not sensibly affect the total electric energy when ε_a is sufficiently small, as is shown at the end of this section.

Clearly, analogous considerations apply also to (4.2) and the field **h**.

Meyer (1969a) has pointed out that the polarization of liquid crystals generally depends on the orientation of the optic axis: it may be different from zero when the optic axis is not uniformly oriented, even if no electric field is applied. The effect discussed by Meyer is vaguely reminiscent of the piezo-electric effect that occurs in solid crystals; here, however, the polarization is to be related to the gradient of the orientation field rather than to an external pressure. This is why De Gennes (1974) (*cf.* p. 97) has suggested calling Meyer's effect **flexo-electric**. In this book we do not pursue the study of this effect any further; equation (4.1) will be employed throughout with the warning that it is less reliable than (4.2), especially when the orientation field exhibits high gradients.

As remarked by Schadt (1972), when (4.1) applies the sign of ε_a may be different for different nematic liquid crystals. For cholesterics also ε_a can be either positive or negative. We learn from p. 186 of De Gennes (1974) that for MBBA near 25°C $\varepsilon_\parallel = 4.7$ and $\varepsilon_\perp = 5.4$.

The free energy densities associated with the fields induced by **e** and **h** are expressed in c.g.s. electromagnetic units by the following equations (*cf. e.g.* pp. 239–240 of Ericksen (1976)):

$$\sigma_e = -\frac{1}{8\pi}\mathbf{d}\cdot\mathbf{e}, \quad \sigma_m = -\tfrac{1}{2}\mathbf{m}\cdot\mathbf{h}, \tag{4.3}$$

which by (4.1) and (4.2) become, respectively,

$$\sigma_e(\mathbf{n}) := -\frac{1}{8\pi}(\varepsilon_\perp \mathbf{e}\cdot\mathbf{e} + \varepsilon_a(\mathbf{e}\cdot\mathbf{n})^2), \tag{4.4}$$

$$\sigma_m(\mathbf{n}) := -\tfrac{1}{2}(\chi_\perp \mathbf{h}\cdot\mathbf{h} + \chi_a(\mathbf{h}\cdot\mathbf{n})^2). \tag{4.5}$$

Thus, electric and magnetic fields contribute to the bulk free energy through the following integrals:

$$\mathscr{F}_e[\mathbf{n}] := \int_{\mathscr{B}} \sigma_e(\mathbf{n})\,dv, \quad \mathscr{F}_m[\mathbf{n}] := \int_{\mathscr{B}} \sigma_m(\mathbf{n})\,dv, \tag{4.6}$$

where $\mathscr{B}$ is the region in space occupied by the liquid crystal. Clearly, by (4.3), formulae (4.6) imply that $\mathscr{F}_e \equiv 0$ and $\mathscr{F}_m \equiv 0$, respectively, whenever $\mathbf{e} \equiv \mathbf{0}$ and $\mathbf{h} \equiv \mathbf{0}$.

Were $\mathscr{F}_e$ the energy functional to be minimized, it would easily follow from (4.4) that the orientation field with minimum energy satisfies either

$$(\mathbf{e}\cdot\mathbf{n})^2 = 1 \quad \text{or} \quad \mathbf{e}\cdot\mathbf{n} = 0 \quad \text{in } \mathscr{B},$$

when $\varepsilon_a > 0$ or $\varepsilon_a < 0$, respectively. In other words, an electric field, in the absence of other distorting causes, tends to orient the optic axis of a liquid crystal in the same direction along which it is applied, if $\varepsilon_a > 0$. Clearly, the same conclusion applies also to a magnetic field, provided that $\chi_a > 0$. When both ε_a and χ_a are negative, as is the case for some cholesterics, a specific alignment of the optic axis can be induced by orthogonal electric and magnetic fields.

In this chapter we take the elastic free energy density as given by Frank's formula (*cf.* equation (3.23) of Chapter 3). Furthermore, the bulk free energy functional defined by (3.2) of Chapter 3 acquires the electric and magnetic contributions expressed by (4.6) above; thus $\mathscr{F}_B$ becomes

$$\mathscr{F}_B[\mathbf{n}] = \int_{\mathscr{B}} (\sigma_F(\mathbf{n}, \nabla\mathbf{n}) + \sigma_e(\mathbf{n}) + \sigma_m(\mathbf{n}))\,dv. \tag{4.7}$$

Here, for simplicity, we separate the effects produced by electric and magnetic fields: in the instabilities that we study below σ_e and σ_m will not simultaneously appear in (4.7).

4.1.1 Approximate energies

As above let $\mathscr{B}$ be a region in space occupied by a liquid crystal. Suppose that its boundary $\partial\mathscr{B}$ is smooth and that electrodes are applied on it so as to produce an electric field $\mathbf{e}$ in $\mathscr{B}$. Formally, this amounts to saying that a smooth scalar-valued function $\bar{\varphi}$ is assigned on $\partial\mathscr{B}$; it represents the electrostatic potential prescribed on the electrodes.

Under the assumption that no free charges are dispersed in the liquid crystal, Maxwell's equations become

$$\operatorname{curl}\mathbf{e} = \mathbf{0}, \tag{4.8}$$

$$\operatorname{div}\mathbf{d} = 0, \tag{4.9}$$

where $\mathbf{d}$ is given by the constitutive equation (4.1). As is customary in electrostatics, we solve equation (4.8) by setting

$$\mathbf{e} = \nabla\varphi, \tag{4.10}$$

where $\varphi: \mathscr{B} \rightarrow \mathbb{R}$ is a function of class C^2 subject to

$$\varphi|_{\partial\mathscr{B}} = \bar{\varphi}; \tag{4.11}$$

it is the electrostatic potential in $\mathscr{B}$. By (4.1) and (4.10) equation (4.9) can easily be given the form

$$\Delta\varphi + \frac{\varepsilon_a}{\varepsilon_\perp}\operatorname{div}((\nabla\varphi\cdot\mathbf{n})\mathbf{n}) = 0. \tag{4.12}$$

For every unit vector field $\mathbf{n}$ of class C^1 on $\mathscr{B}$, (4.12) is a second-order partial differential equation for φ. Clearly, the optic axis in $\mathscr{B}$ affects the electric field. Thus, the electric energy density σ_e depends on $\mathbf{n}$ in a way which is ultimately more intricate than first appears from a glance at (4.4).

The liquid crystal is transparent, as it were, to the electric field whenever (4.12) reads as

$$\Delta\varphi = 0; \tag{4.13}$$

the electric field in $\mathscr{B}$ would then be solely determined by the potential on $\partial\mathscr{B}$, thus being independent of $\mathbf{n}$. Taking $\mathbf{e}$ in σ_e to be such a field even when (4.12) does not reduce to (4.13) is precisely the approximation alluded to above, shortly before we mentioned the flexo-electric effect. Here we explore how such an approximation can alter the total electric energy $\mathscr{F}_e$.

Clearly, (4.12) and (4.13) coincide when ε_a vanishes, which is never the case for liquid crystals, as we have learned above. Here we take ε_a to be small; more precisely, we define

$$\varepsilon := \frac{\varepsilon_a}{\varepsilon_\perp} \tag{4.14}$$

and we treat it as a perturbation parameter. That is, we write φ as a formal series in ε:

$$\varphi = \varphi^{(0)} + \varepsilon\varphi^{(1)} + \varepsilon^2\varphi^{(2)} + o(\varepsilon^2), \tag{4.15}$$

so that the electric field is given by

$$\mathbf{e} = \mathbf{e}^{(0)} + \varepsilon\mathbf{e}^{(1)} + \varepsilon^2\mathbf{e}^{(2)} + o(\varepsilon^2), \tag{4.16}$$

where

$$\mathbf{e}^{(i)} := \nabla \varphi^{(i)} \quad \text{for } i = 0, 1, 2 \tag{4.17}$$

and $\varphi^{(0)}$, $\varphi^{(1)}$ and $\varphi^{(2)}$ are functions of class C^2 on $\mathscr{B}$ such that

$$\varphi^{(0)}|_{\partial\mathscr{B}} = \bar{\varphi} \tag{4.18}$$

and

$$\varphi^{(i)}|_{\partial\mathscr{B}} = 0 \quad \text{for } i = 1, 2. \tag{4.19}$$

Theorem 4.1. Let an orientation field $\mathbf{n}$ of class C^1 on $\mathscr{B}$ be given. If a function φ as in (4.15) solves in $\mathscr{B}$ equation (4.12) subject to (4.18) and (4.19), then the total electric energy associated with it is given by

$$\mathscr{F}_e[\mathbf{n}] = -\frac{\varepsilon_\perp}{8\pi}\int_{\mathscr{B}} \{\mathbf{e}^{(0)}\cdot\mathbf{e}^{(0)} + \varepsilon(\mathbf{e}^{(0)}\cdot\mathbf{n})^2 - \varepsilon^2\mathbf{e}^{(1)}\cdot\mathbf{e}^{(1)}\} + o(\varepsilon^2), \tag{4.20}$$

where $\mathbf{e}^{(0)}$ and $\mathbf{e}^{(1)}$ are defined as in (4.17).

Proof. Inserting (4.15) into (4.12) and bringing together all equal powers of ε we arrive at the following equations in $\mathscr{B}$:

$$\Delta\varphi^{(0)} = 0, \tag{4.21}$$

$$\Delta\varphi^{(1)} = -\operatorname{div}((\nabla\varphi^{(0)}\cdot\mathbf{n})\mathbf{n}), \tag{4.22}$$

$$\Delta\varphi^{(2)} = -\operatorname{div}((\nabla\varphi^{(1)}\cdot\mathbf{n})\mathbf{n}), \tag{4.23}$$

subject to (4.18) and (4.19). Using the solution of (4.21) in (4.22) and the solution of this latter in (4.23), we finally find φ by (4.15), and then $\mathbf{e}$ by (4.16).

Grouping again all equal powers of ε in (4.4) we easily get

$$\begin{aligned}\sigma_e(\mathbf{n}) = -\frac{\varepsilon_\perp}{8\pi}\{&|\nabla\varphi^{(0)}|^2 + \varepsilon(2\nabla\varphi^{(0)}\cdot\nabla\varphi^{(1)} + (\nabla\varphi^{(0)}\cdot\mathbf{n})^2)\\ &+ \varepsilon^2(\nabla\varphi^{(1)}\cdot\nabla\varphi^{(1)} + 2\nabla\varphi^{(0)}\cdot\nabla\varphi^{(2)}\\ &+ 2(\nabla\varphi^{(0)}\cdot\mathbf{n})(\nabla\varphi^{(1)}\cdot\mathbf{n}))\} + o(\varepsilon^2).\end{aligned} \tag{4.24}$$

A formula for $\mathscr{F}_e$ then follows by integrating σ_e over $\mathscr{B}$. Such an integration, however, requires some skill.

We first note that by (4.21)

$$\nabla\varphi^{(0)}\cdot\nabla\varphi^{(1)} = \operatorname{div}(\varphi^{(1)}\nabla\varphi^{(0)}). \tag{4.25}$$

Thus, by the Divergence Theorem (*cf.* Section 2.3), the left-hand side of (4.25) does not contribute to the total electric energy because $\varphi^{(1)}$

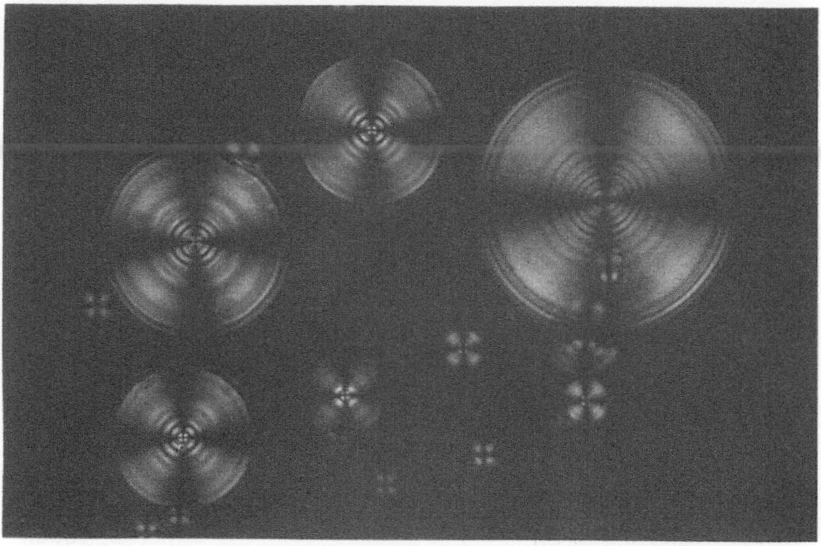

Plate 1.I

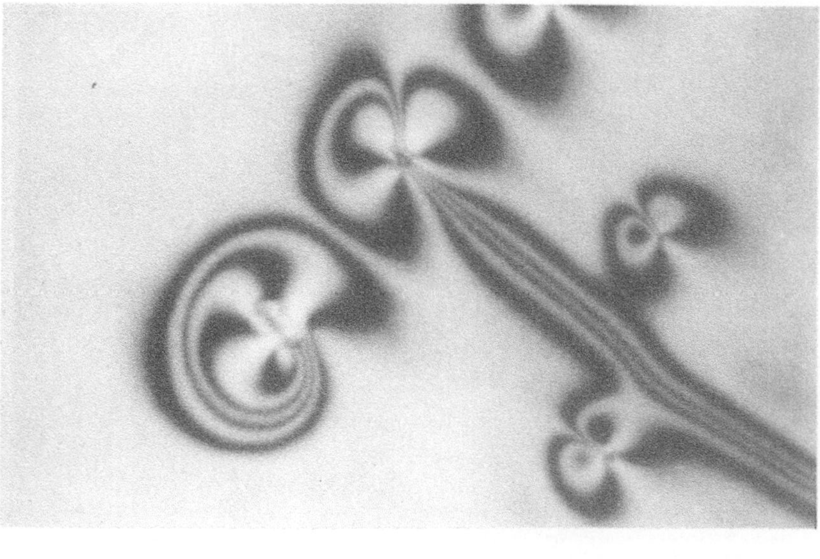

Plate 1.II

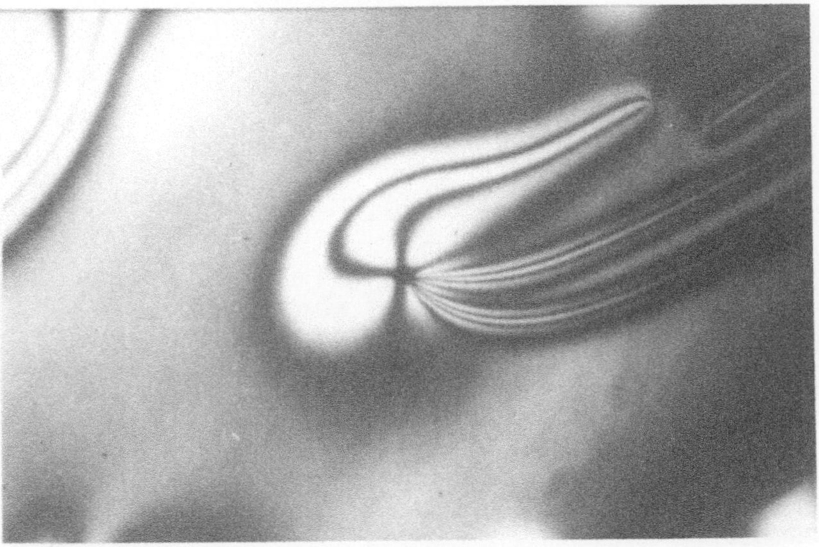

Plate 1.III

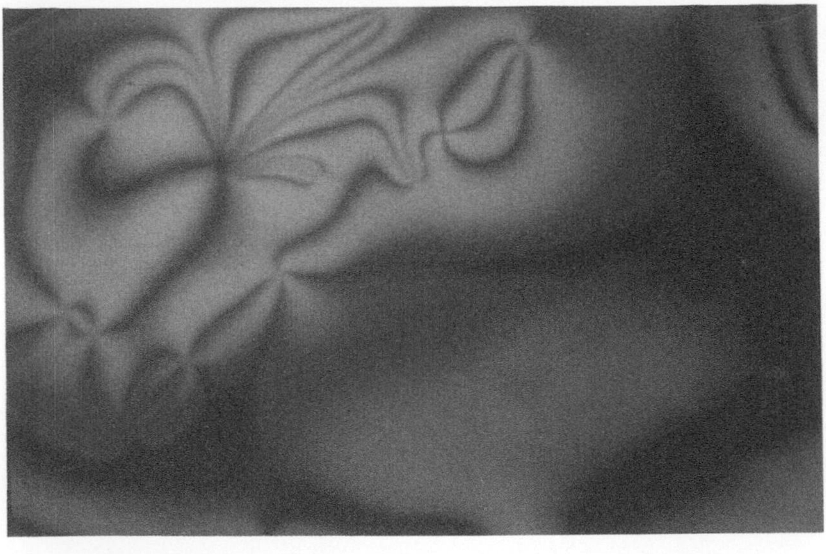

Plate 1.IV

Plate 1.V

Plate 1.VI

Plate 1.VII

Plate 1.VIII

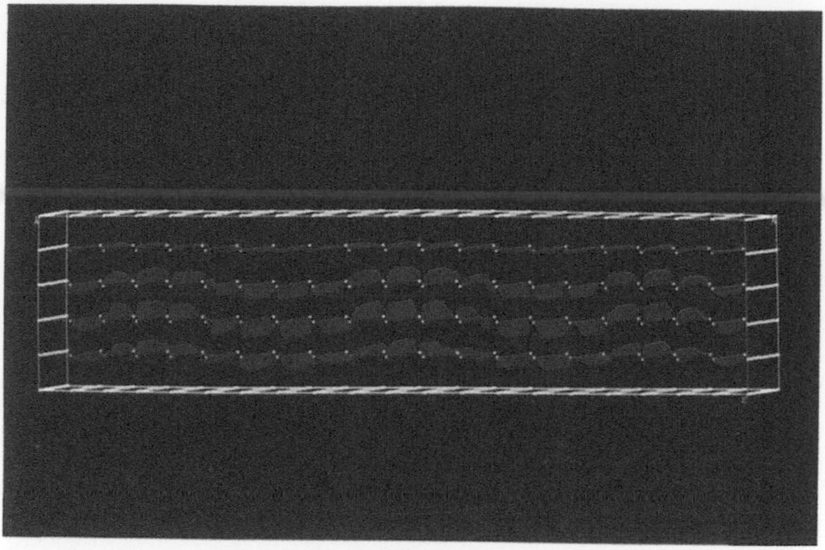

Plate 4.I

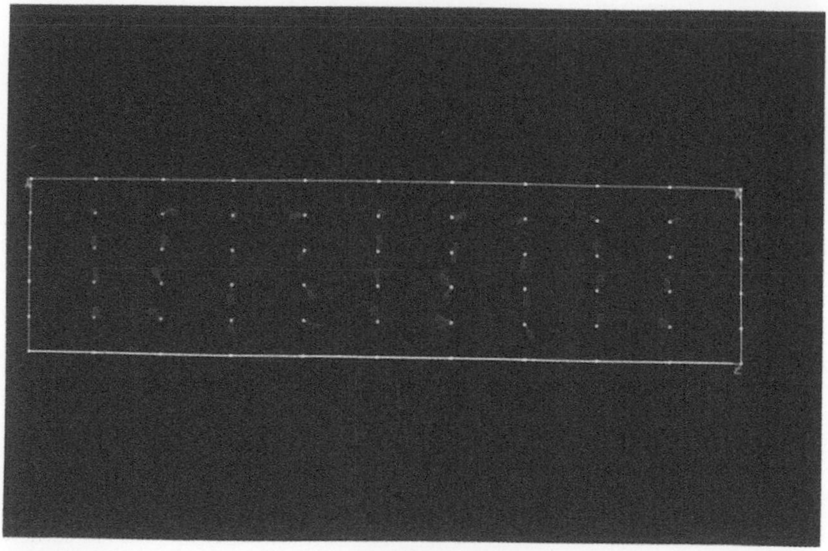

Plate 4.II

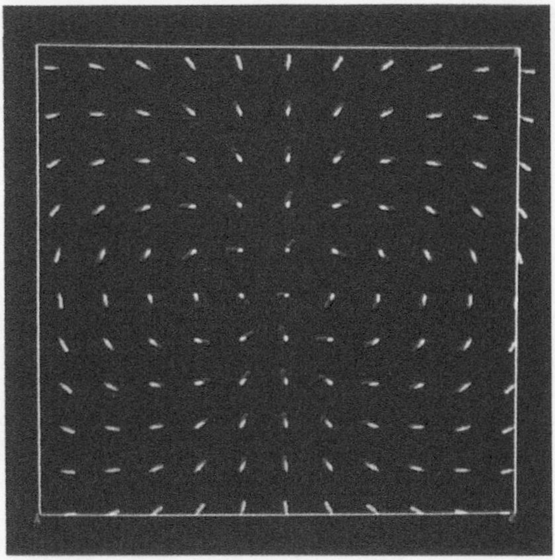

Plate 6.I

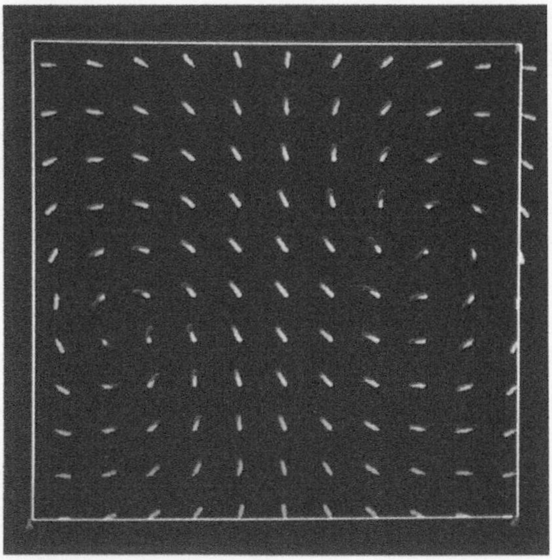

Plate 6.II

vanishes on $\partial\mathscr{B}$. Likewise we show that

$$\int_{\mathscr{B}} \nabla\varphi^{(0)}\cdot\nabla\varphi^{(2)} = 0.$$

Little more labour is needed to show that $(\nabla\varphi^{(0)}\cdot\mathbf{n})(\nabla\varphi^{(1)}\cdot\mathbf{n})$ and $\nabla\varphi^{(1)}\cdot\nabla\varphi^{(1)}$ contribute equal and opposite values to the total electric energy. We easily see that

$$(\nabla\varphi^{(0)}\cdot\mathbf{n})(\nabla\varphi^{(1)}\cdot\mathbf{n}) = \operatorname{div}(\varphi^{(1)}(\nabla\varphi^{(0)}\cdot\mathbf{n})\mathbf{n}) - \varphi^{(1)}\operatorname{div}((\nabla\varphi^{(0)}\cdot\mathbf{n})\mathbf{n}),$$

where, making use of (4.22), we arrive at

$$(\nabla\varphi^{(0)}\cdot\mathbf{n})(\nabla\varphi^{(1)}\cdot\mathbf{n}) = \operatorname{div}(\varphi^{(1)}(\nabla\varphi^{(0)}\cdot\mathbf{n})\mathbf{n}) + \operatorname{div}(\varphi^{(1)}\nabla\varphi^{(1)}) - \nabla\varphi^{(1)}\cdot\nabla\varphi^{(1)}. \qquad (4.26)$$

Integrating both sides of (4.26) over $\mathscr{B}$ we get the desired conclusion, since $\varphi^{(1)}$ vanishes on $\partial\mathscr{B}$.

We now easily conclude the proof of the theorem, making use of (4.17).

It is clear from (4.24) that to the first order in ε the electric energy density also depends on $\mathbf{e}^{(1)}$, the first-order correction to $\mathbf{e}^{(0)}$, but the total electric energy does not; in $\mathscr{F}_e$ the field $\mathbf{e}^{(1)}$ first appears at the second order. Thus, when $\varepsilon_a/\varepsilon_\perp$ is sufficiently small we can take $\mathbf{e}^{(0)}$ for $\mathbf{e}$ in $\mathscr{F}_e$, though $\mathbf{e}$ differs from $\mathbf{e}^{(0)}$.

A formula analogous to (4.20) holds also for the total magnetic energy; the role played here by $\varepsilon_a/\varepsilon_\perp$ is to be taken there by $\chi_a/\chi_\perp$.

In this book we shall always compute both $\mathscr{F}_e$ and $\mathscr{F}_m$ in (4.6) taking $\mathbf{e}$ and $\mathbf{h}$ as the fields excited in $\mathscr{B}$ by either external charges or external currents, as if the liquid crystal were transparent to the action of both. The theorem above indicates when such an approximation may be successful. Generally, it is more sensible to apply it to the magnetic energy, rather than to the electric, because $\varepsilon_a/\varepsilon_\perp$ may be much larger than $\chi_a/\chi_\perp$.

4.2 Freedericks's transition

In this section we consider in detail an instability which was first discovered experimentally by Freedericks (*cf. e.g.* Freedericks and Zolina (1933) for a review of the most relevant experiments) and then explained by Zocher (1933) within his early continuum theory of liquid crystals (*cf.* Section 1.2). Later the theoretical explanation of this phenomenon provided a method, now classic, of measuring all Frank's constants except k_4.

4.2.1 Existence of minimizers

Our analysis will resort here to a fundamental theorem of the calculus of variations, originally due to Tonelli (1914 and 1915), which has been variously rephrased in modern textbooks. This theorem, combined with others, is stated shortly below in a form which is especially expedient to our development.

We need to recall here the definition of a useful class of functions.

Definition 4.1. Let $]\alpha, \beta[$ be an open interval of $\mathbb{R}$. We say that a real function u on $]\alpha, \beta[$ is **absolutely continuous** if its derivative u' exists in $]\alpha, \beta[$, possibly away from a set of measure zero, and it is such that

$$u(y) - u(x) = \int_x^y u'(\zeta)d\zeta \quad \text{for all } x, y \in]\alpha, \beta[. \tag{4.27}$$

Hereafter we denote by $AC(\alpha, \beta)$ the class of all absolutely continuous functions of $]\alpha, \beta[$ into $\mathbb{R}$.

Theorem 4.2. Let k be a real function of class C^1 on $\mathbb{R}$ such that

$$k(u) \geqslant k_0 \tag{4.28}$$

for all $u \in \mathbb{R}$ and for a positive constant k_0. Let h be a real function of class C^1 on $\mathbb{R}$ such that

$$h(u) \geqslant h_0 \tag{4.29}$$

for all $u \in \mathbb{R}$ and for a real constant h_0. Then the functional defined by

$$F[u] := \int_\alpha^\beta \{k(u)u'^p + h(u)\}\, dx \tag{4.30}$$

for $p > 1$ attains its minimum in the class

$$\mathscr{A}_F := \{u \in AC(\alpha, \beta) \mid u(\alpha) = a,\ u(\beta) = b\} \tag{4.31}$$

for any given value of a and b. Furthermore, if both dk/du and dh/du are bounded on $\mathbb{R}$, the minimizers of F in $\mathscr{A}_F$ are of class C^2 and obey the Euler–Lagrange equation

$$p(k(u)u'^{(p-1)})' - \frac{dk}{du}u'^{p} - \frac{dh}{du} = 0 \quad \text{in }]\alpha, \beta[, \tag{4.32}$$

where a prime denotes differentiation with respect to $x \in]\alpha, \beta[$.

Theorem 4.2 relies indeed on several independent results proved by Tonelli (1915) under somewhat more general assumptions (*cf.*, in particular, Sections 12, 27 and 28).

Let $\mathscr{B}$ be the nematic cell already considered in Section 3.9 of Chapter 3. In a Cartesian frame $(o, \mathbf{e}_1, \mathbf{e}_2, \mathbf{e}_3)$ it is represented as follows:

$$\mathscr{B} = \{p = o + x\mathbf{e}_1 + y\mathbf{e}_2 + z\mathbf{e}_3 \,|\, x, y \in]0, \ell[,\ z \in]-d, d[\}.$$

The plates bounding the cell are

$$\mathscr{S}^- := \{p = o + x\mathbf{e}_1 + y\mathbf{e}_2 - d\mathbf{e}_3 \,|\, x, y \in]0, \ell[\}$$

and

$$\mathscr{S}^+ := \{p = o + x\mathbf{e}_1 + y\mathbf{e}_2 + d\mathbf{e}_3 \,|\, x, y \in]0, \ell[\}.$$

We assume that on both $\mathscr{S}^-$ and $\mathscr{S}^+$ $\mathbf{n}$ is taken to be the constant field parallel to one of their edges:

$$\mathbf{n}|_{\mathscr{S}^+} = \mathbf{n}|_{\mathscr{S}^-} \equiv \mathbf{e}_1, \tag{4.33}$$

whereas it is not prescribed at all on the rest of $\partial\mathscr{B}$. Furthermore, we assume that the nematic liquid crystal which occupies $\mathscr{B}$ is subject to a magnetic field $\mathbf{h}$ orthogonal to both $\mathscr{S}^-$ and $\mathscr{S}^+$:

$$\mathbf{h} = H\mathbf{e}_3, \tag{4.34}$$

where H is a positive parameter.

The stable equilibrium configurations of the optic axis in $\mathscr{B}$ are minimizers of the bulk free energy functional $\mathscr{F}_B$ defined by (4.7), where now $\sigma_e \equiv 0$ because $\mathscr{B}$ is not subject to any electric field.

We seek the minimizers of $\mathscr{F}_B$ within the class of orientation fields $\mathbf{n}$ such that

$$\mathbf{n}(p) = \cos\vartheta(z)\mathbf{e}_1 + \sin\vartheta(z)\mathbf{e}_3, \tag{4.35}$$

where ϑ is a function of $]-d, d[$ into $]-\pi, \pi[$ which gives in $\mathscr{B}$ the angle between $\mathbf{n}$ and $\mathbf{e}_1$. All the fields represented by (4.35) satisfy the boundary condition (4.33) if ϑ is such that

$$\vartheta(-d) = \vartheta(d) = 0. \tag{4.36}$$

Here we take ϑ as a member of $AC(-d,d)$, and so

$$\nabla\mathbf{n} = -\vartheta'(\sin\vartheta\mathbf{e}_1\otimes\mathbf{e}_3 - \cos\vartheta\mathbf{e}_3\otimes\mathbf{e}_3); \tag{4.37}$$

this equation holds in $\mathscr{B}$ wherever ϑ' is defined. It follows from (4.37) that

$$\operatorname{div}\mathbf{n} = \vartheta'\cos\vartheta,$$
$$\mathbf{n}\cdot\operatorname{curl}\mathbf{n} = 0,$$
$$(\nabla\mathbf{n})\mathbf{n} = -\vartheta'\sin\vartheta(\sin\vartheta\mathbf{e}_1 - \cos\vartheta\mathbf{e}_3),$$

whence, making use of (3.24) and (3.23) of Chapter 3, we arrive at

$$\sigma_F = (k_1\cos^2\vartheta + k_3\sin^2\vartheta)\vartheta'^2. \tag{4.38}$$

Likewise, from (4.34) and (4.5) above we see that

$$\sigma_m = -\frac{H^2}{2}(\chi_\perp + \chi_a\sin^2\vartheta). \tag{4.39}$$

Finally, inserting (4.38) and (4.39) into (4.7), we compute the total free energy of the orientation fields described by (4.35):

$$\mathscr{F}_B[\mathbf{n}] = \ell^2(F[\vartheta] - \chi_\perp H^2 d);$$

here the functional F is defined on $AC(-d,d)$ by

$$F[\vartheta] := \int_{-d}^{+d}\{k(\vartheta)\vartheta'^2 + h(\vartheta)\}\,dz, \tag{4.40}$$

where

$$k(\vartheta) := k_1\cos^2\vartheta + k_3\sin^2\vartheta, \tag{4.41}$$

$$h(\vartheta) := -\tfrac{1}{2}\chi_a H^2\sin^2\vartheta. \tag{4.42}$$

For any given $H\in\mathbb{R}$, minimizing $\mathscr{F}_B$ among the fields of the form (4.35) amounts to minimizing F within the class $\mathscr{A}_F$ defined by (4.31) with $a=b=0$. If both k_1 and k_3 are positive, the function k defined by (4.41) satisfies (4.28) with $k_0 = \min\{k_1, k_3\}$. Besides, h satisfies (4.29) with $h_0 = \min\{0, -\tfrac{1}{2}\chi_a H^2\}$. Thus, Theorem 4.2 applies, and the variational problem for the functional $\mathscr{F}_B$ arising from a nematic cell in a magnetic field possesses a solution among the orientation fields described by (4.35).

4.2.2 Instability

If $\chi_a > 0$, the magnetic field (4.34) tends to align $\mathbf{n}$ along $\mathbf{e}_3$ inside $\mathscr{B}$, competing with the condition (4.33) on $\mathscr{S}^-$ and $\mathscr{S}^+$, which would induce the uniform field $\mathbf{n} \equiv \mathbf{e}_1$, if it were the case that $H = 0$. Accordingly, we expect the minimizer of F to be the null function, for H sufficiently small, whereas for H sufficiently large it should appreciably deviate from zero in $]-d, d[$, to reflect the distortion produced on $\mathbf{n}$ by the magnetic field. The transition from one of these kinds of minimizers to the other actually occurs at a precise critical value of H. This is **Freedericks's transition**, and it is described in more detail by the following theorem.

Theorem 4.3. Let k_1, k_3 and χ_a be positive constants, and let H_c be defined as

$$H_c := \frac{\pi}{2d}\sqrt{\frac{2k_1}{\chi_a}}. \tag{4.43}$$

If $0 < H \leqslant H_c$, the functional F defined by (4.40) together with both (4.41) and (4.42) attains its minimum in the class

$$\mathscr{A}_F = \{\vartheta \in AC(-d, d) \mid \vartheta(-d) = \vartheta(d) = 0\}$$

when $\vartheta \equiv 0$. If $H > H_c$, F attains its minimum in $\mathscr{A}_F$ on precisely two functions such that one is the opposite of the other and both obey

$$\vartheta(z) \neq 0 \quad \text{for all } z \in]-d, d[. \tag{4.44}$$

The proof of Theorem 4.3 is based upon the qualitative properties of the solutions to the Euler–Lagrange equation for F. Indeed, by Theorem 4.2, the minimizers of F in $\mathscr{A}_F$ are functions of class C^2, which solve the equation

$$2k\vartheta'' + \frac{dk}{d\vartheta}\vartheta'^2 + \chi_a H^2 \sin\vartheta \cos\vartheta = 0 \quad \text{in }]-d, d[, \tag{4.45}$$

where k is as in (4.41) (*cf.* also (4.32) and (4.42)). We collect in the lemmas below the properties of the solutions of (4.45) that will lead us to the proof of Theorem 4.3.

4.2.3 Minimizers

Lemma 4.4. The null function solves equation (4.45). Moreover if ϑ^* solves it, $-\vartheta^*$ also does so.

Proof. A quick glance at equation (4.45) suffices to show that the null function is one of its solutions in $\mathscr{A}_F$. Furthermore, it follows from (4.41) that $\vartheta \mapsto k(\vartheta)$ is an even smooth function, and so its derivative is odd. Thus, an easy computation shows that if $z \mapsto \vartheta^*(z)$ is a function of $\mathscr{A}_F$ which solves (4.45) then the function $z \mapsto -\vartheta^*(z)$ also belongs to $\mathscr{A}_F$ and solves (4.45).

We say that $\vartheta \equiv 0$ is the trivial solution of (4.45). By (4.35) it corresponds to the uniform orientation of the optic axis in $\mathscr{B}$ that obeys the boundary condition (4.33).

We now restrict attention to the solutions of (4.45) in $\mathscr{A}_F$ other than the trivial one. In the following lemma we do not question their existence, but we derive the qualitative properties that they must possess, if they exist.

Lemma 4.5. Let ϑ^* be a non-trivial solution of (4.45) which minimizes F in $\mathscr{A}_F$. Then

(a) the values of ϑ^* range in the interval $[-\pi/2, \pi/2]$;
(b) the sign of ϑ^* is constant in $]-d, d[$;
(c) ϑ^* is an even function;
(d) ϑ^* is either strictly increasing or strictly decreasing in $]-d, 0[$.

Proof. To prove assertion (a), suppose, for a contradiction, that the range of ϑ^* is not included in $[-\pi/2, \pi/2]$. Then let ϑ_C be the function defined by

$$\vartheta_C(z) := \begin{cases} -\dfrac{\pi}{2} & \text{if } \vartheta^*(z) \leqslant -\dfrac{\pi}{2}, \\[2ex] \vartheta^*(z) & \text{if } \vartheta^* \in \left[-\dfrac{\pi}{2}, \dfrac{\pi}{2}\right], \\[2ex] \dfrac{\pi}{2} & \text{if } \vartheta^*(z) \geqslant \dfrac{\pi}{2}; \end{cases}$$

ϑ_C is a function of $\mathscr{A}_F$ whose range is included in $[-\pi/2, \pi/2]$; it can be regarded as ϑ^* confined between two constant functions. A glance at (4.40) and (4.42) will convince the reader that

$$F[\vartheta_C] < F[\vartheta^*],$$

and so ϑ^* cannot be a minimizer of F if its range is not included in $[-\pi/2, \pi/2]$.

To prove (b) we need to look more closely at the consequences of equation (4.45). Let ϑ^* be a non-trivial solution of (4.45) in $\mathscr{A}_F$.

We first observe that the points of $]-d,d[$ where the derivative of ϑ^* vanishes must be isolated. If there were an interval in $]-d,d[$ where $\vartheta^{*\prime}$ vanishes, ϑ^* would indeed be constant on the whole of $[-d,d]$, since the initial-value problem for (4.45) has precisely one solution in all subintervals of $[-d,d]$. This conclusion contradicts the hypothesis that ϑ^* be at the same time in $\mathscr{A}_F$ and non-trivial.

For a non-trivial solution of (4.45) we denote by $S(\vartheta')$ the subset of $[-d,d]$ where ϑ' vanishes:

$$S(\vartheta'):=\{z\in[-d,d] \,|\, \vartheta'(z)=0\}.$$

If in $]-d,d[\backslash S(\vartheta')$ we multiply both sides of (4.45) by ϑ', we easily arrive at

$$\frac{d}{dz}(k(\vartheta)\vartheta'^2+\tfrac{1}{2}\chi_a H^2\sin^2\vartheta)=0 \quad \text{in }]-d,d[\backslash S(\vartheta'),$$

whence it follows that for any interval I in $]-d,d[\backslash S(\vartheta')$ there is a constant c_I such that

$$k(\vartheta)\vartheta'^2+\tfrac{1}{2}\chi_a H^2\sin^2\vartheta=c_I \quad \text{in } I.$$

Since the elements of $S(\vartheta')$ are isolated points of $]-d,d[$ and the solutions of (4.45) are of class C^2 on the whole of $]-d,d[$, we conclude that all constants c_I are equal and the following equation holds in $]-d,d[$ for each solution of (4.45):

$$k(\vartheta)\vartheta'^2+\tfrac{1}{2}\chi_a H^2\sin^2\vartheta=c, \tag{4.46}$$

where c is a constant which depends on the specific solution.

Let ϑ^* be a non-trivial solution of (4.45) which minimizes F in $\mathscr{A}_F$. We show now that ϑ^* does not change sign in $]-d,d[$. Suppose that $S(\vartheta^{*\prime})$ contains more than one point, and let z_1 and z_2 be any two of its elements. Then, by (4.46),

$$\sin^2\vartheta^*(z_1)=\sin^2\vartheta^*(z_2),$$

and so, by assertion (a), either

$$\vartheta^*(z_1)=\vartheta^*(z_2) \quad \text{or} \quad \vartheta^*(z_1)=-\vartheta^*(z_2).$$

This amounts to saying that ϑ^* oscillates between two opposite values in $[-\pi/2,\pi/2]$. Let z' and z'' be the minimum and the maximum of $S(\vartheta^{*\prime})$, respectively. They satisfy

$$z'>-d \quad \text{and} \quad z''<d$$

because if it were the case that $\vartheta^{*\prime}(-d)=0$ or $\vartheta^{*\prime}(d)=0$, ϑ^* would vanish identically on $[-d,d]$ since $\vartheta^*(-d)=\vartheta^*(d)=0$ and (4.45) is an autonomous differential equation for which the initial-value problem possesses only one solution in each subinterval of $[-d,d]$. If $z''>z'$, then both the function defined by

$$\vartheta_M(z):=\begin{cases} |\vartheta^*(z)| & \text{for } z\in[-d,z']\cup[z'',d], \\ \max\limits_{[-d,d]}\vartheta^* & \text{for } z\in[z',z''], \end{cases}$$

and its opposite differ from ϑ^*, and a direct computation shows that

$$F[\vartheta_M]=F[-\vartheta_M]<F[\vartheta^*].$$

Thus, ϑ^* minimizes F only if $z'=z''$. This implies that $S(\vartheta^{*\prime})$ is a singleton, and so the proof of assertion (b) is complete, since ϑ^* vanishes at the end-points of $[-d,d]$.

We now prove assertion (c). Let ϑ^* be a non-trivial minimizer of F in $\mathscr{A}_F$, and let z_0 be the only member of $S(\vartheta^{*\prime})$. Clearly, the restriction of ϑ^* to the interval $[-d,z_0]$ is a solution of (4.45). Moreover, if $z_1:=\min\{d,2z_0+d\}$, ϑ^* must solve in $[z_0,z_1]$ the initial-value problem for (4.45) in which $\vartheta(z_0)=\vartheta^*(z_0)$ and $\vartheta'(z_0)=0$. It is easily seen that the function defined in $[z_0,z_1]$ by

$$\vartheta^*(z):=\vartheta^*(2z_0-z) \tag{4.47}$$

solves this problem, and so the graph of ϑ^* in $[z_0,z_1]$ and that in $[-d,z_0]$ are symmetric with one another. Thus, if $z_0<0$, ϑ^* vanishes at an inner point of $[-d,d]$, which is contrary to assertion (b) above; if $z_0>0$, then $\vartheta^*(d)$ does not vanish, which is contrary to the assumption that $\vartheta^*\in\mathscr{A}_F$. We conclude that $z_0=0$, and so by (4.47) assertions (c) and (d) are proved at once.

The qualitative properties of the minimizers of F in $\mathscr{A}_F$ that we have found above will now lead us to the complete proof of Theorem 4.3.

Proof of Theorem 4.3. Let ϑ^* be a non-trivial minimizer of F in $\mathscr{A}_F$. By Lemma 4.4, its opposite also solves the Euler–Lagrange equation for F. Moreover, since $F[-\vartheta^*]=F[\vartheta^*]$, $-\vartheta^*$ is also a non-trivial minimizer of F. This shows that non-trivial minimizers of F are degenerate: they can be arranged in pairs of opposite functions, which through (4.35) describe different orientation fields with the same energy. Invoking Lemma 4.5, we shall remove such degeneracy from the class of non-trivial minimizers of F by taking them as positive functions on the whole of $]-d,d[$ which attain their maximum at 0. The reader should, however, keep in mind that two different orientation fields correspond in $\mathscr{B}$ to each of the non-trivial minimizers of F that we encounter below.

We denote by ϑ_m the maximum of ϑ^*:

$$\vartheta_m := \vartheta^*(0). \tag{4.48}$$

Thus, for ϑ^* equation (4.46) can be rewritten as

$$k(\vartheta^*)(\vartheta^{*\prime})^2 = \tfrac{1}{2}\chi_a H^2(\sin^2\vartheta_m - \sin^2\vartheta^*), \tag{4.49}$$

whence by separation of variables it follows that

$$z = -d + \frac{1}{H}\sqrt{\frac{2}{\chi_a}}\int_0^{\vartheta^*(z)} \sqrt{\frac{k(\vartheta)}{\sin^2\vartheta_m - \sin^2\vartheta}}\,d\vartheta. \tag{4.50}$$

By the change of variables

$$\sin\lambda := \frac{\sin\vartheta}{\sin\vartheta_m}, \tag{4.51}$$

equation (4.50) can be given the form

$$z = -d + \frac{1}{H}\sqrt{\frac{2}{\chi_a}}\int_0^{\lambda^*(z)} G(\lambda,\vartheta_m)\,d\lambda, \tag{4.52}$$

where

$$\lambda^*(z) := \arcsin\left(\frac{\sin\vartheta^*(z)}{\sin\vartheta_m}\right) \quad \text{for all } z\in[-d,d]$$

and

$$G(\lambda,\vartheta_m) := \sqrt{\frac{k(\arcsin(\sin\vartheta_m\sin\lambda))}{1-(\sin\vartheta_m\sin\lambda)^2}}. \tag{4.53}$$

By (4.48), $\lambda^*(0)=\pi/2$, and so it follows from (4.52) that

$$H = f(\vartheta_m), \tag{4.54}$$

where f is the function defined by

$$f(\vartheta_m) := \frac{1}{d}\sqrt{\frac{2}{\chi_a}}\int_0^{\pi/2} G(\lambda,\vartheta_m)\,d\lambda. \tag{4.55}$$

For each ϑ_m in $]0,\pi/2[$ that solves (4.54), equation (4.50) represents the inverse in $[-d,0]$ of a non-trivial solution of (4.45), a candidate to be a minimizer of F in $\mathscr{A}_F$. Thus, we seek the roots of (4.54). We first observe that f is a function strictly increasing in $]0,\pi/2[$ since $\partial G/\partial\vartheta_m>0$ in $]0,\pi/2[\times]0,\pi/2[$, as is easily checked with the aid of (4.51). Moreover,

$$\lim_{\vartheta_m\to 0^+} f(\vartheta_m)=\frac{1}{d}\sqrt{\frac{2}{\chi_a}}\int_0^{\pi/2}\lim_{\vartheta_m\to 0^+} G(\lambda,\vartheta_m)d\lambda$$

$$=\frac{1}{d}\sqrt{\frac{2}{\chi_a}}\sqrt{k(0)}\frac{\pi}{2} \tag{4.56}$$

and

$$\lim_{\vartheta_m\to(\pi/2)^-} f(\vartheta_m)=\frac{1}{d}\sqrt{\frac{2}{\chi_a}}\int_0^{\pi/2}\frac{\sqrt{k(\lambda)}}{\cos\lambda}d\lambda=+\infty. \tag{4.57}$$

Thus, by (4.56) and (4.41), there is precisely one root of (4.54) if $H\geqslant H_c$, where H_c is as in (4.43), whereas there is none if $H<H_c$. For $H=H_c$, the solution of (4.54) is $\vartheta_m=0$, and so by (4.48) ϑ^* is genuinely non-trivial only for $H>H_c$.

So far we have proved that there is only one non-trivial solution of (4.45) for every $H>H_c$; to see whether it is also a minimizer of F in $\mathscr{A}_F$ we must compare its energy with that of the trivial solution, which exists for all values of H. Since ϑ^* is even, with the aid of (4.49), from (4.40) we arrive at

$$F[\vartheta^*]=\int_{-d}^{0}\chi_a H^2(\sin^2\vartheta_m-2\sin^2\vartheta^*)dz,$$

whence by (4.50) it follows that

$$F[\vartheta^*]=\int_0^{\vartheta_m}\sqrt{2\chi_a}H(\sin^2\vartheta_m-2\sin^2\vartheta)\sqrt{\frac{k(\vartheta)}{\sin^2\vartheta_m-\sin^2\vartheta}}d\vartheta.$$

By use of (4.51), this equation becomes

$$F[\vartheta^*]=\sqrt{2\chi_a}H\sin^2\vartheta_m\int_0^{\pi/2}\cos 2\lambda G(\lambda,\vartheta_m)d\lambda, \tag{4.58}$$

where G is defined as in (4.53). A direct computation, which again employs (4.51), shows that $\partial G/\partial\lambda>0$ in $]0,\pi/2[\times]0,\pi/2[$. Thus, integrating by parts the integral in (4.58), we obtain that

$$F[\vartheta^*]=-\sqrt{\frac{\chi_a}{2}}H\sin^2\vartheta_m\int_0^{\pi/2}\sin 2\lambda\frac{\partial G}{\partial\lambda}d\lambda<0.$$

Since $F[0]=0$, this inequality implies that whenever there is a non-trivial solution of (4.45), it is a minimizer of F. The proof of Theorem 4.3 is thus complete, if we recall that (4.44) is indeed a consequence of Lemma 4.5.

4.2.4 Bifurcation

In Theorem 4.3 above we have shown that Freedericks's transition occurs when $H=H_c$. The following theorem illuminates some mathematical features of this transition and tells us that the mininum of F does not jump when H crosses the critical value H_c.

Theorem 4.6. For any given $H>H_c$ let ϑ_H be the positive minimizer of F in $\mathscr{A}_F$. Then ϑ_H converges uniformly to the trivial solution as $H\to H_c^+$. Moreover,

$$\lim_{H\to H_c^+} F[\vartheta_H]=0. \tag{4.59}$$

Proof. The properties of the function f defined by (4.55) ensure that $\vartheta_m\to 0^+$ when $H\to H_c^+$. Since ϑ_m is the maximum of ϑ_H, this suffices to prove that ϑ_H converges uniformly to the trivial solution as $H\to H_c^+$. Moreover, it follows from (4.49) that for a positive minimizer of F in $\mathscr{A}_F$

$$\vartheta'_H(-d)=\sqrt{\frac{\chi_a}{2k_1}}H\sin\vartheta_m.$$

Since, by Lemma 4.5, ϑ'_H attains its maximum just at $z=-d$, the derivative of ϑ'_H also converges uniformly to the zero function on $[-d,d]$ as $H\to H_c^+$. By the definition of F in (4.40), this immediately implies (4.59).

The transition that we have so far described analytically is illustrated by the bifurcation diagram of Fig. 4.1.

Here ϑ_m is taken as the parameter which represents the positive solutions of the equilibrium equation for F, with the convention that $\vartheta_m=0$ for the trivial solution. In Fig. 4.1 the heavy line represents the **stable** solution, and the dashed line represents the **unstable** one. Thus, for $H\leqslant H_c$ the trivial solution is the only solution of the equilibrium equation, and it is necessarily stable; for $H>H_c$ the trivial solution becomes unstable and a new solution arises, which is stable.

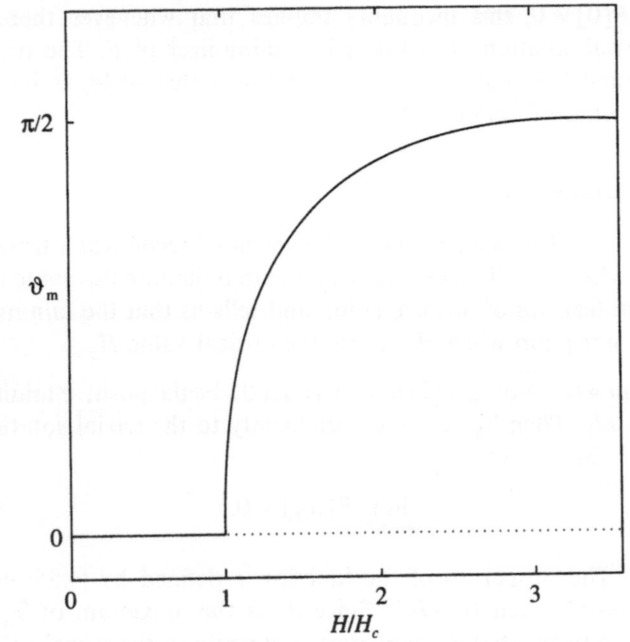

Figure 4.1

In words, we say that a bifurcation with **exchange of stability** occurs in the system at $H = H_c$.

For $H > H_c$ the stable branch of the diagram in Fig. 4.1 represents the graph of the function $\vartheta_m \mapsto f(\vartheta_m)$ (*cf.* equation (4.54)). Though here it is intended to provide just a qualitative picture, we derive one of its features from the asymptotic analysis of f as ϑ_m approaches 0: this will tell us how the stable branch of the diagram in Fig. 4.1 emerges from the H-axis in the vicinity of H_c.

By (4.41), it follows from (4.53) that

$$G(\lambda, \vartheta_m) = \sqrt{k_1}\left(1 + \frac{k_3}{2k_1}\sin^2\lambda\vartheta_m^2\right) + o(\vartheta_m^2),$$

whence, making use of (4.55) and (4.43), we easily arrive at

$$H = H_c\left(1 + \frac{k_3}{4k_1}\vartheta_m^2\right) + o(\vartheta_m^2).$$

Figure 4.2 illustrates the vector lines of the orientation field in $\mathscr{B}$ for both $H < H_c$ and $H > H_c$.

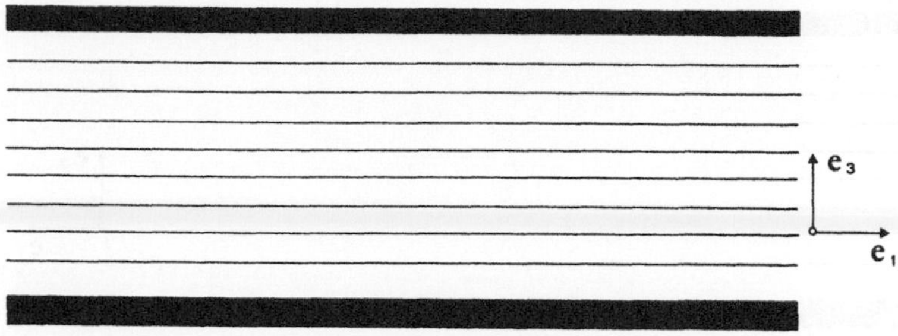

(a) $H < H_c$

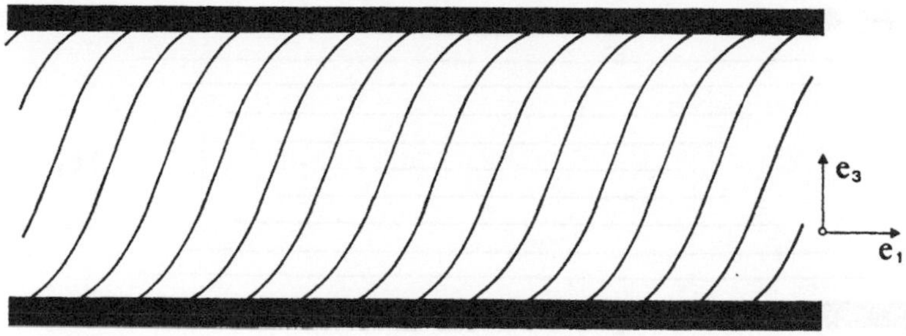

(b) $H > H_c$

Figure 4.2 (*a and b*)

Only positive minimizers of F have been taken into account in our discussion. The vector lines of the orientation fields that would correspond through (4.35) to the negative minimizers of F for $H > H_c$ look like those of Fig. 4.2b when viewed from the opposite side of the page, as though they were drawn on a transparent sheet.

The change in the optical properties of the cell that arises at $H = H_c$ can be detected experimentally. When χ_a is known from other measurements, equation (4.43) allows us to evaluate k_1 by measuring d and H_c.

Likewise, we can measure both k_2 and k_3. Let $\mathbf{n}$ be again prescribed on $\mathscr{S}^-$ and $\mathscr{S}^+$ as in (4.35). Assume that the magnetic field is now applied along $\mathbf{e}_2$, instead of $\mathbf{e}_3$ as in (4.34). Essentially the same analysis carried out above would apply to the minimizers of the bulk free energy functional $\mathscr{F}_B$ within the class of orientation fields

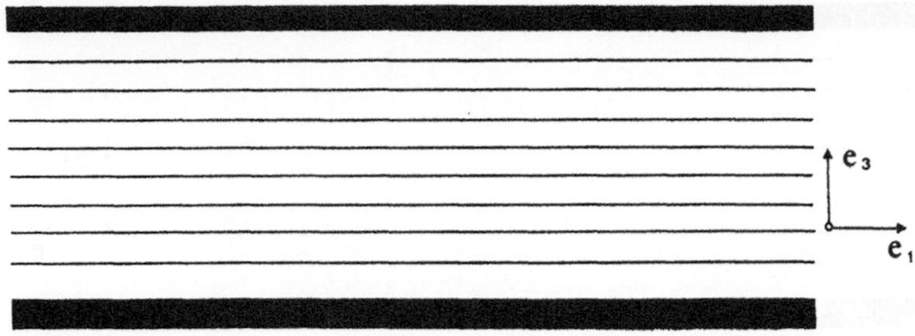

(a) $H < H_c$

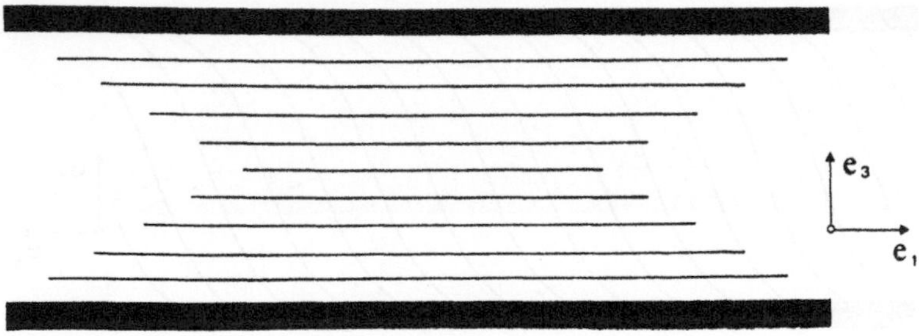

(b) $H > H_c$

Figure 4.3 (*a and b*)

described by

$$\mathbf{n}(p) = \cos\vartheta(z)\mathbf{e}_1 + \sin\vartheta(z)\mathbf{e}_2.$$

We would reach the conclusion illustrated in Fig. 4.3, where H_C is now given by

$$H_c = \frac{\pi}{2d}\sqrt{\frac{2k_2}{\chi_a}}.$$

The graph in Fig. 4.3a describes the vector lines of the uniform orientation field which minimizes the free energy when $H < H_c$, whereas the graph in Fig. 4.3b is a representation of the twisted orientation field that prevails over the uniform field for $H > H_c$.

A similar conclusion holds also when equations (4.33), (4.34) and

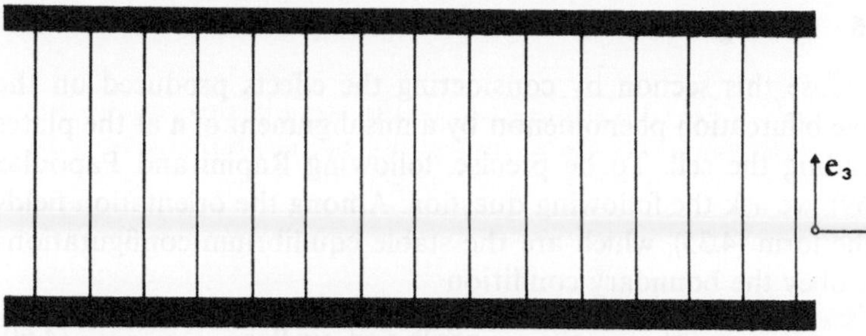

(a) $H < H_c$

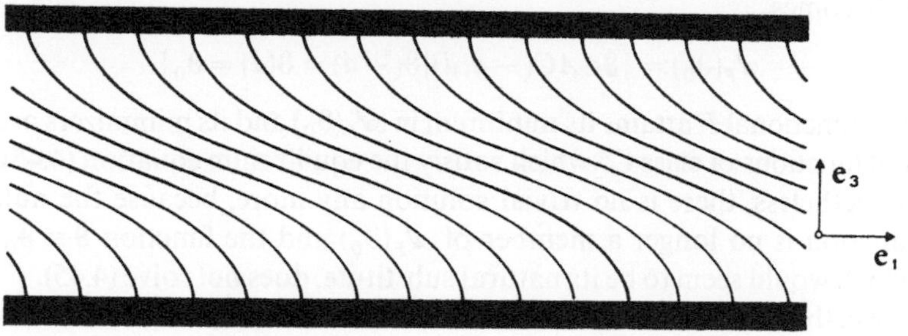

(b) $H > H_c$

Figure 4.4 *(a and b)*

(4.35) are replaced, respectively, by

$$\mathbf{n}|_{\mathscr{S}^-} = \mathbf{n}|_{\mathscr{S}^+} = \mathbf{e}_3,$$
$$\mathbf{h} = H\mathbf{e}_1,$$

and

$$\mathbf{n}(p) = \sin\vartheta(z)\mathbf{e}_1 + \cos\vartheta(z)\mathbf{e}_3.$$

Figure 4.4, where now

$$H_c = \frac{\pi}{2d}\sqrt{\frac{2k_3}{\chi_a}},$$

illustrates the outcomes of the analysis suitable to this case in a form which should by now have become familiar to the reader.

When the distance between the plates of the nematic cell is about 10 μm the critical magnetic field is a few kilogauss.

4.2.5 Misalignments

We close this section by considering the effects produced on the above bifurcation phenomenon by a misalignment of $\mathbf{n}$ at the plates bounding the cell. To be precise, following Rapini and Papoular (1969), we ask the following question: Among the orientation fields of the form (4.35), which are the stable equilibrium configurations that obey the boundary condition

$$\mathbf{n}|_{\mathscr{S}^-} = \mathbf{n}|_{\mathscr{S}^+} = \cos\vartheta_0\mathbf{e}_1 + \sin\vartheta_0\mathbf{e}_3, \tag{4.60}$$

for a given $\vartheta_0 \in]0, \pi/2[$?

As (4.33) is replaced by (4.60), the class of functions admissible for F becomes

$$\mathscr{A}_F(\vartheta_0) := \{\vartheta \in AC(-d, d) | \vartheta(-d) = \vartheta(d) = \vartheta_0\}.$$

The functional F attains its minimum in $\mathscr{A}_F(\vartheta_0)$ and its minimizers are still functions of class C^2 which satisfy the equilibrium equation (4.45). Nonetheless, there is no trivial solution any more, because the null function is no longer a member of $\mathscr{A}_F(\vartheta_0)$ and the function $\vartheta \equiv \vartheta_0$, which would seem to be its natural substitute, does not solve (4.45).

Furthermore, Lemma 4.4 above does not apply here, whereas assertions (a), (b), and (c) of Lemma 4.5 do, as the reader will easily verify by going through their proofs again. The new boundary conditions imposed on ϑ remove completely the degeneracy of the minimizers of F that we have encountered above, because they *break the symmetry* of $\mathscr{A}_F$. Accordingly, assertion (d) of Lemma 4.5 becomes more stringent.

Lemma 4.7. Every function which minimizes F in $\mathscr{A}_F(\vartheta_0)$ is strictly increasing in $]-d, 0[$.

Proof. Let ϑ^* be a minimizer of F in $\mathscr{A}_F(\vartheta_0)$. The qualitative properties of ϑ^* shown in the proof of assertions (b) and (c) of Lemma 4.5 would indeed apply also to a positive function strictly decreasing in $]-d, 0[$, which attains its minimum at 0. To show that this cannot be the case it suffices to observe that wherever ϑ^* attains an isolated minimum in $]-d, d[$, by (4.45) it must be negative, since k_1, k_3 and χ_a are all strictly positive, which is contrary to assertion (b) of Lemma 4.5.

The bifurcation described by Theorem 4.3 above no longer exists in the class $\mathscr{A}_F(\vartheta_0)$, however small we take ϑ_0.

Theorem 4.8. Let ϑ_0 be given in $]0, \pi/2[$. If k_1, k_2, and χ_a are positive constants, for every $H > 0$ there is precisely one minimizer of F in $\mathscr{A}_F(\vartheta_0)$. It satisfies

$$\vartheta^*(z) > \vartheta_0 \quad \text{for all } z \in]-d, d[.$$

Proof. As in the proof of Theorem 4.3, here also equation (4.46) plays a central role. Combining it with the qualitative features of the minimizers of F stated in assertions (a), (b), and (c) of Lemma 4.5 and in Lemma 4.7 above, we easily arrive at the formula

$$z = -d + \frac{1}{H}\sqrt{\frac{2}{\chi_a}}\int_{\lambda_0(\vartheta_m)}^{\lambda^*(z)} G(\lambda, \vartheta_m)\,d\lambda \quad \text{for all } z \in [-d, 0], \tag{4.61}$$

which is to replace (4.52) here. In (4.61) the function G is defined as in (4.53) and

$$\lambda_0(\vartheta_m) := \arcsin\left(\frac{\sin\vartheta_0}{\sin\vartheta_m}\right). \tag{4.62}$$

There is only one solution of (4.45) in $\mathscr{A}_F(\vartheta_0)$, which is related to the function λ^* in (4.61) through

$$\vartheta^*(z) := \arcsin(\sin\vartheta_m \sin\lambda^*(z)) \quad \text{for all } z \in [-d, 0],$$

provided the following equation holds

$$d = \frac{1}{H}\sqrt{\frac{2}{\chi_a}}\int_{\lambda_0(\vartheta_m)}^{\pi/2} G(\lambda, \vartheta_m)\,d\lambda. \tag{4.63}$$

It is easily seen that the function defined by (4.62), which maps the interval $[\vartheta_0, \pi/2]$ into itself, is strictly decreasing, and so, since $\partial G/\partial\vartheta_m > 0$, the function g defined on $]\vartheta_0, \pi/2[$ as

$$g(\vartheta_m) := \frac{1}{d}\sqrt{\frac{2}{\chi_a}}\int_{\lambda_0(\vartheta_m)}^{\pi/2} G(\lambda, \vartheta_m)\,d\lambda$$

is strictly increasing. Thus, equation (4.63), which can now be rewritten as

$$H = g(\vartheta_m), \tag{4.64}$$

possesses only one root in $]\vartheta_0, \pi/2[$ for every $H > 0$, because

$$\lim_{\vartheta_m \to \vartheta_0^+} g(\vartheta_m) = 0 \quad \text{and} \quad \lim_{\vartheta_m \to \pi/2^-} g(\vartheta_m) = +\infty.$$

This completes the proof of the theorem.

What makes the conclusion of this theorem different from that of Theorem 4.3 is that the function g defined above satisfies

$$\min_{[\vartheta_0,\pi/2]} g = 0,$$

while the function f in (4.55) is such that

$$\min_{[\vartheta_0,\pi/2]} f > 0.$$

It is worth noting that just a slight perturbation of the boundary conditions (4.36) is enough to destroy the bifurcation diagram drawn in Fig. 4.1. Thus, the method of measuring Frank's constants that is based on (4.43) and the other formulae akin to it appears to be very sensitive to parameters hard to control experimentally. A small misalignment of the optic axis at the two plates of the nematic cell causes the critical phenomenon outlined in Theorem 4.3 to disappear; the orienting effect of the magnetic field starts at the onset of the field and gradually proceeds further as its intensity increases. Nevertheless, when ϑ_0 is sufficiently small the graph of g is very similar to the

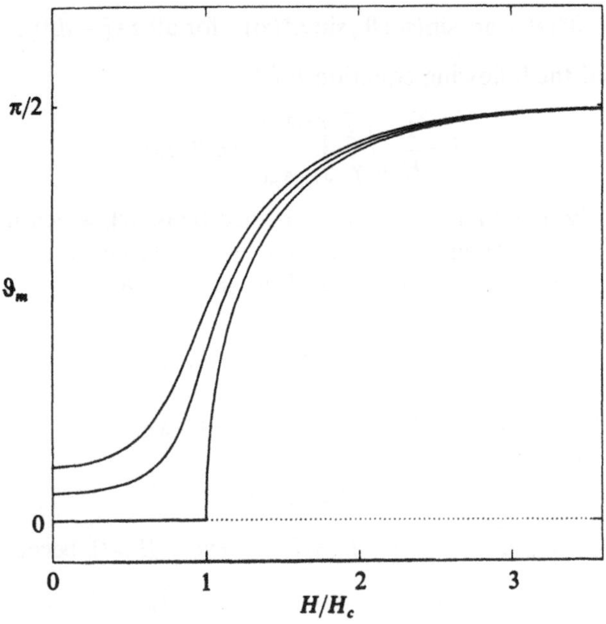

Figure 4.5

heavy line of Fig. 4.1, though it cannot be regarded by any means as a bifurcation diagram. Figure 4.5 illustrates equation (4.64) for different values of ϑ_0: the corresponding graphs accumulate on one side of the bifurcation diagram of Fig. 4.1 as ϑ_0 converges to 0.

In Fig. 4.5 H_c is defined as in (4.43); analogous conclusions hold for the other critical phenomena considered above.

It is clear from the graphs of Fig. 4.5 that in experiments affected by a slight misalignment of the optic axis at the plates of the cell, the *measured* critical values of H are smaller than the theoretical value in (4.43); this leads to measurements of Frank's constants smaller than their actual values. Rapini and Papoular (1969) have shown that this error is not negligible: for example, they have estimated that $\Delta H_c/H_c \approx -0.1$ when $\vartheta_0 \approx 2°$, and so the error which affects Frank's constants is about 20%.

It is worth noting that our discussion about misalignments of the optic axis at the plates of the nematic cell applies *verbatim* to misalignments of the magnetic field, which thus represent another source of uncertainty in the measure of Frank's constants. Dafermos (1968) first studied the variational problem that arises when the magnetic field is not orthogonal to the plates of the cell. His analysis applies also when both elastic and magnetic energies are given by formulae far more general than Frank's and (4.5).

4.3 Periodic Freedericks's transition

If we were content with evaluating only the critical value of the magnetic intensity H above which Freedericks's transition occurs, the analysis carried out in the preceding section could be made considerably simpler. Here we illustrate such a simpler argument because it leads to the prediction of a new instability, which within a class richer in orientations than that represented by (4.35) may occur before Freedericks's when H is steadily increased from zero.

Suppose that the orientation $\mathbf{n}$ in the nematic cell $\mathscr{B}$ considered in the preceding section is given by

$$\mathbf{n} = \mathbf{e}_1 + \varepsilon(f(z)\mathbf{e}_2 + g(z)\mathbf{e}_3), \qquad (4.65)$$

where ε is a given small parameter and both f and g are scalar-valued functions defined on the interval $[-d, d]$. Equation (4.65) defines a unit vector field to within terms of order higher than one in ε:

$$\mathbf{n}\cdot\mathbf{n} = 1 + o(\varepsilon).$$

Let $\mathbf{n}$ be prescribed on the faces $\mathscr{S}^+$ and $\mathscr{S}^-$ of $\mathscr{B}$ precisely as in (4.33) above; thus, both f and g must vanish at the end-points of the interval $[-d, d]$:

$$f(-d) = f(d) = 0, \quad g(-d) = g(d) = 0. \tag{4.66}$$

The fields described by (4.65) represent small perturbations of the constant orientation that satisfies (4.33). The elastic free energy associated with these perturbations can easily be computed by the use of the following formula:

$$\nabla\mathbf{n} = \varepsilon(f'\mathbf{e}_2 \otimes \mathbf{e}_3 + g'\mathbf{e}_3 \otimes \mathbf{e}_3), \tag{4.67}$$

which implies that

$$\operatorname{div}\mathbf{n} = \varepsilon g', \quad \operatorname{curl}\mathbf{n} = -\varepsilon f'\mathbf{e}_1.$$

If the cell is subject to a magnetic field as in (4.34), the functional F defined by (4.40) above is to be replaced by

$$F[f, g] = \varepsilon^2 F_2[f, g] + o(\varepsilon^2), \tag{4.68}$$

where

$$F_2[f, g] := \int_{-d}^{+d} \{k_1 g'^2 + k_2 f'^2 - \tfrac{1}{2}\chi_a H^2 g^2\}\, dz. \tag{4.69}$$

The Euler–Lagrange equations of F_2 are

$$k_2 f'' = 0, \tag{4.70}$$

$$k_1 g'' = -\tfrac{1}{2}\chi_a H^2 g, \tag{4.71}$$

whence it follows that $f = 0$ and $g = 0$ describe an equilibrium configuration for F_2. If $k_2 > 0$, $f = 0$ is the only solution of (4.70) which satisfies $(4.66)_1$. Were $g = 0$ also the only solution of (4.71) which satisfies $(4.66)_2$, we should conclude that the field $\mathbf{n} \equiv \mathbf{e}_1$ is the only equilibrium configuration for the free energy functional subject to (4.33) within the class represented by (4.65).

The qualitative properties we have learned in the preceding section about the equilibrium configurations which are approximated by (4.65) suggest that only solutions of (4.71) even in $[-d, d]$ be considered. We shall direct attention only to these solutions.

The appearence of a solution of (4.71) and $(4.66)_2$ other than $g = 0$

as H is progressively increased from zero is to be interpreted as the occurrence of another equilibrium configuration for the liquid crystal in the cell, which is close to the unperturbed one. This is a sign that a branch of new equilibrium solutions is beginning to bifurcate from the trivial solution: the least value of H for which this happens is to be equal to H_c, the critical value obtained through the fully nonlinear analysis performed in Section 4.2.

The general even solution of (4.71) is

$$g(z) = g_0 \cos \sqrt{\frac{\chi_a H^2}{2k_1}}\, z, \tag{4.72}$$

where g_0 is an arbitrary constant. This function obeys $(4.66)_2$ whenever

$$\sqrt{\frac{\chi_a H^2}{2k_1}} = m\frac{\pi}{2} \tag{4.73}$$

for any $m \in \mathbb{N}$. The least value of H for which (4.73) is satisfied corresponds to $m = 1$, and so it is precisely H_c as given in (4.43). Though here we have determined the critical value of H at which Freedericks's transition occurs, the nonlinear analysis of Section 4.2 tells us that the new branch of equilibrium solutions remains stable away from the bifurcation point. Below, following Lonberg and Meyer (1985), we apply this method again to discover in the same nematic cell $\mathscr{B}$ considered so far an instability which may occur before Freedericks's.

We represent the orientation field in $\mathscr{B}$ through the formula

$$\mathbf{n} = \mathbf{e}_1 + \varepsilon(f(y,z)\mathbf{e}_2 + g(y,z)\mathbf{e}_3), \tag{4.74}$$

where f and g are now scalar-valued functions defined on $[0,\ell] \times [-d,d]$. The field $\mathbf{n}$ described by (4.74) obeys the boundary conditions (4.33) whenever f and g are such that

$$f(y,d) = f(y,-d) = 0 \quad \text{and} \quad g(y,d) = g(y,-d) = 0 \tag{4.75}$$

for all $y \in [0,\ell]$. It is easily seen from (4.74) that

$$\nabla\mathbf{n} = \varepsilon(f_{,y}\mathbf{e}_2 \otimes \mathbf{e}_2 + f_{,z}\mathbf{e}_2 \otimes \mathbf{e}_3 + g_{,y}\mathbf{e}_3 \otimes \mathbf{e}_2 + g_{,z}\mathbf{e}_3 \otimes \mathbf{e}_3),$$

and hence

$$\operatorname{div}\mathbf{n} = \varepsilon(f_{,y} + g_{,z}), \quad \operatorname{curl}\mathbf{n} = \varepsilon(g_{,y} - f_{,z})\mathbf{e}_1.$$

Thus, the functional F still has the form (4.68), but F_2 is to be replaced here by

$$F_2[f,g]:=\frac{1}{\ell}\int_0^\ell dy\int_{-d}^{+d} dz\{k_1(g_{,z}+f_{,y})^2+k_2(g_{,y}-f_{,z})^2-\tfrac{1}{2}\chi_a H^2 g^2\}. \tag{4.76}$$

The Euler–Lagrange equations of this functional read as follows (*cf. e.g.* p. 192 of Courant and Hilbert (1989), Vol. 1):

$$k_1(g_{,z}+f_{,y})_{,y}-k_2(g_{,y}-f_{,z})_{,z}=0, \tag{4.77}$$

$$k_2(g_{,y}-f_{,z})_{,y}+k_1(g_{,z}+f_{,y})_{,z}=-\tfrac{1}{2}\chi_a H^2 g. \tag{4.78}$$

We assume that both k_1 and k_2 are positive and we set

$$\rho:=\frac{k_2}{k_1},\quad h:=\sqrt{\frac{\chi_a H^2}{2k_1}}. \tag{4.79}$$

It is easily seen from (4.43) and (4.79) that if $H=H_c$ then $h=h_c$, where

$$h_c:=\frac{\pi}{2d}; \tag{4.80}$$

moreover, h increases as H does. We further look for solutions to (4.77) and (4.78) of the form

$$f(y,z)=\varphi(z)\sin qy,\quad g(y,z)=\gamma(z)\cos qy, \tag{4.81}$$

where q is a positive parameter and both φ and γ are functions of class C^2 on $[-d,d]$ which satisfy the conditions

$$\varphi(-d)=\varphi(d)=0,\quad \gamma(-d)=\gamma(d)=0. \tag{4.82}$$

These solutions represent equilibrium configurations for the orientation of the liquid crystal in the cell, which are close to the uniform field $\mathbf{n}\equiv\mathbf{e}_1$, but periodic along the direction of $\mathbf{e}_2$.

It is worth noting that, by (4.81),

$$(g_{,z}+f_{,y})=(\gamma'-q\varphi)\cos qy,\quad (g_{,y}-f_{,z})=-(q\gamma+\varphi')\sin qy,$$

and so the contribution of the splay distortion to the functional in (4.76) may vanish, while that of the twist does not. This shows how the class of distortions we consider here is different from that of the preceding section, though unlike this latter it includes only small perturbations of a uniform alignment.

Inserting (4.81) into both (4.77) and (4.78), by (4.79) we arrive at the following ordinary differential equations for φ and γ:

$$\rho\varphi'' - (1-\rho)q\gamma' - q^2\varphi = 0, \tag{4.83}$$

$$\gamma'' + (1-\rho)q\varphi' + (h^2 - \rho q^2)\gamma = 0. \tag{4.84}$$

Here we consider only solutions φ and γ of (4.83) and (4.84) which are, respectively, odd and even on $[-d, d]$.

Whenever there are a $q > 0$ and two functions of class $C^2(-d, d)$, φ and γ, which solve both (4.83) and (4.84) and satisfy (4.82), but do not vanish identically in $[-d, d]$, we say that there is a **periodic equilibrium configuration** of F_2.

Clearly, periodic equilibrium configurations of F_2 may fail to exist if h is not given appropriate values. We show below that when ρ is sufficiently small, they do exist for values of the magnetic strength smaller than the threshold of Freedericks's transition.

Theorem 4.9. Let ρ_c be the positive root of the equation

$$\rho^2 - 2a\rho - a = 0, \tag{4.85}$$

where $a := (\pi^2/8) - 1$. For all $\rho \in]0, \rho_c[$ there is a periodic equilibrium configuration of F_2 such that

$$h < h_c.$$

Proof. For $\rho \neq 1$ equation (4.83) can also be given the form

$$\gamma' = \frac{\rho}{(1-\rho)q}\varphi'' - \frac{q}{1-\rho}\varphi. \tag{4.86}$$

Differentiating both sides of (4.84) and making use of (4.86), we obtain the following differential equation for φ only:

$$\rho\varphi^{(iv)} + \rho(h^2 - 2q^2)\varphi'' - q^2(h^2 - \rho q^2)\varphi = 0; \tag{4.87}$$

the characteristic equation associated with it is

$$\rho\lambda^4 + \rho(h^2 - 2q^2)\lambda^2 - q^2(h^2 - \rho q^2) = 0.$$

Let η be the dimensionless parameter defined by

$$\eta := \frac{q}{h}. \tag{4.88}$$

It is easily seen that if

$$\frac{1}{\rho} > \eta^2, \tag{4.89}$$

then the characteristic equation of (4.87) possesses two real roots and two imaginary roots, and the real and imaginary roots are each the opposite of the other:

$$\lambda_1 = \mu_1, \quad \lambda_2 = -\mu_1, \quad \lambda_3 = i\mu_2, \quad \lambda_4 = -i\mu_2,$$

where

$$\begin{aligned} \mu_1 &:= \frac{h}{\sqrt{2}}\sqrt{-1 + 2\eta^2 + \sqrt{1 + 4\eta^2\left(\frac{1}{\rho} - 1\right)}}, \\ \mu_2 &:= \frac{h}{\sqrt{2}}\sqrt{1 - 2\eta^2 + \sqrt{1 + 4\eta^2\left(\frac{1}{\rho} - 1\right)}}. \end{aligned} \tag{4.90}$$

Thus, all odd solutions of (4.87) are represented through the formula

$$\varphi(z) = b_1 \sinh \mu_1 z + b_2 \sin \mu_2 z. \tag{4.91}$$

Inserting (4.91) into (4.86), we arrive at the following representation formula for γ:

$$\gamma(z) = b_1 \frac{\rho\bar{\mu}_1^2 - \eta^2}{(1-\rho)\eta\bar{\mu}_1} \cosh \bar{\mu}_1 hz + b_2 \frac{\rho\bar{\mu}_2^2 + \eta^2}{(1-\rho)\eta\bar{\mu}_2} \cos \bar{\mu}_2 hz + b, \tag{4.92}$$

where

$$\bar{\mu}_i := \frac{\mu_i}{h} \quad \text{for } i = 1, 2,$$

and b is an arbitrary constant. Making use of both (4.91) and (4.92) in (4.84), we easily see that this equation is satisfied only if

$$b = 0.$$

By (4.91) and (4.92), conditions (4.82) require b_1 and b_2 to satisfy the equations

$$b_1 \sinh \bar{\mu}_1 hd + b_2 \sin \bar{\mu}_2 hd = 0, \tag{4.93}$$

$$b_1 \frac{\rho\bar{\mu}_1^2 - \eta^2}{(1-\rho)\eta\bar{\mu}_1} \cosh \bar{\mu}_1 hd + b_2 \frac{\rho\bar{\mu}_2^2 + \eta^2}{(1-\rho)\eta\bar{\mu}_2} \cos \bar{\mu}_2 hd = 0. \tag{4.94}$$

Solutions of (4.93) and (4.94) exist, provided the following condition holds:

$$\frac{\bar{\mu}_1}{\rho\bar{\mu}_1^2 - \eta^2} \tanh\left(\bar{\mu}_1 \frac{\pi}{2}\frac{h}{h_c}\right) = \frac{\bar{\mu}_2}{\rho\bar{\mu}_2^2 + \eta^2} \tan\left(\bar{\mu}_2 \frac{\pi}{2}\frac{h}{h_c}\right), \tag{4.95}$$

where use has also been made of (4.80).

For each given positive η which satisfies (4.89), (4.95) becomes an equation for h only. One value of q then corresponds through (4.88) to every h that solves (4.95). For each of these values of q, equations (4.91) and (4.92) satisfy (4.82), and so they represent a periodic equilibrium configuration of F_2. Thus, we find an entire class of such configurations, if we find all h which solve (4.95) as η varies in $]0, 1/\sqrt{\rho}[$. To do so we first remark that by (4.90) both $\bar{\mu}_1$ and $\bar{\mu}_2$ do not depend on h, and so both sides of (4.95) are indeed elementary functions of h. Furthermore, if

$$\rho < 1, \tag{4.96}$$

it is easily seen that

$$\rho\bar{\mu}_1^2 - \eta^2 < 0.$$

Thus, for given η, the roots of (4.95) correspond to the intersections between the graphs sketched in Fig. 4.6 where $\xi := \bar{\mu}_2 h/h_c$, and so there are infinitely many.

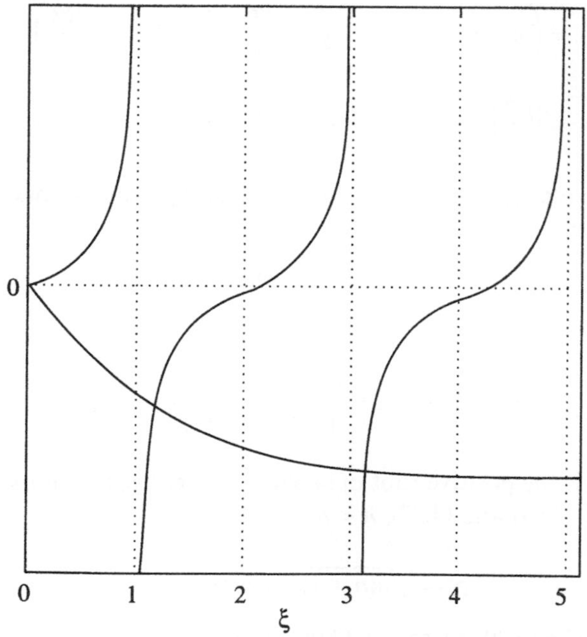

Figure 4.6

Here, for a given value of η, we shall confine attention to the least of these roots, which we denote by $h(\eta)$.

Inspired by a paper of Oldano (1986), we now study the asymptotic behaviour of $h(\eta)$ as $\eta \to 0$. We take $h(\eta)$ as

$$h(\eta) = h_c(\alpha + \beta\eta^{\omega}), \tag{4.97}$$

where α, β and ω are parameters to be determined. Straightforward computations lead us to the following formulae

$$\bar{\mu}_1 = \frac{\eta}{\sqrt{\rho}}\left(1 - \frac{1}{2}\rho\left(\frac{1}{\rho} - 1\right)^2 \eta^2\right) + o(\eta^3),$$

$$\bar{\mu}_2 = 1 + \frac{1}{2}\left(\frac{1}{\rho} - 2\right)\eta^2 + o(\eta^2),$$

whence, by (4.97), it follows that (4.95) becomes

$$\begin{aligned}
&-\frac{2\rho^2}{\pi\alpha}\left(\frac{1}{\rho} - 1\right)^2 \eta^2 + o(\eta^2)\\
&= \left(\rho + \left(\frac{3}{2} - \rho\right)\eta^2\right)\left(\cot\frac{\pi\alpha}{2} - \frac{\pi}{2}\left(1 + \cot^2\frac{\pi\alpha}{2}\right)\right.\\
&\quad\left.\cdot\left(\frac{\alpha}{2}\left(\frac{1}{\rho} - 2\right)\eta^2 + \beta\eta^{\omega}\right)\right) + o(\eta^{\bar{\omega}}),
\end{aligned} \tag{4.98}$$

where $\bar{\omega} := \max\{2, \omega\}$. A close look at this equation shows that it is not contradictory only if

$$\alpha = 1 \quad \text{and} \quad \omega = 2. \tag{4.99}$$

Inserting (4.99) into (4.98), we find that

$$\beta = \frac{4}{\pi^2\rho}\left(\rho^2 + 2\left(\frac{\pi^2}{8} - 1\right)\rho - \left(\frac{\pi^2}{8} - 1\right)\right).$$

Thus, if ρ_c is the positive root of (4.85), β is negative for all $\rho \in]0, \rho_c[$, and so, by (4.99) and (4.97), $h < h_c$.

Since

$$\rho_c = \sqrt{a(a+1)} - a \approx 0.303,$$

all values of ρ in $]0, \rho_c[$ satisfy (4.96), a restriction which has already been imposed.

Clearly, the theorem above just shows that there is a family of periodic equilibrium configurations of F_2 when the magnetic strength is less than the threshold at which Freedericks's transition should occur. It does not prove that this branch of solutions is stable nor that when $\rho < \rho_c$ a transition to a periodic equilibrium configuration does indeed take place. There is, however, evidence that this is the case. Lonberg and Meyer (1986) have performed an experiment where a polymeric liquid crystal confined in a cell was subject to a sufficiently strong magnetic field. They observed a stable periodic pattern of the orientation field, which closely resembles the periodic equilibrium configurations that we have encountered above.

In their paper the reader can also find the outcome of a numerical analysis of the compatibility condition (4.95), which shows that for all $\rho < \rho_c$ the least value of h for which there is a periodic equilibrium configuration in the nematic cell is always less than h_c, even when η is far greater than zero.

We do not attempt here to evaluate the energy of the above periodic equilibrium configurations, even for small values of η. Nonetheless, a paper by Cohen and Luskin (1991) should easily convince us that a **periodic Freedericks's transition** actually occurs whenever $\rho < \rho_c$.

Cohen and Luskin studied a nematic cell with the orientation prescribed on the *whole* of its boundary so as to be parallel to a given edge. Their analysis shows that when the ratio k_2/k_1 is sufficiently small the critical magnetic strength at which the uniform alignment becomes unstable with respect to generic perturbations is strictly smaller than Freedericks's threshold. They have also computed the stable equilibrium configurations of the orientation in the cell corresponding to several values of k_1 and k_2 which agree with the experimental data available for some polymeric liquid crystals: they all exhibit a periodic roll pattern, where the axes of the rolls are parallel to the direction of the unperturbed orientation. Plates 4.I and 4.II, which are reproduced here by courtesy of M. Luskin, illustrate two views of one periodic pattern which minimizes the energy stored in the cell.

It is worth noting that no periodic Freedericks's transition can occur when $k_1 = k_2$, as is proved in Lemma 4.3 of Cohen and Luskin (1991). On the other hand, since $\rho_c \approx 0.303$, the anisotropy required for Frank's constants to make a periodic Freederick's transition possible is not too big. This explains why roll patterns have been observed only in polymeric liquid crystals: their elastic constants must be just right to sustain a periodic Freedericks's transition.

4.4 De Gennes's transition

In this section we treat an instability which occurs in cholesteric liquid crystals. It actually consists in a transition which changes cholesterics into nematics with no change in the temperature, but in the presence of a magnetic field, whose strength exceeds a critical value.

Consider the natural orientation of a cholesteric liquid crystal described, for example, by equation (3.16) of Chapter 3. If the material possesses a positive diamagnetic anisotropy, one would expect that a magnetic field applied at right angles to the cholesteric axis would tend to align the orientation field. Intuitively, the magnetic field should unwind the helical structure characteristic of cholesterics and change their chiral natural orientation into the uniform orientation pertaining to nematics.

We wonder whether the cholesteric helix can be unwound by a field of finite strength. Following a celebrated paper by De Gennes (1968), we shall learn below that this is indeed the case and we shall compute the critical value of the magnetic field at which the transition occurs. An approximate analysis of the same problem was also proposed by Meyer (1968 and 1969b).

Let $\mathbf{e}_3$ be the cholesteric axis of a natural orientation. We consider the orientation fields of the liquid crystal that are expressed in the form

$$\mathbf{n} = \cos\vartheta\mathbf{e}_1 + \sin\vartheta\mathbf{e}_2, \tag{4.100}$$

where ϑ is a function which depends only on z, the co-ordinate along $\mathbf{e}_3$, and $\mathbf{e}_1$ and $\mathbf{e}_2$ are unit vectors which together with $\mathbf{e}_3$ constitute the same frame. Clearly, the undistorted configuration of the liquid crystal is among those represented by (4.100); it occurs when ϑ is taken to be

$$\vartheta(z) = \tau z + \vartheta_0, \tag{4.101}$$

where τ is the characteristic twist of the material. In general, if ϑ is a function of class C^1, it follows from (4.100) that

$$\nabla\mathbf{n} = -\vartheta'(\sin\vartheta\mathbf{e}_1 \otimes \mathbf{e}_3 - \cos\vartheta\mathbf{e}_2 \otimes \mathbf{e}_3), \tag{4.102}$$

where a prime denotes differentiation with respect to z. Hence, for the orientation fields described by (4.100) Frank's formula with the aid of $(3.79)_1$ of Chapter 3 leads to the following expression for the energy density:

$$\sigma_F = k_2(\vartheta' - \tau)^2. \tag{4.103}$$

Remark. Equation $(3.79)_2$ of Chapter 3 need not be invoked to arrive at (4.103) because, by (4.102), in the class of orientations described by (4.100) both $\operatorname{tr} \nabla\mathbf{n}$ and $\operatorname{tr}(\nabla\mathbf{n})^2$ vanish.

We suppose that the cholesteric liquid crystal occupies the whole space. We formulate a variational problem in a variable set, which represents an elementary cell periodically repeated in space, where the orientation of the optic axis exhibits a single twist. We now make this idea more precise. Let $\mathscr{B}$ be the cell confined between two parallel plates orthogonal to $\mathbf{e}_3$, which in the Cartesian co-ordinates (x, y, z) is defined by

$$\mathscr{B} = \{(x, y, z) \mid x, y \in]0, \ell[,\ z \in]0, d_\vartheta[\}.$$

For the fields in (4.100) the height of $\mathscr{B}$ depends on the orientation of the optic axis in $\mathscr{B}$, and it is defined so that

$$\vartheta(d_\vartheta) - \vartheta(0) = \pi. \tag{4.104}$$

Thus, at the plates bounding $\mathscr{B}$ the two orientations of $\mathbf{n}$ are opposite to each other, and they represent the same optical state (*cf.* Section 1.4 of Chapter 1).

In the absence of any external field the total elastic free energy stored in $\mathscr{B}$ reads as

$$\mathscr{F}_B[\mathbf{n}] = \ell^2 F[\vartheta]$$

with

$$F[\vartheta] := \int_0^{d_\vartheta} k_2(\vartheta' - \tau)^2 dz, \tag{4.105}$$

where d_ϑ is also to be regarded as unknown. If $\mathscr{F}_B$ measures properly the extent at which the elastic energy of the liquid crystal is dispersed in the whole space, determining the stable equilibrium configuration of the optic axis amounts to finding a positive d_ϑ and a function ϑ of class C^1 on $[0, d_\vartheta]$ that minimize F subject to (4.104).[1] This is indeed an easy task, because when $k_2 > 0$ the minimum of F is zero

[1] The reader should compare Theorem 4.2 to reassure himself that the minimizing ϑ is a smooth function.

and it is actually attained only when $\vartheta' \equiv \tau$. Thus, as was to be expected, (4.101) represents all minimizers of the free energy when no distorting field is acting on $\mathscr{B}$, and by (4.104) the corresponding value of d_ϑ is

$$d_0 := \frac{\pi}{\tau}. \tag{4.106}$$

When ϑ is as in (4.101) we denote by L_0 the **pitch** of the helix described by $\mathbf{n}$ when (4.100) is evaluated along a line parallel to the cholesteric axis. L_0 is the distance between two planes which enclose a complete revolution of $\mathbf{n}$; clearly, it is equal to $2d_0$.

We now consider the distorting effect of a magnetic field applied transversely to $\mathbf{e}_3$. We still represent $\mathbf{n}$ in $\mathscr{B}$ through (4.100) and we take the magnetic field in the form

$$\mathbf{h} = H\mathbf{e}_2 \tag{4.107}$$

By (4.5) and (4.6) a straightforward computation shows that, *modulo* an additive constant, the functional F defined by (4.105) is to be replaced now by

$$F[\vartheta] = \int_0^{d_\vartheta} \{k_2(\vartheta' - \tau)^2 - \tfrac{1}{2}\chi_a H^2 \sin^2 \vartheta\}\, dz, \tag{4.108}$$

where χ_a is taken as a positive constant.

For any given value of d_ϑ, the minimizers of F must satisfy the Euler–Lagrange equation

$$k_2 \vartheta'' = -\tfrac{1}{2}\chi_a H^2 \sin\vartheta \cos\vartheta. \tag{4.109}$$

Since no anchoring condition is prescribed for $\mathbf{n}$ on the plates that bound $\mathscr{B}$, this equation will not be subject to any condition at the end-points of the interval $[0, d_\vartheta]$. Thus, among the equilibrium configurations for $\mathbf{n}$ in $\mathscr{B}$ there are the uniform fields corresponding to the following solutions of (4.109):

$$\vartheta \equiv 0 \quad \text{and} \quad \vartheta \equiv \frac{\pi}{2}, \tag{4.110}$$

apart from the others which differ from these by a multiple of $\pi/2$ and do not bear further information. The solutions in (4.110) cannot both be minimizers of F since the energy associated with the former is higher:

$$F[0] - F[\tfrac{1}{2}\pi] = \chi_a H^2 d_\vartheta > 0.$$

Thus, searching for minimizers of F, we shall hereafter disregard $(4.110)_1$ and consider only $(4.110)_2$ as the prototype of all constant solutions of (4.109).

The analysis illustrated below parallels closely that which led us to complete the proof of Theorem 4.3 in Section 4.2, and so we skip here several details which would sound merely repetitious.

Let d_ϑ be given. We consider solutions of (4.109) whose first derivative does not vanish in $]0, d_\vartheta[$. Multiplying both sides of (4.109) by ϑ' we arrive at the equation

$$2k_2\vartheta'^2 = -\chi_a H^2 \sin^2\vartheta + c, \tag{4.111}$$

where c is an arbitrary positive constant. We set

$$c := \frac{\chi_a H^2}{k^2} \tag{4.112}$$

with k in $\mathbb{R}$. Thus, (4.111) can be rewritten in the form

$$2k_2\vartheta'^2 = \frac{\chi_a H^2}{k^2}(1 - k^2 \sin^2\vartheta). \tag{4.113}$$

This equation requires that $\sin^2\vartheta \leqslant (1/k^2)$, and so only if $k^2 < 1$ do its solutions describe orientation fields which wind completely around the cholesteric axis as when the distorting magnetic field is absent. This explains why hereafter we take $k \in]0, 1[$. The limiting case when $k = 1$ plays a special role in our development: this is the only value of k for which a constant solution of (4.113) is also a solution of (4.109); such a solution is precisely $(4.110)_2$.

For each solution of (4.113) there is another which differs from it only in sign: if one is an increasing function, the other is decreasing. We shall confine attention to the increasing solutions of (4.113), and so this equation can be given the form

$$\vartheta' = \frac{1}{\xi k}\sqrt{1 - k^2\sin^2\vartheta}, \tag{4.114}$$

where

$$\xi := \frac{1}{H}\sqrt{\frac{2k_2}{\chi_a}} \tag{4.115}$$

is the **magnetic coherence length**. It is easily seen that for all $k \in]0, 1[$ the solution of (4.114) is invertible and a straightforward integration yields the following formula for its inverse:

$$z(\vartheta) = k\xi \int_0^{\vartheta} \frac{d\eta}{\sqrt{1-k^2\sin^2\eta}}. \tag{4.116}$$

Since this function is increasing, equation (4.104) is satisfied if

$$d_\vartheta = k\xi \int_0^{\pi} \frac{d\eta}{\sqrt{1-k^2\sin^2\eta}}. \tag{4.117}$$

Comparing (4.116) and (4.101) we see that for a cholesteric liquid crystal in the presence of a magnetic field transverse to the cholesteric axis the equilibrium configurations of **n** suffer a distortion which alters the natural chiral structure of the substance, but does not completely destroy its periodicity. The twist of the orientation field is no longer uniform, but the field is still periodic and so a **distorted pitch** can be defined as

$$L(k) := 2d_\vartheta = 4k\xi \int_0^{\pi/2} \frac{d\eta}{\sqrt{1-k^2\sin^2\eta}}. \tag{4.118}$$

We have found a family of equilibrium configurations of **n** in the class (4.100). As k varies in $]0,1[$, (4.116) represents all of them and (4.118) tells us which is their distorted pitch. Finding the equilibrium configuration which is indeed stable is a rather delicate issue which demands some knowledge of elliptic integrals. For the reader who is not already familiar with this matter I collect here the results that are relevant to our development (*cf.* Sections 8.11 and 8.12 of Gradshteyn and Ryzhik (1980) for further information).

Definition 4.2. For any given $k \in]0,1[$, the integrals defined by

$$K(k) := \int_0^{\pi/2} \frac{d\eta}{\sqrt{1-k^2\sin^2\eta}}, \tag{4.119}$$

$$E(k) := \int_0^{\pi/2} \sqrt{1-k^2\sin^2\eta}\,d\eta \tag{4.120}$$

are called **complete elliptic integrals** of the **first** and **second kind**, respectively.

These integrals possess the following properties, which will often be employed below.

Theorem 4.10. Both K and E are differentiable functions, and the following relations hold in $]0,1[$:

$$\frac{dK}{dk}=\frac{1}{k}\left(\frac{E(k)}{1-k^2}-K(k)\right), \tag{4.121}$$

$$\frac{dE}{dk}=\frac{1}{k}(E(k)-K(k)). \tag{4.122}$$

Furthermore, the asymptotic behaviour of both K and E near the end-points of the interval where they are defined is illustrated by the formulae below:

$$K(k)=\frac{\pi}{2}\left(1+\frac{1}{4}k^2+\frac{9}{64}k^4\right)+o(k^4), \tag{4.123}$$

$$K(k)=\ln\frac{4}{\sqrt{1-k^2}}+\frac{1}{4}\left(\ln\frac{4}{\sqrt{1-k^2}}-1\right)(1-k^2)+o(1-k^2), \tag{4.124}$$

$$E(k)=\frac{\pi}{2}\left(1-\frac{1}{4}k^2-\frac{3}{64}k^4\right)+o(k^4), \tag{4.125}$$

$$E(k)=1+\frac{1}{2}\left(\ln\frac{4}{\sqrt{1-k^2}}-1\right)(1-k^2)+o(1-k^2). \tag{4.126}$$

These formulae show that K diverges logarithmically as k approaches 1, while E tends to 1 and that both functions remain bounded when k approaches 0.

A glance at (4.119) will suffice to rewrite (4.118) in the form

$$L(k)=4k\xi K(k). \tag{4.127}$$

It is now a simple matter to evaluate the functional F on the equilibrium configurations described by (4.116). Inserting (4.114) into (4.108) and interpreting (4.116) as a change of variables, with the aid of (4.117) we express the energy of the solutions to the Euler–Lagrange equation of F as a function of k:

$$F(k):=\frac{k_2}{k\xi}\{2E(k)-\tau k\xi\pi+(\tau^2k^2\xi^2-1)K(k)\}, \tag{4.128}$$

where use has also been made of (4.119) and (4.120).

One would imagine that finding the stable equilibrium configuration of the cholesteric liquid crystal in the whole space should amount to finding the minimizer of $F(k)$ in $]0,1[$. We now show that this view is too naïve and that the functional F fails to measure the total energy distributed in space when the complete unwinding of the cholesteric helix is about to happen.

By Theorem 4.10, $F(k)$ is a differentiable function and an easy computation shows that its derivative can be given the form

$$F'(k) = \frac{E(k)}{1-k^2}\left(\tau^2\xi^2 - \frac{1}{k^2}\right). \tag{4.129}$$

Thus, $F(k)$ attains its minimum at $k_0 := 1/\tau\xi$, provided $\tau\xi > 1$, whereas it is everywhere decreasing in $]0,1[$ if $\tau\xi < 1$. Since the case when $k = 1$ corresponds to the solution of the equilibrium equation that describes $\mathbf{n}$ perfectly aligned to the magnetic field, one would conclude that this is indeed the energy minimizer when $\tau\xi < 1$. Such a conclusion is, however, incorrect because, by (4.124), when $\tau\xi < 1$ the function defined by (4.128) diverges to $-\infty$ as $k \to 1^-$.

The above analysis provides only a scant indication that the chiral configuration of the cholesteric liquid crystal is completely unwound when ξ is sufficiently small, that is, when H is sufficiently large. Nevertheless, the critical value of H at which this transition should occur cannot be computed through the analysis of the minimizers of F.

This just shows that choosing F as the energy functional to be minimized was not wise. Since the unwinding of the cholesteric helix is attained through an extreme stretching of the distorted pitch, a divergence of the energy stored between two plates separated by a pitch is to be expected. Thus, we replace the functional F by its average on $[0, d_\vartheta]$:

$$\bar{F}[\vartheta] := \frac{1}{d_\vartheta} F[\vartheta]. \tag{4.130}$$

The functional so defined turns out to be suitable to express a measure of the elastic energy distributed periodically in the distorted helix of the cholesteric liquid crystal.

Theorem 4.11. Let H_c be given by

$$H_c := \frac{\pi\tau}{2}\sqrt{\frac{2k_2}{\chi_a}}. \tag{4.131}$$

If $H > H_c$ the minimizer of $\bar{F}$ is the function $\vartheta \equiv \pi/2$.

Proof. For any given d_9 the Euler–Lagrange equation of $\bar{F}$ is again (4.109), and so the analysis carried out above for the minimizers of F still applies to the minimizers of $\bar{F}$. In particular, (4.117) gives the equilibrium value of d_9 for $\bar{F}$ also. What actually changes is the function of k which expresses the energy stored in a single spiral of the distorted helix: by $(4.118)_2$ and (4.119), it is now defined by

$$\bar{F}(k) := \frac{F(k)}{2k\xi K(k)},$$

where $F(k)$ is as in (4.128). It is easily seen that $\bar{F}(k)$ can be given the form

$$\bar{F}(k) = \frac{k_2\tau^2}{2} \frac{(k^2 - h^2)K(k) + h(2hE(k) - \pi k)}{k^2 K(k)}, \tag{4.132}$$

where h, the **reduced field**, is the dimensionless parameter defined as

$$h := \frac{1}{\tau\xi} = \frac{H}{\tau}\sqrt{\frac{\chi_a}{2k_2}}. \tag{4.133}$$

A tedious computation based on (4.121) and (4.122) leads us to the formula

$$\bar{F}'(k) = \frac{k_2\tau^2 h}{k^3 K^2(k)} \frac{E(k)}{1 - k^2}\left(\frac{\pi k}{2h} - E(k)\right).$$

As the reader will easily verify by making use of (4.122), (4.125) and (4.126), the function E decreases in $]0,1[$ from $\pi/2$ to 1, and so $\bar{F}'(k)$ vanishes only at one point in $]0,1[$ if $h < \pi/2$, whereas it is everywhere negatve if $h > \pi/2$. This reproduces one qualitative feature of the function $F'(k)$ studied above.

If $h < \pi/2$ there is precisely one k_0 in $]0,1[$ which satisfies the equation

$$E(k_0) = \frac{\pi k_0}{2h}, \tag{4.134}$$

and it is the minimizer of $\bar{F}(k)$. On the other hand, if $h > \pi/2$ the function $\bar{F}(k)$ decreases in $]0,1[$ and attains its minimum at $k = 1$ since

$$\lim_{k \to 1^-} \bar{F}(k) = \frac{k_2\tau^2}{2}(1 - h^2).$$

When $h = \pi/2$ equation (4.133) implies that $H = H_c$, as in (4.131), and so the proof of the theorem is complete, since $k = 1$ corresponds to the constant alignment where $\vartheta \equiv \pi/2$.

We now illustrate how the pitch of the cholesteric helix is distorted by the magnetic field. For each value of $H < H_c$ there is precisely one value of L which is given by (4.127) when we substitute for k the root of (4.134). Since k_0 spans $]0, 1[$ when H varies in $]0, H_c[$, the following formulae give us a parametric representation of the mapping of $]0, 1[$ into $]1, \infty[$, which assigns L/L_0 to H/H_c:

$$\frac{L}{L_0} = \frac{4}{\pi^2} E(k_0) K(k_0), \quad \frac{H}{H_c} = \frac{2h}{\pi} = \frac{k_0}{E(k_0)}. \tag{4.135}$$

Thus, taking the limit as $k_0 \to 0^+$ and that as $k_0 \to 1^-$, we learn that L converges to L_0 as $H \to 0^+$, whereas L diverges logarithmically as $H \to H_c^-$. The following theorem shows some other qualitative features of the function represented by (4.135).

Theorem 4.12. The function $H/H_c \mapsto L/L_0$ is increasing on $]0, 1[$. Moreover, it exhibits the following asymptotic behaviour as $H/H_c \to 0^+$:

$$\frac{L}{L_0} = 1 + \frac{1}{2}\left(\frac{\pi}{4}\right)^4 \left(\frac{H}{H_c}\right)^4 + o\left(\left(\frac{H}{H_c}\right)^4\right). \tag{4.136}$$

Proof. To prove that L/L_0 is an increasing function of H/H_c, we differentiate with respect to k_0 both functions in (4.135). By Theorem 4.10, we arrive at

$$\frac{d}{dk_0}\left(\frac{H}{H_c}\right) = \frac{K(k_0)}{E(k_0)}, \quad \frac{d}{dk_0}\left(\frac{L}{L_0}\right) = \frac{4}{\pi^2} \frac{1}{k_0(1 - k_0^2)} \lambda(k_0),$$

where the function λ is defined by

$$\lambda(k_0) := E^2(k_0) - (1 - k_0^2) K^2(k_0).$$

While it is clear that H/H_c is an increasing function of k_0, to show that L/L_0 is also one, we study the sign of λ in $]0, 1[$. Making use again of Theorem 4.10, we get

$$\lambda'(k_0) = \frac{2}{k_0}(E(k_0) - K(k_0))^2,$$

and so $\lambda(k_0) > 0$ for all $k_0 \in]0, 1[$ since

$$\lim_{k_0 \to 0^+} \lambda(k_0) = 0.$$

This proves the first assertion of the theorem. To prove (4.136), we make use of both (4.123) and (4.125) in (4.135), so arriving at

$$\frac{L}{L_0} = 1 + \frac{1}{32} k_0^4 + o(k_0^4), \quad \frac{H}{H_c} = \frac{2}{\pi} k_0 + o(k_0),$$

whence we easily draw the desired conclusion.

Thus, the distorted pitch of the cholesteric helix grows slowly when H is far from its critical value, whereas it diverges when it gets closer to it. Figure 4.7 illustrates the function described by (4.135).

In the experiment that Durand, Leger, Rondalez and Yeyssie (1969) performed with a mixture of an ordinary nematic (PAA) and a tiny quantity of an optically active substance, the graph of Fig. 4.7 was found to be in good agreement with the observed

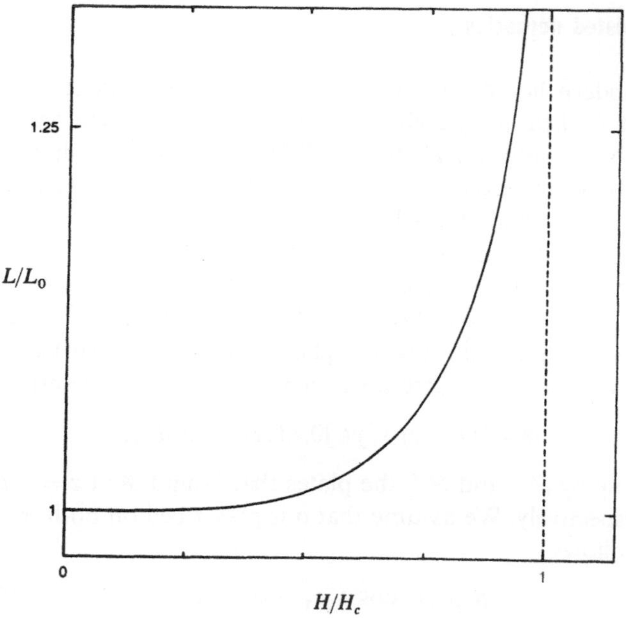

Figure 4.7

data. They measured $H_c = 9.6\,\text{kG}$, and beyond this value of H the sample became optically nematic.

The development that has led us to Theorem 4.11 may seem slightly artificial to a critical reader, who might object that the difficulty offered by the functional F could also be overcome by defining functionals other than $\bar{F}$, which in principle would lead to critical values of H different from (4.131).

The best way to dispel this doubt is by reading a paper of Dreher (1973). He considers a fixed cell bounded by two parallel plates on which $\mathbf{n}$ is strongly anchored. The cholesteric axis is orthogonal to both plates, and a magnetic field is applied transversely to it, as above. The main difference between our analysis and Dreher's is that the distorted pitch no longer increases continuously as H increases, but it suffers jumps at several critical values, whose supremum is precisely H_c as in (4.131); here again the complete unwinding of the cholesteric helix occurs. The number of jumps in the function that now delivers L depends on the distance between the bounding plates.

4.5 Twisted nematics

Most modern liquid-crystal displays make use of an effect which was first studied analytically by Leslie (1970) and then applied to devices by Schadt and Helfrich (1971). Such an effect involves an instability, which Leslie thought of as driven by a magnetic field, whereas Schadt and Helfrich found it easier to experiment through an electric field. Here we first illustrate Leslie's analysis and we determine the critical value of the magnetic field which marks the onset of the transition. We also describe Schadt and Helfrich's experiment and explain how a liquid crystal display works.

Let $(o, \mathbf{e}_1, \mathbf{e}_2, \mathbf{e}_3)$ be a Cartesian frame in which a cell is described by

$$\mathscr{B} = \{(x, y, z) | x, y \in]0, \ell[, z \in]-d, d[\}.$$

We denote by $\mathscr{S}^-$ and $\mathscr{S}^+$ the plates that bound $\mathscr{B}$ at $z = -d$ and $z = d$, respectively. We assume that $\mathbf{n}$ is prescribed on both $\mathscr{S}^-$ and $\mathscr{S}^+$ as follows:

$$\mathbf{n}|_{\mathscr{S}^-} = \cos\varphi_0 \mathbf{e}_1 - \sin\varphi_0 \mathbf{e}_2, \tag{4.137}$$

$$\mathbf{n}|_{\mathscr{S}^+} = \cos\varphi_0 \mathbf{e}_1 + \sin\varphi_0 \mathbf{e}_2, \tag{4.138}$$

so that $\mathbf{n}$ is constant on both plates and the angle between the two boundary orientations is $2\varphi_0$. Here we take φ_0 in the interval $[0, \pi]$.

We consider a class of orientation fields in $\mathscr{B}$ which allows $\mathbf{n}$ to flip out of the plane of the bounding plates. That is, we represent $\mathbf{n}$ in the form

$$\mathbf{n} = \cos\vartheta(z)\cos\varphi(z)\mathbf{e}_1 + \cos\vartheta(z)\sin\varphi(z)\mathbf{e}_2 + \sin\vartheta(z)\mathbf{e}_3, \quad (4.139)$$

where ϑ is a function of class C^1 of $]-d, d[$ into $[0, \pi]$ and φ is a function of class C^1 of $]-d, d[$ into $[-\pi, \pi]$. By (4.137) and (4.138), these functions are subject to

$$\vartheta(-d) = \vartheta(d) = 0, \quad (4.140)$$

$$\varphi(-d) = -\varphi_0, \quad \varphi(d) = \varphi_0. \quad (4.141)$$

From (4.139) we arrive at

$$\begin{aligned}\nabla\mathbf{n} = &-(\vartheta'\sin\vartheta\cos\varphi + \varphi'\cos\vartheta\sin\varphi)\mathbf{e}_1 \otimes \mathbf{e}_3\\ &+ (\varphi'\cos\vartheta\cos\varphi - \vartheta'\sin\vartheta\sin\varphi)\mathbf{e}_2 \otimes \mathbf{e}_3 + \vartheta'\cos\vartheta\mathbf{e}_3 \otimes \mathbf{e}_3,\end{aligned} \quad (4.142)$$

whence it follows that

$$\operatorname{div}\mathbf{n} = \vartheta'\cos\vartheta, \quad (4.143)$$

$$\begin{aligned}\operatorname{curl}\mathbf{n} = &(\vartheta'\sin\vartheta\sin\varphi - \varphi'\cos\vartheta\cos\varphi)\mathbf{e}_1\\ &-(\vartheta'\sin\vartheta\cos\varphi + \varphi'\cos\vartheta\sin\varphi)\mathbf{e}_2.\end{aligned} \quad (4.144)$$

Moreover, a straightforward computation shows that

$$\operatorname{tr}(\nabla\mathbf{n})^2 = (\operatorname{tr}\nabla\mathbf{n})^2. \quad (4.145)$$

Thus, when $\mathbf{n}$ is as in (4.139) Frank's formula gives the following expression for the elastic energy density

$$\sigma_F = f(\vartheta)\vartheta'^2 + g(\vartheta)\varphi'^2 \quad (4.146)$$

where the functions f and g are defined by

$$f(\vartheta) := k_1\cos^2\vartheta + k_3\sin^2\vartheta, \quad (4.147)$$

$$g(\vartheta) := \cos^2\vartheta(k_2\cos^2\vartheta + k_3\sin^2\vartheta). \quad (4.148)$$

Let a magnetic field be applied in the direction orthogonal to both $\mathscr{S}^-$ and $\mathscr{S}^+$:

$$\mathbf{h} = H\mathbf{e}_3.$$

By (4.5) and (4.139), the energy density associated with it reads as

$$\sigma_m = -\tfrac{1}{2}(\chi_\perp + \chi_a \sin^2 \vartheta)H^2, \tag{4.149}$$

where χ_a is taken to be positive. Thus, *modulo* an additive constant, the free energy stored in the cell is given by

$$\mathscr{F}_B[\mathbf{n}] = \ell^2 F[\vartheta, \varphi],$$

where F is the functional defined by

$$F[\vartheta, \varphi] := \int_{-d}^{+d} \{f(\vartheta)\vartheta'^2 + g(\vartheta)\varphi'^2 - \tfrac{1}{2}\chi_a H^2 \sin^2 \vartheta\}\, dz. \tag{4.150}$$

Following Leslie (1970), we illustrate the analysis of the equilibrium configurations for F, which closely resembles that performed in Section 4.2 to describe Freedericks's transition. Nevertheless, here the fully nonlinear analysis of the equilibrium equations cannot be pursued as far as to get all the qualitative features of the bifurcated configuration.

4.5.1 Nonlinear analysis

The Euler–Lagrange equations associated with F are

$$2(f(\vartheta)\vartheta')' = \frac{df}{d\vartheta}\vartheta'^2 + \frac{dg}{d\vartheta}\varphi'^2 - \chi_a H^2 \sin\vartheta \cos\vartheta, \tag{4.151}$$

$$(g(\vartheta)\varphi')' = 0, \tag{4.152}$$

where a prime denotes differentiation with respect to z. These equations, which are subject to (4.140) and (4.141), represent all equilibrium configurations for $\mathbf{n}$ in the class of orientations described by (4.139). Since both $df/d\vartheta$ and $dg/d\vartheta$ vanish at $\vartheta = 0$, there is one equilibrium configuration that is independent of H; it represents a uniform twist of $\mathbf{n}$ between the plates $\mathscr{S}^-$ and $\mathscr{S}^+$:

$$\vartheta^* \equiv 0, \quad \varphi^*(z) = \frac{\varphi_0}{d} z. \tag{4.153}$$

We now look for equilibrium configurations such that neither ϑ' nor φ' vanish in $]-d, d[$. Equation (4.152) can be integrated at once to give

$$g(\vartheta)\varphi' = k, \tag{4.154}$$

where k is an arbitrary constant; but it can also be written in a form which is soon to become expedient to our development, that is

$$2\frac{dg}{d\vartheta}\vartheta'\varphi' + 2g(\vartheta)\varphi'' = 0. \tag{4.155}$$

If we multiply both sides of (4.151) by ϑ' and both sides of (4.155) by φ' and add them together, we arrive at the equation

$$f(\vartheta)\vartheta'^2 + g(\vartheta)\varphi'^2 + \tfrac{1}{2}\chi_a H^2 \sin^2\vartheta = c, \tag{4.156}$$

where c is any positive constant. Since we take both k_2 and k_3 to be strictly positive, g never vanishes as long as $\vartheta \neq \pi/2$, and use of (4.154) in (4.156) leads us to

$$f(\vartheta)\vartheta'^2 + \frac{k^2}{g(\vartheta)} + \tfrac{1}{2}\chi_a H^2 \sin^2\vartheta = c. \tag{4.157}$$

Just as we did in Section 4.2, we now seek solutions of (4.154) and (4.157) such that ϑ is even and φ is odd on $]-d, d[$. Thus, it is sufficient to study these functions in $]-d, 0[$, and so conditions (4.140) and (4.141) can be replaced, respectively, by

$$\vartheta'(0) = 0, \quad \vartheta(-d) = 0, \tag{4.158}$$

$$\varphi(0) = 0, \quad \varphi(-d) = -\varphi_0. \tag{4.159}$$

We also restrict attention to positive functions ϑ, bearing in mind that because both f and g are even functions, if ϑ is a solution of both (4.151) and (4.152), its opposite is too. We set

$$\vartheta_m := \vartheta(0), \tag{4.160}$$

and we use this parameter to label the members of a family of solutions to (4.154) and (4.157). More precisely, by (4.158) and (4.160) we write c as

$$c = \frac{k^2}{g(\vartheta_m)} + \tfrac{1}{2}\chi_a H^2 \sin^2\vartheta_m,$$

and so we give (4.157) the following form:

$$f(\vartheta)\vartheta'^2 = k^2\left(\frac{1}{g(\vartheta_m)} - \frac{1}{g(\vartheta)}\right) + \tfrac{1}{2}\chi_a H^2(\sin^2\vartheta_m - \sin^2\vartheta). \tag{4.161}$$

Furthermore, we take both k and H as functions of ϑ_m to be determined so that there is a solution of both (4.154) and (4.161) other than (4.153).

For a given $\vartheta_m \in]0, \pi[$ let $k(\vartheta_m)$ and $H(\vartheta_m)$ be such that the right-hand side of (4.161) is a positive function of ϑ in $[0, \vartheta_m]$. Integrating (4.161) by separation of variables and making use of $(4.158)_2$ we arrive at the following formula for the inverse of the function ϑ:

$$z = \int_0^{\vartheta} \sqrt{\frac{f(\eta)}{k^2(\vartheta_m)\left(\frac{1}{g(\vartheta_m)} - \frac{1}{g(\eta)}\right) + \frac{1}{2}\chi_a H^2(\vartheta_m)(\sin^2\vartheta_m - \sin^2\eta)}}\, d\eta - d. \tag{4.162}$$

With the aid of (4.162) we can integrate (4.154) also:

$$\varphi(z) = \int_0^{\vartheta(z)} \sqrt{\frac{f(\eta)}{k^2(\vartheta_m)\left(\frac{1}{g(\vartheta_m)} - \frac{1}{g(\eta)}\right) + \frac{1}{2}\chi_a H^2(\vartheta_m)(\sin^2\vartheta_m - \sin^2\eta)}} \cdot \frac{k(\vartheta_m)}{g(\eta)}\, d\eta - \varphi_0, \tag{4.163}$$

where use has also been made of $(4.159)_2$.

For each value of ϑ_m equations (4.162) and (4.163) represent a solution to both (4.154) and (4.157), provided that k and H can be chosen so as to satisfy the following conditions:

$$d = \int_0^{\vartheta_m} \sqrt{\frac{f(\eta)}{k^2(\vartheta_m)\left(\frac{1}{g(\vartheta_m)} - \frac{1}{g(\eta)}\right) + \frac{1}{2}\chi_a H^2(\vartheta_m)(\sin^2\vartheta_m - \sin^2\vartheta)}}\, d\vartheta,$$

$$\varphi_0 = \int_0^{\vartheta_m} \sqrt{\frac{f(\eta)}{k^2(\vartheta_m)\left(\frac{1}{g(\vartheta_m)} - \frac{1}{g(\eta)}\right) + \frac{1}{2}\chi_a H^2(\vartheta_m)(\sin^2\vartheta_m - \sin^2\vartheta)}} \cdot \frac{k(\vartheta_m)}{g(\vartheta)}\, d\vartheta.$$

By the change of variables

$$\lambda := \arcsin\left(\frac{\sin\vartheta}{\sin\vartheta_m}\right)$$

these equations can be given a neater form:

$$d = \int_0^{\pi/2} G(\vartheta_m, \lambda)\, d\lambda, \tag{4.164}$$

$$\varphi_0 = \int_0^{\pi/2} G(\vartheta_m, \lambda) \frac{k(\vartheta_m)}{g^*(\vartheta_m, \lambda)}\, d\lambda, \tag{4.165}$$

where

$G(\vartheta_m, \lambda) :=$

$$= \sqrt{\frac{f^*(\vartheta_m, \lambda)}{k^2(\vartheta_m)\left(\dfrac{1}{g^*(\vartheta_m, \pi/2)} - \dfrac{1}{g^*(\vartheta_m, \lambda)}\right) + \frac{1}{2}\chi_a H^2(\vartheta_m) \sin^2\vartheta_m \cos^2\lambda}}$$

$$\cdot \frac{\sin\vartheta_m \cos\lambda}{\sqrt{1 - \sin\vartheta_m \sin\lambda}}\, d\lambda$$

and

$$f^*(\vartheta_m, \lambda) := k_1 + (k_3 - k_1)(\sin\vartheta_m \sin\lambda)^2,$$

$$g^*(\vartheta_m, \lambda) := k_2 + (k_3 - 2k_2)(\sin\vartheta_m \sin\lambda)^2 - (k_3 - k_2)(\sin\vartheta_m \sin\lambda)^4.$$

The functions $\vartheta_m \mapsto k(\vartheta_m)$ and $\vartheta_m \mapsto H(\vartheta_m)$ that are implicitly defined by (4.164) and (4.165) represent a family of equilibrium configurations for F, whose members are given by (4.162) and (4.163). If these functions fail to be made explicit, the above analysis simply indicates that (4.153) is the only equilibrium configuration of F. Equations (4.164) and (4.165) play here the same role as equation (4.54) in the analysis of Freedericks's transition, though they are far more complicated than that.

For small ϑ_m they become, however, a little easier to inspect. Leslie (1970) has carried out this asymptotic analysis and has arrived at the following explicit formulae for k and H^2 in terms of ϑ_m:

$$k(\vartheta_m) = k_2 \frac{\varphi_0}{d} + (k_3 - 2k_2)\frac{\varphi_0}{2d}\vartheta_m^2 + o(\vartheta_m^2), \tag{4.166}$$

$$H^2(\vartheta_m) = \frac{2}{\chi_a d^2}\left(k_1\left(\frac{\pi}{2}\right)^2 + (k_3 - 2k_2)\varphi_0^2\right)$$
$$+ \frac{1}{\chi_a d^2}\left(k_3\left(\frac{\pi}{2}\right)^2 + (k_3^2 - k_3 k_2 + k_2^2)\frac{\varphi_0^2}{k_2}\right)\vartheta_m^2 + o(\vartheta_m^2). \tag{4.167}$$

It follows from (4.166) that when $\vartheta_m \to 0^+$, k approaches the value corresponding through (4.154) to the solution (4.153), which persists for all values of H. Correspondingly, $H(\vartheta_m)$ tends to a limiting value, which should be interpreted as the critical strength of the magnetic field, above which (4.153) is no longer the only equilibrium configuration for F. Thus, equation (4.167) gives the following formula for H_c:

$$H_c := \frac{1}{d}\sqrt{\frac{2}{\chi_a}}\sqrt{k_1\left(\frac{\pi}{2}\right)^2 + (k_3 - 2k_2)\varphi_0^2}, \tag{4.168}$$

provided that

$$k_3 - 2k_2 \geqslant 0 \quad \text{or} \quad \varphi_0 \leqslant \frac{\pi}{2}\sqrt{\frac{k_1}{2k_2 - k_3}}. \tag{4.169}$$

Furthermore, equation (4.167) represents a bifurcation like that depicted in Fig. 4.1 whenever the mapping $\vartheta_m \mapsto H(\vartheta_m)$ is increasing in the vicinity of $\vartheta_m = 0$, that is whenever

$$\varphi_0 < \frac{\pi}{2}\sqrt{\frac{k_2 k_3}{k_2^2 + k_3^2 - k_2 k_3}} =: \varphi_m. \tag{4.170}$$

Remark 1. It is worth noting that when, for example, the total twist imposed between the two plates is equal to $\pi/2$, namely when $\varphi_0 = \pi/4$, conditions $(4.169)_2$ and (4.170) become, respectively,

$$\frac{1}{4} \leqslant \frac{k_1}{2k_2 - k_3}, \quad \frac{k_2 k_3}{k_2^2 + k_3^2 - k_2 k_3} > \frac{1}{4},$$

which define a domain of admissible values for Frank's constants.

To explore the stability of the solution branching off the one that always exists, Leslie (1970) has evaluated for small values of ϑ_m the energy associated with the functions described by (4.162) and (4.163). He arrived at the following formula:

$$F[\vartheta, \varphi] = F[\vartheta^*, \varphi^*] - \frac{1}{8d}\frac{k_2^2 + k_3^2 - k_2 k_3}{k_2}(\varphi_m^2 - \varphi_0^2)\vartheta_m^4 + o(\vartheta_m^4),$$

which shows that the new branch is stable whenever φ_0 obeys (4.170).

Thus, we have learned that when φ_0 is not too large there is a critical value of H above which the uniform planar twist of $\mathbf{n}$ in $\mathscr{B}$ described by (4.153) no longer minimizes the free energy and a new stable configuration of $\mathbf{n}$ arises which fails to be planar and exhibits a slight alignment in the direction of the magnetic field. Such a conclusion, however, stands on firm ground only if we assume that both (4.169) and (4.170) hold and, moreover, we take H just slightly above H_c as given by (4.168). We expect also that when H grows larger than H_c the uniform twist remains unstable, whereas the stable orientation field in the cell becomes closer and closer to the uniform alignment in the direction of the magnetic field, at least away from the bounding plates. Though this reasoning would be supported by the analogous conclusion that we drew in Section 4.2, the fact remains that we have not reached here the complete proof of it.

The reader with some skill in numerical analysis might examine equations (4.164) and (4.165) for various values of ϑ_m to find out whether the energy of the bifurcating branch remains below that of the fundamental solution also when ϑ_m is far from zero.

We shall see later in this section how the experiment of Schadt and Helfrich (1971) has indeed shown that an almost completely aligned state prevails when the strength of the applied field is nearly twice as much as the critical strength. To strengthen this experimental evidence with a mathematical argument we show below that the uniform twist described by (4.153) cannot be stable when $H > H_c$. This conclusion is based on the analysis of the second variation of F at (ϑ^*, φ^*).

4.5.2 Instability of the uniform twist

Definition 4.3. Let $\mathscr{A}_F$ be the class of admissible pairs defined by

$$\mathscr{A}_F := \{(\vartheta, \varphi) \mid \vartheta, \varphi \text{ are of class } C^1(-d, d) \text{ and obey (4.140), (4.141)}\}.$$

Theorem 4.13. Let both k_2 and k_3 be positive and let φ_0 be chosen in $]0, \pi/2[$. If condition (4.169) is satisfied, then F cannnot attain its minimum in $\mathscr{A}_F$ on the pair (ϑ^*, φ^*) when $H > H_c$. However, when $H < H_c$, (ϑ^*, φ^*) is a local minimizer of F in $\mathscr{A}_F$.

Proof. Let ε_0 be a given positive number. If u and v are functions of class $C^1(-d, d)$ which vanish at the end-points of the interval $[-d, d]$, then for every $\varepsilon \in [-\varepsilon_0, \varepsilon_0]$ the functions defined by

$$\vartheta_\varepsilon := \vartheta^* + \varepsilon u, \quad \varphi_\varepsilon := \varphi^* + \varepsilon v$$

constitute a pair of $\mathscr{A}_F$. The mapping $\varepsilon \mapsto (\vartheta_\varepsilon, \varphi_\varepsilon)$ describes a curve in $\mathscr{A}_F$ which passes through (ϑ^*, φ^*). We evaluate below the functional F along each of these curves in the vicinity of (ϑ^*, φ^*).

It follows from (4.147), (4.148) and (4.153) that

$$f(\vartheta_\varepsilon) = k_1 + (k_3 - k_1)\varepsilon^2 u^2 + o(\varepsilon^2),$$

$$g(\vartheta_\varepsilon) = k_2 + (k_3 - 2k_2)\varepsilon^2 u^2 + o(\varepsilon^2),$$

$$\vartheta' = \varepsilon u', \quad \varphi'_\varepsilon = \frac{\varphi_0}{d} + \varepsilon v',$$

whence we readily arrive at

$$F[\vartheta_\varepsilon, \varphi_\varepsilon] = F[\vartheta^*, \varphi^*] + \varepsilon^2 \delta^2 F(u, v) + o(\varepsilon^2),$$

where $\delta^2 F(u, v)$, the second variation of F, is given by

$$\delta^2 F(u, v) := \int_{-d}^{+d} \left\{ k_2 v'^2 + k_1 u'^2 + \left(\left(\frac{\varphi_0}{d} \right)^2 (k_3 - 2k_2) - \tfrac{1}{2}\chi_a H^2 \right) u^2 \right\} dz. \tag{4.171}$$

When $H > H_c$, it follows from (4.168) that

$$\left(\frac{\varphi_0}{d} \right)^2 (k_3 - 2k_2) - \tfrac{1}{2}\chi_a H^2 < -k_1 \left(\frac{\pi}{2d} \right)^2,$$

and so (4.171) implies that

$$\delta^2 F(u, v) < \int_{-d}^{+d} \left\{ k_2 v'^2 + k_1 \left(u'^2 - \left(\frac{\pi}{2d} \right)^2 u^2 \right) \right\} dz. \tag{4.172}$$

Now, the following inequality applies to every function u of class C^1 on $[-d, d]$ which vanishes at the end-points of this interval (*cf. e.g.* p. 185 of Hardy, Littlewood and Pólya (1952)):

$$\int_{-d}^{+d} u^2 dz \leqslant \left(\frac{2d}{\pi} \right)^2 \int_{-d}^{+d} u'^2 dz, \tag{4.173}$$

where the equality is attained if, and only if,

$$u(z) = c \cos \frac{\pi z}{2d} \tag{4.174}$$

for any constant c.

Thus, by (4.172), choosing $v \equiv 0$ and u as in (4.174), we make $\delta^2 F(u, v)$ negative: in $\mathscr{A}_F$ there are curves along which the energy functional is not minimum at (ϑ^*, φ^*), and so the uniform twist of **n** described by this pair cannot be a stable equilibrium configuration.

To prove that, instead, for $H < H_c$ this is indeed the case, at least when only small perturbations of (ϑ^*, φ^*) are considered, we make use of (4.173) in (4.171). By (4.168), we obtain that

$$\delta^2 F(u, v) \geqslant \int_{-d}^{+d} \{k_2 v'^2 + \tfrac{1}{2}\chi_a(H_c^2 - H^2)u^2\}\,dz, \qquad (4.175)$$

and so we see that, when $H < H_c$, $\delta^2 F(u, v)$ is never negative for all admissible u and v. This amounts to saying that (ϑ^*, φ^*) is a local minimizer.

Remark 2. The critical reader might find the above proof reminiscent of the reasoning which at the beginning of Section 4.3 led us to a simpler evaluation of the critical magnetic strength at which the classical Freedericks's transition occurs. The functions ϑ_ε and φ_ε are indeed small alterations of ϑ^* and φ^*. Just as in Section 4.3, the linearized equilibrium equations for u and v would have no solution for $H < H_c$ and infinitely many for $H = H_c$, namely, u as in (4.174) for any constant c, and $v \equiv 0$.

Remark 3. The proof of the above theorem has made no use of inequality (4.170). In the light of the preceding remark, this should not surprise us. The theorem essentially refers to the critical value of H at which a new branch of equilibrium configurations comes into being, whereas (4.167), whence (4.170) was derived, tells us how the new branch behaves near the bifurcation point. More precisely, when (4.170) is satisfied, the bifurcation takes place to the right of H_c; if (4.170) were replaced by its reverse, the bifurcation would instead develop to the left of H_c.

4.5.3 Electro-optical effect

Let the nematic liquid crystal which fills the cell $\mathscr{B}$ possess a positive dielectric anisotropy ε_a. If we apply a voltage V to the plates that bound $\mathscr{B}$ we excite in it the electric field

$$\mathbf{e} = E\mathbf{e}_3,$$

where

$$E = \frac{V}{2d}. \qquad (4.176)$$

Such a field would have precisely the same aligning effect on the optic axis as the magnetic field considered above. Comparing (4.4) and (4.5) we readily learn that the conclusions of the preceding analysis would still hold, provided we replace H by E and χ_a by $(1/4\pi)\varepsilon_a$. Thus, we see from (4.168) that the critical value of E is

$$E_c := \frac{1}{d}\sqrt{\frac{8\pi}{\varepsilon_a}}\sqrt{k_1\left(\frac{\pi}{2}\right)^2 + (k_3 - 2k_2)\varphi_0^2},$$

whence, by (4.176), we get the critical voltage

$$V_c := 4\sqrt{\frac{2\pi}{\varepsilon_a}}\sqrt{k_1\left(\frac{\pi}{2}\right)^2 + (k_3 - 2k_2)\varphi_0^2}. \tag{4.177}$$

Clearly, we still regard condition (4.169) as valid.

Schadt and Helfrich (1971) put a thin layer of nematic liquid crystal between two glass plates which were provided with a conductive coating and treated so as to induce, respectively, two uniform orientations of **n**, one orthogonal to the other. Thus, the total twist imposed on the nematic was $\pi/2$, and so φ_0 is to be taken as $\pi/4$. They also put the nematic cell between polarizers, whose axes were oriented so as to be parallel to the direction of **n** prescribed on the adjacent plate.

The thickness of the cell was about 10^{-3} cm, and so in the absence of any applied voltage the equilibrium configuration of the optic axis in $\mathscr{B}$ would resemble the natural orientation of a cholesteric liquid crystal whose pitch is 4×10^{-3} cm, which is much larger than the wavelengths in the visible spectrum. Thus, as we learned in Section 1.4 of Chapter 1, as the light travels orthogonally to the plates, the polarization vector follows **n**, and so it suffers a total rotation of $\pi/2$ in the cell. Since the polarizers are also rotated by the same angle, light propagates through the cell when **n** is in its unperturbed state, which is to say that the cell is transparent when no voltage is applied to the plates. According to our analysis, this should be the case as long as the applied voltage V is below V_c.

On the other hand, when $V > V_c$, **n** no longer exhibits a uniform twist in the cell, and we expect that the more the voltage exceeds V_c the more **n** tends to be aligned to the electric field in the middle of the cell. When the voltage is sufficiently high we can imagine **n** to be almost perpendicular to the plates everywhere in the cell, apart from two narrow boundary layers, where it exhibits high splay and bend distortions, but virtually no twist. If this orientation of **n** can indeed be achieved with voltages not extremely high, we expect the cell to act as an optical shutter. The polarization vector would not be much affected by an optic axis almost parallel to the direction of propagation, and so it would be orthogonal to the polarizer axis at the plate opposite to that through which it entered the cell.

Figure 4.8 illustrates the optical behaviour of the cell when it transmits light. Of course, if the axes of the polarizers were parallel, the cell would be opaque for small voltages and transparent for large voltages.

Schadt and Helfrich used two different substances in their experiment, namely *n*(4′-ethoxybenzylidene)4-amino-benzonitride (PEBAB), which is nematic between 106 and 128°C, and a racemic mixture, which is nematic between 20 and 94°C. For both materials they had a cell of thickness 10 μm and put it between parallel polarizers, after having properly treated the surfaces of the plates bounding it.

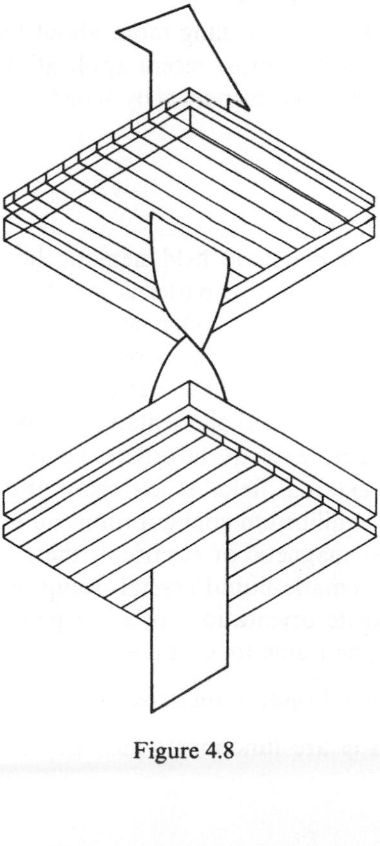

Figure 4.8

Applying both dc and ac voltages, they indeed observed the cell to become brighter above a critical voltage, which was higher for dc fields than for ac fields. For example, the cell filled with the mixed nematic liquid crystal at room temperature exhibited a threshold of just 3 V at 1 kHz and the brightening was 90% complete at 6 V. This effect showed no hysteresis: the cell became opaque again when the voltage was decreased.

The critical voltages measured by Schadt and Helfrich (1971) were surprisingly smaller than the voltages needed to induce other electro-optical effects known at that time. A glance at (4.177) suggests that this may be easily explained if the Frank's constants of the liquid crystal are such that $k_3 < 2k_2$.

This effect has come to be the one most used in modern liquid crystal displays. The technological applications of it have much improved since it was patented by M. Schadt and W. Helfrich (Swiss Patent No. 532261 of 4 December 1970) and J. Fergason (U.S. Patent No. 3818796 of 9 February 1971).

The reader interested in learning more about the electro-optical effect outlined here and its most recent applications to commercial devices is referred to two review papers by Schadt (1988 and 1989).

4.6 Inversion walls

In the presence of a magnetic field nematic liquid crystals may exhibit equilibrium orientation patterns which are reminiscent of Bloch and Néell walls, structures often encountered in ferromagnetic systems (for these latter the reader is referred *e.g.* to Chapter 9 of Kléman (1983)). One or other of these structures is found to possess less energy within a special class of orientations, depending on whether one of Frank's constants exceeds the others. Though this instability is not driven by the external field, as are those above, we discuss it here because these equilibrium configurations do not exist unless an external field, either magnetic or electric, is there to sustain them.

Suppose that a nematic liquid crystal occupies the whole space and that two opposite orientations of $\mathbf{n}$ are prescribed at infinity. More precisely, in the frame $(o, \mathbf{e}_1, \mathbf{e}_2, \mathbf{e}_3)$ we represent $\mathbf{n}$ as

$$\mathbf{n} = \sin\vartheta \sin\varphi\, \mathbf{e}_1 + \sin\vartheta \cos\varphi\, \mathbf{e}_2 + \cos\vartheta\, \mathbf{e}_3, \qquad (4.178)$$

where both ϑ and φ are functions which depend only on x, the spatial co-ordinate along $\mathbf{e}_1$, and whose ranges are $[0, \pi]$ and $[0, 2\pi]$, respectively. We further assume that both ϑ and φ are of class $C^1(-\infty, \infty)$ and that

$$\lim_{x\to+\infty} \vartheta(x) = \pi, \quad \lim_{x\to-\infty} \vartheta(x) = 0, \tag{4.179}$$

$$\lim_{x\to+\infty} \vartheta'(x) = \lim_{x\to-\infty} \vartheta'(x) = 0. \tag{4.180}$$

Thus, $\mathbf{n}$ is asymptotically constant and parallel to $\mathbf{e}_3$, but with opposite limiting values in the two half-spaces bounded by the plane

$$\mathscr{P} := \{o + y\mathbf{e}_2 + z\mathbf{e}_3 \mid y, z \in \mathbb{R}\}.$$

An easy computation leads us from (4.178) to the following formula:

$$\begin{aligned}\nabla\mathbf{n} &= (\cos\vartheta\sin\varphi\vartheta' + \sin\vartheta\cos\varphi\varphi')\mathbf{e}_1\otimes\mathbf{e}_1 \\ &\quad + (\cos\vartheta\cos\varphi\vartheta' - \sin\vartheta\sin\varphi\varphi')\mathbf{e}_2\otimes\mathbf{e}_1 - \sin\vartheta\vartheta'\mathbf{e}_3\otimes\mathbf{e}_1,\end{aligned} \tag{4.181}$$

and hence to

$$\operatorname{div}\mathbf{n} = \cos\vartheta\sin\varphi\varphi' + \sin\vartheta\cos\varphi\varphi', \tag{4.182}$$

$$\operatorname{curl}\mathbf{n} = \sin\vartheta\vartheta'\mathbf{e}_2 + (\cos\vartheta\cos\varphi\vartheta' - \sin\vartheta\sin\varphi\varphi')\mathbf{e}_3, \tag{4.183}$$

$$|(\nabla\mathbf{n})\mathbf{n}|^2 = \sin^2\vartheta\sin^2\varphi(\vartheta'^2 + \sin^2\vartheta\varphi'^2), \tag{4.184}$$

where a prime denotes differentiation with respect to x. Furthermore, it follows from (4.181) that $\operatorname{tr}(\nabla\mathbf{n})^2 = (\operatorname{div}\mathbf{n})^2$, and so making use also of equation (3.24) of Chapter 3, we see that when $\mathbf{n}$ is as in (4.178) the elastic energy density becomes

$$\begin{aligned}\sigma_F &= k_1(\cos\vartheta\sin\varphi\vartheta' + \sin\vartheta\cos\varphi\varphi')^2 \\ &\quad + k_2(\cos\varphi\vartheta' - \sin\vartheta\cos\vartheta\sin\varphi\varphi')^2 \\ &\quad + k_3\sin^2\vartheta\sin^2\varphi(\vartheta'^2 + \sin^2\vartheta\varphi'^2).\end{aligned} \tag{4.185}$$

Let a uniform magnetic field be applied in the direction of $\mathbf{e}_3$:

$$\mathbf{h} = H\mathbf{e}_3.$$

If $\mathscr{B}$ is any infinite cylinder orthogonal to $\mathscr{P}$ with cross sectional area A, the total free energy stored in it is given by

$$\mathscr{F}_B[\mathbf{n}] = AF[\vartheta, \varphi],$$

where F is the functional defined by

$$F[\vartheta, \varphi] := \int_{-\infty}^{+\infty} \{\sigma_F - \tfrac{1}{2}\chi_a H^2 \cos^2 \vartheta\} dx, \qquad (4.186)$$

and σ_F is as in (4.185).

Because the integral in (4.186) is defined on the whole real line, we must be sure that it converges at least when $\mathbf{n}$ is constant in the whole space and agrees with each of the asymptotic conditions in (4.179). Since no constant added to the free energy density alters the equilibrium configurations of the orientation field, we rewrite F in the form

$$F[\vartheta, \varphi] = \int_{-\infty}^{+\infty} \{\sigma_F - \tfrac{1}{2}\chi_a H^2 \cos^2 \vartheta + \sigma_0\} dx, \qquad (4.187)$$

and determine the constant σ_0 so as to make (4.187) compatible with the requirement of integrability. Clearly, F converges for both $\vartheta \equiv 0$ and $\vartheta \equiv \pi$ only if $\sigma_0 = \frac{1}{2}\chi_a H^2$, and so (4.187) finally becomes

$$F(\vartheta, \varphi] = \int_{-\infty}^{+\infty} \{\sigma_F + \tfrac{1}{2}\chi_a H^2 \sin^2 \vartheta\} dx. \qquad (4.188)$$

Helfrich (1968) has considered special equilibrium configurations of this functional: namely, those for which

$$\varphi \equiv \varphi_0, \qquad (4.189)$$

where $\varphi_0 \in [0, 2\pi]$. Inserting (4.189) into (4.188), by (4.185) we arrive at

$$F[\vartheta, \varphi_0] = \int_{-\infty}^{+\infty} \{(k_1 \sin^2 \varphi_0 \cos^2 \vartheta + k_2 \cos^2 \varphi_0 + k_3 \sin^2 \varphi_0 \sin^2 \vartheta)\vartheta'^2 + \tfrac{1}{2}\chi_a H^2 \sin^2 \vartheta\} dx. \qquad (4.190)$$

The values of φ_0 that make $F[\vartheta, \varphi_0]$ stationary must satisfy

$$\frac{\partial}{\partial \varphi_0} F[\vartheta, \varphi_0] = 0 \qquad (4.191)$$

whenever ϑ solves the following Euler–Lagrange equation:

$$((k_1 \sin^2 \varphi_0 \cos^2 \vartheta + k_2 \cos^2 \varphi_0 + k_3 \sin^2 \varphi_0 \sin^2 \vartheta)\vartheta')' = ((k_3 - k_1) \sin^2 \varphi_0 \vartheta'^2 + \tfrac{1}{2}\chi_a H^2) \sin \vartheta \cos \vartheta, \qquad (4.192)$$

subject to (4.179) and (4.180).

A straightforward computation shows that

$$\frac{\partial}{\partial \varphi_0} F[\vartheta, \varphi_0] = \sin 2\varphi_0 \int_{-\infty}^{+\infty} (k_1 \cos^2 \vartheta + k_3 \sin^2 \vartheta - k_2)\vartheta'^2 dx,$$

and so equation (4.191) is satisfied whenever

$$\sin 2\varphi_0 = 0.$$

Two solutions of this equation represent all, namely $\varphi_0 = 0$ and $\varphi_0 = \pi/2$; they will correspond to two different inversion walls, which are separately studied below.

4.6.1 Twist wall

Let $\varphi_0 = 0$. Equation (4.192) then reduces to

$$k_2 \vartheta'' = \tfrac{1}{2}\chi_a H^2 \sin\vartheta \cos\vartheta, \tag{4.193}$$

for which we seek solutions that satisfy (4.179) and (4.180), and are such that

$$\vartheta(x) - \frac{\pi}{2} = \frac{\pi}{2} - \vartheta(-x) \quad \text{for all } x \in \mathbb{R}. \tag{4.194}$$

In particular, (4.194) requires that

$$\vartheta(0) = \frac{\pi}{2}. \tag{4.195}$$

Multiplying both sides of (4.193) by ϑ' and integrating, we arrive at the equation

$$k_2 \vartheta'^2 = \tfrac{1}{2}\chi_a H^2 \sin^2 \vartheta + c, \tag{4.196}$$

where c is an arbitrary constant. By (4.179) and (4.180), it follows from (4.196) that $c = 0$. Thus, the equilibrium equation reduces to

$$\vartheta' = \sqrt{\frac{\chi_a}{2k_2}} H \sin\vartheta, \tag{4.197}$$

which can be integrated at once by separation of variables. There is precisely one solution of (4.197) that satisfies (4.195); its inverse is given by

$$x = \frac{1}{H}\sqrt{\frac{2k_2}{\chi_a}} \ln \tan\frac{\vartheta}{2}. \tag{4.198}$$

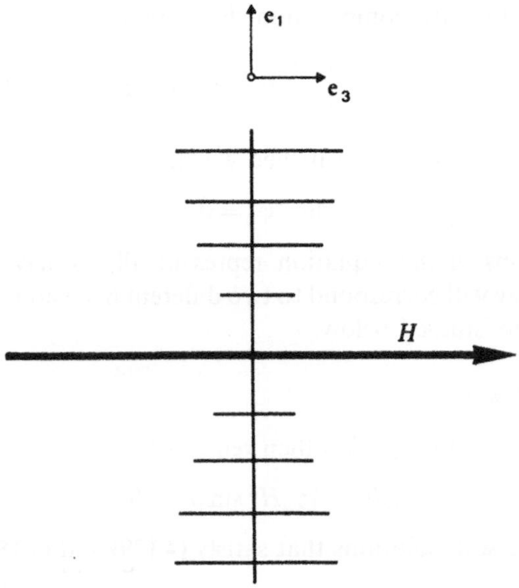

Figure 4.9

When ϑ is as in (4.198) and $\varphi \equiv 0$ the orientation field represented by (4.178) is like that illustrated in Fig. 4.9. It is a twisted field, which gets reversed only once along any straight line orthogonal to $\mathscr{P}$. We call it a **twist wall**. It closely resembles the pattern of the magnetization in a Bloch wall, and so hereafter we denote by ϑ_B the function defined by (4.198).

The energy of such an equilibrium configuration can easily be evaluated with the aid of (4.197). Inserting this equation into (4.190) leads us to

$$F[\vartheta_B, 0] = \chi_a H^2 \int_{-\infty}^{+\infty} \sin^2 \vartheta_B dx,$$

which, again by (4.197), can be converted into an easier integral:

$$F[\vartheta_B, 0] = H\sqrt{2k_2\chi_a} \int_0^{\pi} \sin \vartheta d\vartheta = 2H\sqrt{2k_2\chi_a}. \tag{4.199}$$

We agree to describe as the **width** of the twist wall that distance w_B between two planes parallel to $\mathscr{P}$ on which the values of ϑ_B differ by $\pi/2$, which is just half the complete twist angle. It follows from

(4.198) that

$$w_B = \left(2\ln\tan\frac{3\pi}{8}\right)\frac{1}{H}\sqrt{\frac{2k_2}{\chi_a}} \approx 1.76\frac{1}{H}\sqrt{\frac{2k_2}{\chi_a}}. \tag{4.200}$$

Thus, w_B decreases as H increases. This is not surprising because we expect that higher magnetic fields make more stable both asymptotic orientations, squeezing the wall.

4.6.2 Splay–bend wall

Let $\varphi_0 = \pi/2$. The energy functional now becomes

$$F[\vartheta, \tfrac{1}{2}\pi] = \int_{-\infty}^{+\infty} \{(k_1\cos^2\vartheta + k_3\sin^2\vartheta)\vartheta'^2 + \tfrac{1}{2}\chi_a H^2 \sin^2\vartheta\}\, dx, \tag{4.201}$$

and the Euler–Lagrange equation associated with it reads as

$$(k_1\cos^2\vartheta + k_3\sin^2\vartheta)\vartheta'' + (k_3 - k_1)\sin\vartheta\cos\vartheta\,\vartheta'^2 = \tfrac{1}{2}\chi_a H^2\sin\vartheta\cos\vartheta, \tag{4.202}$$

whence it follows that

$$(k_1\cos^2\vartheta + k_3\sin^2\vartheta)\vartheta'^2 = \tfrac{1}{2}\chi_a H^2\sin^2\vartheta + c, \tag{4.203}$$

for any constant c. By (4.179) and (4.180), c must vanish, and (4.203) can be given the form

$$\vartheta' = H\sqrt{\frac{\chi_a}{2}}\frac{\sin\vartheta}{\sqrt{k_1\cos^2\vartheta + k_3\sin^2\vartheta}}. \tag{4.204}$$

The solution of (4.204) that satisfies (4.195) is represented by the following integral:

$$x = \frac{1}{H}\sqrt{\frac{2}{\chi_a}}\int_{\pi/2}^{\vartheta}\frac{\sqrt{k_1\cos^2\vartheta + k_3\sin^2\vartheta}}{\sin\vartheta}\,d\eta. \tag{4.205}$$

When $k_1 = k_3$, (4.205) reduces to (4.198), but the orientation field that it now represents is quite different: $\mathbf{n}$ is still a planar field, but it lies everywhere in a plane orthogonal to $\mathscr{P}$; Fig. 4.10 illustrates its vector lines. No twist distortion is associated with this orientation field, which we call the **splay–bend wall**.

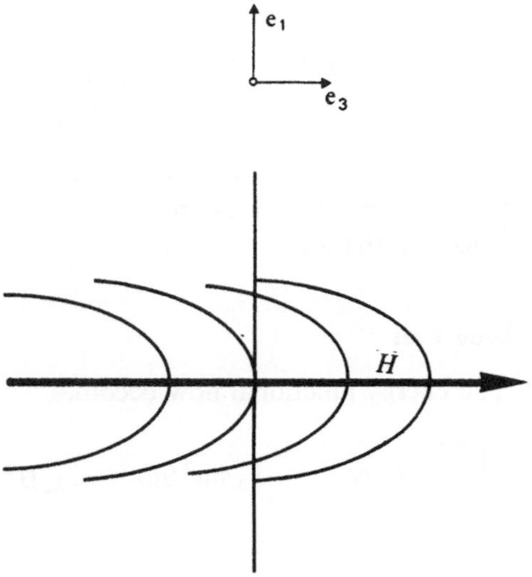

Figure 4.10

Also when k_1 and k_3 are not equal the main qualitative features of the orientation field remain those depicted in Fig. 4.10: they are reminiscent of Néel walls in ferromagnetism, and so in the following we shall denote by ϑ_N the function defined by (4.205). Making use of (4.204), we easily evaluate its energy:

$$F[\vartheta_N, \tfrac{1}{2}\pi] = H\sqrt{2\chi_a}\int_0^\pi \sqrt{k_1\cos^2\vartheta + k_3\sin^2\vartheta}\,d\vartheta. \qquad (4.206)$$

Precisely as we did above for the twist wall, we define here the width w_N of the splay–bend wall; by (4.205) it is given by the formula

$$w_N = \frac{2}{H}\sqrt{\frac{2}{\chi_a}}\int_{\pi/4}^{\pi/2} \frac{\sqrt{k_1\cos^2\vartheta + k_3\sin^2\vartheta}}{\sin\vartheta}\,d\vartheta, \qquad (4.207)$$

and so, like w_B, it decreases as H increases.

4.6.3 Stability

We have found above two special equilibrium configurations of the energy functional defined by (4.188), but so far we have not explored their stability. Here we find out whether they are stable with respect to different perturbations.

We denote by k_m and k_M the minimum and the maximum of $\{k_1, k_3\}$.

Furthermore, we introduce two classes of admissible functions for F, namely

$$\mathscr{A}_F := \{(\vartheta, \varphi) \mid \vartheta, \varphi \in C^1(-\infty, \infty), (4.179), (4.180) \text{ and } (4.194) \text{ are satisfied}\}$$

$$\bar{\mathscr{A}}_F := \{(\vartheta, \varphi) \in \mathscr{A}_F \mid \varphi' \equiv 0\}.$$

Theorem 4.14. If $k_2 < k_m$, then the minimizer of F in $\bar{\mathscr{A}}_F$ is $(\vartheta_B, 0)$, whereas it is $(\vartheta_N, \pi/2)$, if $k_2 > k_M$.

Proof. We have already seen that F is stationary in $\bar{\mathscr{A}}_F$ only at $(\vartheta_B, 0)$ and $(\vartheta_N, \pi/2)$. Moreover, (4.206) implies that

$$2H\sqrt{2k_m\chi_a} < F[\vartheta_N, \tfrac{1}{2}\pi] < 2H\sqrt{2k_M\chi_a},$$

and so, by (4.199), the desired conclusion easily follows.

Remark. It follows from (4.207) that

$$\sqrt{k_m} < \frac{w_N}{w_B}\sqrt{k_2} < \sqrt{k_M}.$$

We now study the stability of both twist and splay–bend walls in the whole of $\mathscr{A}_F$. To make our development a little easier, hereafter we take

$$k_1 = k_3 =: k, \tag{4.208}$$

though the outcome of our analysis would not be much different if (4.208) were not valid.

Theorem 4.15. The absolute minimizer of F in $\bar{\mathscr{A}}_F$ is a local minimizer of F in $\mathscr{A}_F$.

Proof. We first compute the second variation of F at the pair $(\vartheta_B, 0)$. Let $\varepsilon_0 > 0$ be given. For each v of class $C^1(-\infty, \infty)$ and each u of class $C^1(-\infty, \infty)$ such that

$$\lim_{x\to+\infty} u(x) = \lim_{x\to-\infty} u(x) = 0 \quad \text{and} \quad \lim_{x\to+\infty} u'(x) = \lim_{x\to-\infty} u'(x) = 0,$$

we define on the interval $[-\varepsilon_0, \varepsilon_0]$ a curve in $\mathscr{A}_F$ which passes through the pair $(\vartheta_B, 0)$:

$$\varepsilon \mapsto (\vartheta_\varepsilon, \varphi_\varepsilon), \quad \vartheta_\varepsilon := \vartheta_B + \varepsilon u, \quad \varphi_\varepsilon := \varepsilon v.$$

By (4.208), (4.185) and (4.188) lead us to

$$F[\vartheta_\varepsilon, \varphi_\varepsilon] = F[\vartheta_B, 0] + \varepsilon\delta F(u, v) + \varepsilon^2 \tfrac{1}{2}\delta^2 F(u, v) + o(\varepsilon^2);$$

$\delta F(u, v)$ and $\delta^2 F(u, v)$ are the first and second variation of F, respectively:

$$\tfrac{1}{2}\delta F(u, v) := \int_{-\infty}^{+\infty} \{k_2 \vartheta_B' u' + \tfrac{1}{2}\chi_a H^2 \sin\vartheta_B \cos\vartheta_B\} dx,$$

$$\tfrac{1}{2}\delta^2 F(u, v) := U[u] + V[v], \tag{4.209}$$

where U and V are the functionals defined by

$$U[u] := \int_{-\infty}^{+\infty} \{k_2 u'^2 + \tfrac{1}{2}\chi_a H^2(\cos^2\vartheta_B - \sin^2\vartheta_B)u^2\} dx, \tag{4.210}$$

$$V[v] := \int_{-\infty}^{+\infty} \{k \sin^2\vartheta_B v'^2 + (k - k_2)\vartheta_B'^2 v^2 + 2(k - k_2)\sin\vartheta_B \cos\vartheta_B v v'\} dx. \tag{4.211}$$

Since ϑ_B is a solution of the equilibrium equation (4.193), the first variation of F vanishes identically. We now show that $U[u]$ is always positive, whereas $V[v]$ is strictly positive whenever $(\vartheta_B, 0)$ is the minimizer of F in $\bar{\mathscr{A}}_F$, that is whenever $k_2 < k$.

It is clear from (4.197) that the function ϑ_B is strictly increasing on $\mathbb{R}$, and so it induces a change of variables which maps $\mathbb{R}$ onto $[0, \pi]$. In the new variable both u and v become functions on $[0, \pi]$, which are defined by

$$u_B(\vartheta_B(x)) := u(x), \quad v_B(\vartheta_B(x)) := v(x) \quad \text{for all } x \in \mathbb{R}. \tag{4.212}$$

Inserting (4.212) into both (4.210) and (4.211), by use of (4.197) we give U and V the following form:

$$U[u_B] = \sqrt{\frac{\chi_a k_2}{2}} H \int_0^\pi \left\{ u_B'^2 + \left(\frac{\cos^2\vartheta}{\sin^2\vartheta} - 1 \right) u_B^2 \right\} \sin\vartheta \, d\vartheta, \tag{4.213}$$

$$V[v_B] = \sqrt{\frac{\chi_a}{2k_2}} H \int_0^\pi \{k \sin^2\vartheta \, v_B'^2 + (k - k_2) v_B^2 + 2(k - k_2)\sin\vartheta \cos\vartheta \, v_B v_B'\} \sin\vartheta \, d\vartheta, \tag{4.214}$$

where a prime now denotes differentiation with respect to the variable ϑ.

Since u_B vanishes at both end-points of the interval $[0, \pi]$, integrating by parts the third addend on the right-hand side of (4.213), we arrive at

$$U[u_B] = \sqrt{\frac{\chi_a k_2}{2}} H \int_0^\pi \left(u_B' - \frac{\cos\vartheta}{\sin\vartheta} u_B \right)^2 \sin\vartheta \, d\vartheta.$$

Thus, $U[u_B] > 0$ unless $u_B = u_0 \sin\vartheta$, for any constant u_0. Furthermore, when $k_2 < k$ we easily obtain from (4.214) that

$$V[v_B] > \sqrt{\frac{\chi_a}{2k_2}} (k_2 - k) H \int_0^\pi \{(\sin^2\vartheta) v_B'^2 + v_B^2$$

$$+ 2\sin\vartheta\cos\vartheta \, v_B v_B'\} \sin\vartheta \, d\vartheta$$

$$\geqslant \sqrt{\frac{\chi_a}{2k_2}} (k_2 - k) H \int_0^\pi (\sin\vartheta \, v_B' + \cos\vartheta \, v_B)^2 \sin\vartheta \, d\vartheta \geqslant 0,$$

and so $V[v_B] > 0$ unless $v_B \equiv 0$. We conclude that the second variation of F at $(\vartheta_B, 0)$ is strictly positive if the perturbation of φ is not constant. If, on the other hand, the pair $(\vartheta_B, 0)$ is perturbed so as to remain in $\bar{\mathscr{A}}_F$, we know from Theorem 4.14 above that the energy of the twist wall cannot decrease even if the second variation of F vanishes.

Let $k_2 > k$, so that $(\vartheta_N, \pi/2)$ is the minimizer of F in $\bar{\mathscr{A}}_F$. For any u and v as above we now define the following curve in $\mathscr{A}_F$:

$$\varepsilon \mapsto (\vartheta_\varepsilon, \varphi_\varepsilon), \quad \vartheta_\varepsilon := \vartheta_N + \varepsilon u, \quad \varphi_\varepsilon := \frac{\pi}{2} + \varepsilon v.$$

An easy but tedious computation still leads us to express the second variation of F at $(\vartheta_N, \pi/2)$ in the form (4.209), where now

$$U[u] := \int_{-\infty}^{+\infty} \{k u'^2 + \tfrac{1}{2}\chi_a H^2 (\cos^2\vartheta_N - \sin^2\vartheta_N) u^2\} dx, \tag{4.215}$$

$$V[v] := \int_{-\infty}^{+\infty} \{(k\sin^2\vartheta_N + k_2\cos^2\vartheta_N)\sin^2\vartheta_N v'^2$$

$$+ (k_2 - k)\vartheta_N'^2 + 2(k_2 - k)\sin\vartheta_N\cos\vartheta_N\vartheta_N' v v'\} \, dx. \tag{4.216}$$

When (4.208) applies, (4.204) reduces to

$$\vartheta' = \sqrt{\frac{\chi_a}{2k}} H \sin\vartheta;$$

hence ϑ_N, which solves this equation, defines a change of variables similar to (4.212):

$$u_N(\vartheta_N(x)) := u(x), \quad v_N(\vartheta_N(x)) := v(x) \quad \text{for all } x \in \mathbb{R}.$$

Thus, the functional defined by (4.215) can be given precisely the form (4.213) with k in place of k_2, while (4.216) becomes

$$V[v_N] = \sqrt{\frac{\chi_a}{2k}} H \int_0^\pi \{(k \sin^2 \vartheta + k_2 \cos^2 \vartheta) \sin^2 \vartheta v_N'^2 + (k_2 - k) v_N^2 + 2(k_2 - k) \sin \vartheta \cos \vartheta v_N v_N'\} \sin \vartheta d\vartheta. \tag{4.217}$$

If $k < k_2$, (4.217) implies that

$$V[v_N] > \sqrt{\frac{\chi_a}{2k}} H \int_0^\pi \{k_2 \sin^2 \vartheta v_N'^2 + (k_2 - k) v_N^2 + 2(k_2 - k) \sin \vartheta \cos \vartheta v_N v_N'\} \sin \vartheta d\vartheta$$

$$> \sqrt{\frac{\chi_a}{2k}} (k_2 - k) H \int_0^\pi (\sin \vartheta v_N' + \cos \vartheta v_N)^2 \sin \vartheta d\vartheta \geqslant 0.$$

This completes the proof of the theorem.

To evaluate the order of magnitude of both w_B and w_N Helfrich (1968) considered PAA, which is nematic between 118 and 135°C; he took the values of the material moduli at 120°C:

$$2k_1 = 0.7 \times 10^{-6} \text{dyne}, \quad 2k_2 = 0.43 \times 10^{-6} \text{dyne},$$
$$2k_3 = 1.7 \times 10^{-6} \text{dyne}, \quad \chi_a = 1.3 \times 10^{-7} \text{c.g.s. units}.$$

For $H = 0.5\,\text{kG}$ he found:

$$w_B = 0.64 \times 10^{-2}\,\text{cm} \quad \text{and} \quad 0.81 \times 10^{-2}\,\text{cm} < w_N < 1.3 \times 10^{-2}\,\text{cm}.$$

Thus for PAA the width of both twist and splay–bend walls is about 1×10^{-2} cm. This result led Helfrich to interpret the experiments of Williams (1968) as supporting the evidence of alignment walls in nematics. Williams put a thin layer of PAA at 120°C on a glass substrate and applied to it a magnetic field parallel to the substrate. The layer was illuminated from below the glass and observed through a microscope from above. When the strength of the magnetic field was larger than 0.5 kG, Williams saw stripes, not wider than 1×10^{-2} cm, forming in the sample, Helfrich suggested interpreting these stripes as alignment inversion walls set upright on the substrate. The stripes disappeared when the applied field was below 0.5 kG. According to Helfrich this should be regarded as a dynamical instability driven by convection currents able to prevail over the aligning effect of weak fields.

4.7 References

Cohen, R. and Luskin, M. (1991) Field-induced instabilities in nematic liquid crystals, in: *Nematics. Mathematical and physical aspects*, J.-M. Coron, J.-M. Ghidaglia and F. Hélein eds., Kluwer Academic Publishers, Dordrecht, 261–278.

Courant, R. and Hilbert, D. (1989) *Methods of mathematical physics*, Volumes 1 and 2, J. Wiley & Sons, New York. Republication in the Wiley Classics Library Series of the first English edition (1953), Interscience Publishers.

Dafermos, C.M. (1968) Stability of orientation patterns of liquid crystals subject to magnetic fields, *SIAM J. Appl. Math.*, **16**, 1305–1318.

De Gennes, P.G. (1968) Calcul de la distorsion d'une structure cholesterique par un champ magnétique, *Solid State Commun.*, **6**, 163–165.

De Gennes, P.G. (1974) *The physics of liquid crystals*, Clarendon Press, Oxford.

Dreher, R. (1973) Remarks on the distortion of a cholesteric structure by a magnetic field, *Solid State Commun.*, **13**, 1517–1574.

Durand, G., Leger, L., Rondalez, F. and Veyssie, M. (1969) Magnetically induced cholesteric-to-nematic phase transition in liquid crystals, *Phys. Rev. Lett.*, **22**, 227–228.

Ericksen, J.L. (1976) Equilibrium theory of liquid crystals, in: *Advances in liquid crystals*, Vol. 2, G.H. Brown ed., Academic Press, New York.

Freedericks, V. and Zolina, V. (1933) Forces causing the orientation of an anisotropic liquid. *Trans. Faraday Soc.*, **29**, 919–930.

Gradshteyn, I.S. and Ryzhik, I.M. (1980) *Tables of integrals, series, and products*. Corrected and enlarged edition prepared by A. Jeffrey, Academic Press, New York.

Hardy, G.H., Littlewood, J.E. and Pólya, G. (1952) *Inequalities*, second edition, Cambridge University Press, Cambridge.

Helfrich, W. (1968) Alignment-inversion walls in nematic liquid crystals in the presence of a magnetic field, *Phys. Rev. Lett.*, **21**, 1518–1521.

Kléman, M. (1983) *Points, lines and walls*, J. Wiley and Sons, Chichester.

Leslie, F.M. (1970) Distortion of twisted orientation patterns in liquid crystals by magnetic fields, *Mol. Cryst. Liq. Cryst.*, **12**, 57–72.

Lonberg, F. and Meyer, R.B. (1985) New ground state for the splay-Freedericks transition in a polymer nematic liquid crystal, *Phys. Rev. Lett.*, **55**, 718–721.

Meyer, R.B. (1968) Effects of electric and magnetic fields on the structure of cholesteric liquid crystals, *Appl. Phys. Lett.*, **12**, 281–282.

Meyer, R.B. (1969a) Piezoelectric effects in liquid crystals, *Phys. Rev. Lett.*, **18**, 918–921.

Meyer, R.B. (1969b) Distortion of a cholesteric structure by a magnetic field, *Appl. Phys. Lett.*, **14**, 208–209.

Oldano, C. (1986) Comment on 'New ground state for the splay–Freedericks transition in a polymer nematic liquid crystal', *Phys. Rev. Lett.*, **56**, 1098.

Rapini, A. and Papoular, M. (1969) Distorsion d'une lamelle nématique sous champ magnétique. Conditions d'ancrage aux parois, *J. Phys. (Paris), Colloque* **C4**, 54–56.

Schadt, M. (1972) Dielectric properties of some nematic liquid crystals with strong positive dielectric anisotropy, *J. Chem. Phys.*, **56**, 1494–1497.

Schadt, M. (1988) The twisted nematic effect: liquid crystal displays and liquid crystal materials, *Mol. Cryst. Liq. Cryst.*, **165**, 405–438.

Schadt, M. (1989) The history of the liquid crystal display and liquid crystal material technology, *Liq. Crystals*, **5**, 57–71.

Schadt, M. and Helfrich, W. (1971) Voltage-dependent optical activity of a twisted nematic liquid crystal, *Appl. Phys. Lett.*, **18**, 127–128.

Tonelli, L. (1914) Sur une méthode directe du calcul des variations, *C.R. Acad. Sc. Paris*, **158**, 1776–1778, 1983–1985. Reprinted in Leonida Tonelli, *Opere Scelte*, Vol. II, *Calcolo delle Variazioni*, Cremonese, Roma, 1961, 177–182.

Tonelli, L. (1915) Sur une méthode directe du calcul des variations, *Rend. Matem. Palermo*, **39**, 233–264. Reprinted in Leonida Tonelli, *Opere Scelte*, Vol. II, *Calcolo delle Variazioni*, Cremonese, Roma, 1961, 289–333.

Williams, R. (1968) Optical rotatory effect in the nematic liquid phase of *p*-azoxyanisole, *Phys. Rev. Lett.*, **21**, 342–344.

Zocher, H. (1933) The effect of magnetic field on the nematic state, *Trans. Faraday Soc.*, **29**, 945–957.

4.8 Further reading

Electric and magnetic energies

Dubois-Violette, E. and Parodi, O. (1969) Émulsions nématiques. Effets de champs magnétiques et effets piézoélectriques, *J. Phys. (Paris), Colloque* **C4**, 57–64.

Freedericks's transition

Dubois-Violette, E. (1982) Instabilités dans les nématiques, in: Cristalli liquidi, *Atti della I Scuola Nazionale del GNCL, Rende*, 10–21 Settembre 1981, CLU, Torino, 587–601.

Fraser, C. (1978) Theoretical investigation of Freedericks transitions in twisted nematics with surface tilt, *J. Phys. A*, **11**, 1439–1448.

Leslie, F.M. (1987) Some topics in equilibrium theory of liquid crystals, in: *Theory and applications of liquid crystals*, J.L. Ericksen and D. Kinderlehrer eds., *IMA Volumes in Mathematics and its Applications*, **5**, 211–234.

Periodic Freedericks's transition

Allender, D.W., Frisken, B.J. and Palffy-Muhoray, P. (1989) Theory of an electric field induced periodic phase in a nematic film, *Liq. Crystals*, **5**, 735–738.

Barbero, G. and Miraldi, E. (1989) Forbidden periodic Freedericks transition: an example, *Nuovo Cimento*, **11 D**, 1265–1272.

Schiller, P. (1989) Perturbation theory for planar nematic twisted layers, *Liq. Crystals*, **4**, 69–79.

Sparavigna, A., Komotov, L., Stebler, B. and Strigazzi, A. (1991) Static splay-stripes in a hybrid aligned nematic layer, *Mol. Cryst. Liq. Cryst.*, **207**, 265–280.

Sparavigna, A., Komotov, L. and Strigazzi, A. (1992) Influence of a nematic field on the periodic splay-stripes in hybrid aligned nematics, *Mol. Cryst. Liq. Cryst.*, **212**, 289–303.

De Gennes's transition

De Gennes, P.G. (1971) Mouvements de parois dans un nématique sous champ tournant, *J. Phys.* (*Paris*), **32**, 789–792.
Helfrich, W. (1970) Deformation of cholesteric liquid crystals with low threshold voltage, *Appl. Phys. Lett.*, **17**, 531–532.
Helfrich, W. (1971) Electrohydrodynamic and dielectric instabilities of cholesteric liquid crystals, *J. Chem. Phys.*, **55**, 839–842.
Leslie, F.M. (1968) Some thermal effects in cholesteric liquid crystals, *Proc. Roy. Soc. A*, **307**, 358–372.
Mioskowski, C., Bourguignon, J., Candau, S. and Solladie, G. (1976) Photochemically induced cholesteric–nematic transition in liquid crystals, *Chem. Phys. Letts.*, **38**, 456–459.
Niggemann, E. and Stregemeyer, H. (1989) Magnetic field-induced instabilities in cholesteric liquid crystals: periodic deformations of the Grandjean texture, *Liq. Crystals*, **5**, 739–747.
Schiller, P. and Schiller, K. (1990) Phase diagrams of cholesteric films in electric fields, *Liq. Crystals*, **3**, 553–564.
Wysocki, J.J., Adams, J. and Haas, W. (1968), Electric-field-induced phase change in cholesteric liquid crystals, *Phys. Rev. Lett.*, **20**, 1024–1025.

Twisted nematics

Berreman, D.W. (1973) Optics in smoothly varying anisotropic planar structures: Applications to liquid-crystal twist cells, *J. Opt. Soc. Am.*, **63**, 1374–1380.
Berreman, D.W. (1974) Dynamics of liquid-crystal twist cells, *Appl. Phys. Lett.*, **25**, 12–15.
Berreman, D.W. (1984) Domain-wall tension allows bistability in imperfect laminar cholesteric twist cell, *J. Appl. Phys.*, **55**, 806–809.
Berreman, D.W. and Heffner, W.R. (1980) New bistable cholesteric liquid crystal display, *Appl. Phys. Lett.*, **37**, 109–111.
Blinov, L.M. (1983) *Electro-optical and magneto-optical principles of liquid crystals*, J. Wiley & Sons, Chichester.
Leslie, F.M. (1987) Some topics in equilibrium theory of liquid crystals, in: *Theory and applications of liquid crystals*, J.L. Ericksen and D. Kinderlehrer eds., *IMA Volumes in Mathematics and its Applications*, **5**, 211–234.

Raynes, E.P. (1986) The theory of supertwist transitions, *Mol. Cryst. Liq. Cryst. Letters*, **4**, 1–8.

Thurston, R.N. (1981a) Unit sphere description of liquid-crystal configurations, *J. Appl. Phys.*, **52**, 3040–3052.

Thurston, R.N. (1981b) Stability of nematic liquid crystal configurations, *J. Phys. (Paris)*, **42**, 419–425.

Thurston, R.N. (1982) Bistabilities in twisted nematics with a holding voltage, *J. Phys. (Paris)*, **43**, 117–128.

Thurston, R.N. (1984) Liquid crystal bistable displays, Invited lecture at the Tenth International Liquid Crystal Conference, York, U.K.

Thurston, R.N. and Berreman, D.W. (1981) Equilibrium and stability of liquid-crystal configurations in an electric field, *J. Appl. Phys.*, **52**, 508–509.

Inversion walls

Ericksen, J.L. (1969) Twisting of liquid crystals by magnetic fields, *ZAMP*, **20**, 383–388.

Helfrich, W. (1969) Erratum to 'Alignment-inversion walls in nematic liquid crystals in the presence of a magnetic field', *Phys. Rev. Lett.*, **22**, 1342.

5

Drops

This chapter mostly concerns free-boundary problems for liquid crystals. The literature about them is so huge as to demand a whole book to describe it. Here, apart from the last section, I will mostly follow three papers, namely Virga (1989 and 1990) and Roccato and Virga (1990), which deal with the equilibrium shapes of a drop of liquid crystal surrounded by an isotropic fluid. The prototype of this problem was first considered by Oseen (1933). The list of further reading appended to this chapter attempts to indicate other problems akin to this, which have also received attention in recent times.

5.1 Surface free energy

We have learnt in Section 1.1 of Chapter 1 that liquid crystals are to be regarded as incompressible fluids. Thus, when the region in space occupied by any such fluid is not prescribed, but is allowed to vary like the orientation field within it, a new constraint arises, which has not come into play so far, and which assigns the volume of its shape:

$$v(\mathscr{B}) = \beta, \tag{5.1}$$

where $\mathscr{B}$ is any admissible region occupied by the liquid crystal, β is a given positive constant, and v denotes the volume-measure as defined in Chapter 2 (*cf*. Definition 2.8 of Section 2.3).

A **configuration** for a drop of liquid crystal is described here as a pair $(\mathscr{B}, \mathbf{n})$, where $\mathscr{B}$ is a fit region subject to (5.1), and $\mathbf{n}$ is a field of class C^2 of $\mathscr{B}$ into $\mathbb{S}^2$. Thus, the interface between a drop and the surrounding environment is to be identified with the reduced boundary $\partial^*\mathscr{B}$ (*cf*. Theorem 2.13 in Section 2.3 of Chapter 2).

In this chapter we consider drops of nematic liquid crystals surrounded by isotropic fluids.

As is well known, the free energy associated with the interface between two isotropic fluids is proportional to the area of the surface of contact by the **surface tension**, a positive constant depending on both the temperature and the nature of the fluids in contact. On the other hand, when one of the fluids is not isotropic, as is the case here, such a representation of the surface free energy is too simplistic. As liquid crystals are transversely isotropic about the optic axis **n**, the simplest formula to represent the surface free energy, already suggested by the pioneers (*cf.* Friedel (1922) and Oseen (1931)), is

$$\mathscr{F}_S[\mathscr{B}, \mathbf{n}] := \int_{\partial^*\mathscr{B}} \varphi(\mathbf{n}, \boldsymbol{\nu})\, da, \tag{5.2}$$

where $\boldsymbol{\nu}$ is the outer unit normal to $\mathscr{B}$ and φ is a function of $\mathbb{S}^2 \times \mathbb{S}^2$ into the positive real line $\mathbb{R}^+$ such that

$$\varphi(-\mathbf{n}, \boldsymbol{\nu}) = \varphi(\mathbf{n}, \boldsymbol{\nu}) \quad \text{for all } \mathbf{n}, \boldsymbol{\nu} \in \mathbb{S}^2. \tag{5.3}$$

As φ depends on the orientation of both the optic axis and the normal at the reduced boundary of the drop, it is often called anisotropic surface tension.

Requiring φ to be frame-indifferent, as was the elastic energy density σ in Section 3.1 of Chapter 3, amounts to imposing the following condition:

$$\varphi(\mathbf{Q}\mathbf{n}, \mathbf{Q}\boldsymbol{\nu}) = \varphi(\mathbf{n}, \boldsymbol{\nu}) \quad \text{for all } \mathbf{Q} \in SO(\mathscr{V}) \text{ and } \mathbf{n}, \boldsymbol{\nu} \in \mathbb{S}^2. \tag{5.4}$$

By Theorem 2.26 in Section 2.6 of Chapter 2, (5.3) and (5.4) together are equivalent to

$$\varphi(\mathbf{n}, \boldsymbol{\nu}) = \tau(\mathbf{n} \cdot \boldsymbol{\nu}), \tag{5.5}$$

where τ is an even function of $[-1, 1]$ into $\mathbb{R}^+$. We assume that τ is of class C^2 and that there is only one point ξ_0 in the interval $[0, 1]$ such that

$$\tau(\xi) > \tau(\xi_0) \quad \text{for all } \xi \in [0, 1] \setminus \{\xi_0\}; \tag{5.6}$$

that is, ξ_0 is the unique minimizer of τ in $[0, 1]$.

It is customary to call the angle between **n** and $\boldsymbol{\nu}$ at the reduced boundary of the drop the **tilt angle** and to denote it by ϑ, so that the argument of τ can also be written as

$$\xi = \cos \vartheta. \tag{5.7}$$

Sluckin and Poniewierski (1986) suggest employing the following formula for τ:

$$\tau(\cos \vartheta) = \tau_0 + \tau_1 \cos^2 \vartheta + \tau_2 \cos^4 \vartheta, \tag{5.8}$$

where τ_0, τ_1 and τ_2 are material moduli depending on both the temperature and the two fluids in contact. For simplicity, we set

$$\tau_2 = 0 \quad \text{and} \quad \tau_1 = \omega\tau_0, \tag{5.9}$$

and so (5.8) reduces to

$$\tau(\cos \vartheta) = \tau_0(1 + \omega \cos^2 \vartheta), \tag{5.10}$$

with $\tau_0 > 0$ and $\omega > -1$. This is the well-known formula of Rapini and Papoular, which has been confirmed by several experiments, as is shown *e.g.* in Naemura (1978 and 1979), and Rivière, Lévy and Guyon (1979).

Clearly, (5.10) satisfies (5.6) with $\xi_0 = 0$ for $\omega > 0$, and $\xi_0 = 1$ for $-1 < \omega < 0$. Thus, when $\omega > 0$ the surface energy favours a tangential orientation of the optic axis at the free surface of the drop, whereas when $-1 < \omega < 0$ it favours the homeotropic alignment.

The bulk free energy of the drop will still be expressed by formula (3.2) of Chapter 3, but here also the region occupied by the liquid crystal, being unknown, is to be regarded as a variable:

$$\mathscr{F}_B[\mathscr{B}, \mathbf{n}] = \int_{\mathscr{B}} \sigma(\mathbf{n}, \nabla \mathbf{n}) dv. \tag{5.11}$$

The elastic free energy density σ obeys (3.10) of Chapter 3 and may be given by Frank's formula for nematics (*cf.* equation (3.23) of Chapter 3):

$$\begin{aligned} \sigma(\mathbf{n}, \nabla \mathbf{n}) = {} & k_1(\operatorname{div} \mathbf{n})^2 + k_2(\mathbf{n} \cdot \operatorname{curl} \mathbf{n})^2 + k_3 |\mathbf{n} \wedge \operatorname{curl} \mathbf{n}|^2 \\ & + (k_2 + k_4)(\operatorname{tr}(\nabla \mathbf{n})^2 - (\operatorname{div} \mathbf{n})^2). \end{aligned} \tag{5.12}$$

Here we seldom employ (5.12) for σ, rather we shall quite often exploit two qualitative features of the function which gives it, namely that there exists $\nu > 1$ such that

$$\sigma(\mathbf{n}, \mu \nabla \mathbf{n}) = \mu^\nu \sigma(\mathbf{n}, \nabla \mathbf{n}) \quad \text{for all } \mu \in \mathbb{R}, \tag{5.13}$$

and

$$\sigma(\mathbf{n}, \nabla\mathbf{n}) > 0 \quad \text{unless } \mathbf{n} \text{ is constant.} \tag{5.14}$$

Clearly, Frank's formula satisfies (5.13) with $\nu = 2$. Moreover, by (3.65) of Chapter 3, (5.14) holds for it if Frank's constants satisfy the following inequalities:

$$2k_1 > k_2 + k_4, \quad k_2 > |k_4|, \quad k_3 > 0, \tag{5.15}$$

which are taken as valid whenever in this chapter we make use of (5.12).

Neglecting the work done by body forces, we write the total energy of a drop of nematic liquid crystal surrounded by an isotropic fluid as

$$\mathscr{F}[\mathscr{B}, \mathbf{n}] := \mathscr{F}_B[\mathscr{B}, \mathbf{n}] + \mathscr{F}_S[\mathscr{B}, \mathbf{n}], \tag{5.16}$$

where $\mathscr{F}_B$ and $\mathscr{F}_S$ are as in (5.11) and (5.2), with the anisotropic surface tension given by (5.5).

For a configuration $(\mathscr{B}, \mathbf{n})$ of the drop we say that $\mathscr{B}$ is its **shape**. All shapes must obey (5.1) with a given constant β. An **equilibrium configuration** is a configuration at which $\mathscr{F}$ attains an extremum; an equilibrium configuration is **stable** if it minimizes $\mathscr{F}$. When $(\mathscr{B}, \mathbf{n})$ is an equilibrium configuration we say that $\mathscr{B}$ is an **equilibrium shape**; when $(\mathscr{B}, \mathbf{n})$ is a stable equilibrium configuration we say that $\mathscr{B}$ is a **stable shape**.

The problem of finding the stable equilibrium configurations of a drop of liquid crystal is fascinating, but formidable; it has not yet been completely solved. In the next section I open the way to the approximate and numerical solutions that form the main object of this chapter.

An exhaustive analytical treatment is still lacking. There is some indication that it might be based on a new class of functions with bounded variation which has recently been introduced by De Giorgi and Ambrosio (1988). An application to liquid crystals of such a class was first pursued by Ambrosio (1989a and 1990); an enlightening introduction to this matter, oriented towards another successful application, can be found in Ambrosio (1989b).

I shall come back to this issue in Section 5.3 below, where I touch upon some promising new findings of Lin and Poon (1992).

5.2 Equilibrium configurations

Here we shall not keep the level of generality held in the preceding section. The admissible shapes of a drop of liquid crystal are to be taken as regular regions.

As we learned in Section 2.3 of Chapter 2, a regular region $\mathcal{B}$ admits a regular partition whose members are $\mathcal{B}$ itself and the sides, edges and vertices of its boundary. Each side of a regular partition is a regular orientable surface in the three-dimensional Euclidean space $\mathcal{E}$ (*cf*. Definitions 2.10 and 2.11 in Section 2.3 of Chapter 2).

We agreed above to represent the shape of a drop of nematic liquid crystal as a fit region. Here, to make our development a little easier, we look for the equilibrium shapes of a drop among some special regular regions: those whose boundary possesses at most one edge, which is a closed curve in $\mathcal{E}$; that is, we represent $\partial\mathcal{B}$ in the form

$$\partial\mathcal{B} = \mathcal{S}_1 \cup \mathcal{S}_2 \cup \mathcal{C}, \tag{5.17}$$

where $\mathcal{S}_1$ and $\mathcal{S}_2$ are regular orientable surfaces, and $\mathcal{C}$ is either the empty set or the only edge in the border of both $\mathcal{S}_1$ and $\mathcal{S}_2$. When $\mathcal{C}$ is not empty it is also the reduced border of both $\mathcal{S}_1$ and $\mathcal{S}_2$, so that they possess no vertex. When $\mathcal{C}$ is the empty set, $\mathcal{S}_1$ and $\mathcal{S}_2$ cannot be disjoint, because $\partial\mathcal{B}$ is a closed set, but both $\mathcal{S}_1$ and $\mathcal{S}_2$ are open.

The border and the reduced border of a regular orientable surface are defined in the closing pages of Section 2.3 in Chapter 2; the reader is referred there for these and other details concerning the geometric properties invoked here.

When $\mathcal{C}$ is the empty set, the reduced boundary $\partial^*\mathcal{B}$ is just the same as the boundary of $\mathcal{B}$. On the other hand, when $\mathcal{C}$ is not empty, $\partial^*\mathcal{B} = \mathcal{S}_1 \cup \mathcal{S}_2$. Figure 5.1 illustrates a region $\mathcal{B}$ for which $\mathcal{C}$ is not empty. It is to be noted that two pairs of mutually orthogonal unit vectors can be defined along it, one relative to each surface in the reduced boundary (see Fig. 5.1). We denote by $(\mathbf{v}_1, \mathbf{v}_{1,2})$ and $(\mathbf{v}_2, \mathbf{v}_{2,1})$ such pairs: $\mathbf{v}_1$ and $\mathbf{v}_2$ are normal to $\mathcal{S}_1$ and $\mathcal{S}_2$, respectively, whereas $\mathbf{v}_{1,2}$ and $\mathbf{v}_{2,1}$ satisfy

$$\mathbf{v}_1 \cdot \mathbf{v}_{1,2} = 0, \quad \mathbf{v}_2 \cdot \mathbf{v}_{2,1} = 0$$

In the notation introduced in Section 2.3 of Chapter 2 for regular

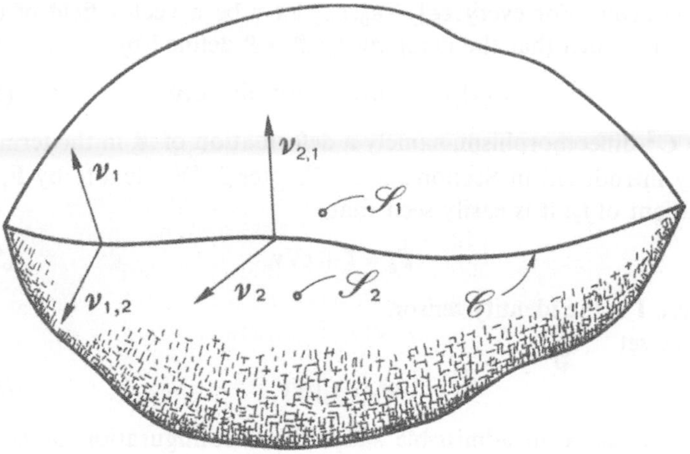

Figure 5.1

orientable surfaces $\mathbf{v}_{1,2}$ and $\mathbf{v}_{2,1}$ would read as $\mathbf{v}_{\mathscr{S}_1}$ and $\mathbf{v}_{\mathscr{S}_2}$, respectively; such a notation is not used here to avoid clutter.

The equilibrium equations that we shall derive in this section apply also when the boundary of the regular region occupied by the drop is represented by a formula more general than (5.17). As soon as we master the easiest case, the harder ones will become manageable (*cf.* Noll and Virga (1990) for a systematic treatment of integrals over the boundary of general regular regions).

We shall employ here the following class of admissible configurations for $\mathscr{F}$:

$$\mathscr{A}_{\mathscr{F}} := \{(\mathscr{B}, \mathbf{n}) | \mathscr{B} \text{ is a regular region, } \partial\mathscr{B} \text{ is as in (5.17),} \\ \text{and } \mathbf{n}: \mathscr{B} \to \mathbb{S}^2 \text{ is of class } C^2\}.$$

Let $\mathscr{F}^*$ be the functional defined by

$$\mathscr{F}^*[\mathscr{B}, \mathbf{n}] := \mathscr{F}[\mathscr{B}, \mathbf{n}] + \pi_0 v(\mathscr{B}), \tag{5.18}$$

where π_0 is a Lagrange multiplier.

All configurations admissible for $\mathscr{F}$ are also admissible for $\mathscr{F}^*$, and so we shall employ the class $\mathscr{A}_{\mathscr{F}}$ also for $\mathscr{F}^*$. The configurations of $\mathscr{F}^*$ are not subject to (5.1). The equilibrium configurations of $\mathscr{F}$ are those equilibrium configurations of $\mathscr{F}^*$ whose shapes satisfy (5.1).

Let a configuration $(\mathscr{B}, \mathbf{n})$ be given in $\mathscr{A}_{\mathscr{F}}$. We now construct two paths of configurations in $\mathscr{A}_{\mathscr{F}}$ close to $(\mathscr{B}, \mathbf{n})$. Let a positive scalar

ε_0 be given. For every $\varepsilon \in [-\varepsilon_0, \varepsilon_0]$ let $\mathbf{v}$ be a vector field of class C^2 on $\mathscr{B}$ such that the mapping $f_\varepsilon : \mathscr{B} \to \mathscr{E}$ defined by

$$f_\varepsilon(x) := x + \varepsilon \mathbf{v}(x) \quad \text{for all } x \in \mathscr{B} \tag{5.19}$$

is a C^2-diffeomorphism, namely a deformation of $\mathscr{B}$, in the terminology introduced in Section 2.3 of Chapter 2. We denote by $\mathbf{F}_\varepsilon$ the gradient of f_ε; it is easily seen that

$$\mathbf{F}_\varepsilon = \mathbf{I} + \varepsilon \nabla \mathbf{v}, \tag{5.20}$$

where $\mathbf{I}$ is the identity tensor.

We set

$$\mathscr{B}_\varepsilon := f_\varepsilon(\mathscr{B}). \tag{5.21}$$

Clearly, $\mathscr{B}_\varepsilon$ is an admissible shape for a configuration of $\mathscr{A}_{\mathscr{F}}$. It follows from (5.20) that

$$\det \mathbf{F}_\varepsilon = 1 + \varepsilon \operatorname{div} \mathbf{v} + o(\varepsilon). \tag{5.22}$$

Hence, by the Divergence Theorem,

$$v(\mathscr{B}_\varepsilon) = v(\mathscr{B}) + \varepsilon \int_{\partial^* \mathscr{B}} \mathbf{v} \cdot \boldsymbol{\nu} \, da + o(\varepsilon), \tag{5.23}$$

where $\partial^* \mathscr{B} = \mathscr{S}_1 \cup \mathscr{S}_2$ by (5.17), and $\boldsymbol{\nu}$ is the outer unit normal to $\mathscr{B}$.

It requires a little more labour to relate the unit normal of $\mathscr{B}_\varepsilon$ to that of $\mathscr{B}$. It is shown *e.g.* on p. 53 of Gurtin (1981) that $|\mathbf{F}_\varepsilon^* \boldsymbol{\nu}|$, where $\mathbf{F}_\varepsilon^*$ is the adjugate of $\mathbf{F}_\varepsilon$ according to formula (2.40) of Chapter 2, expresses at all points of $\partial^* \mathscr{B}$ the surface dilation factor, precisely as $\det \mathbf{F}_\varepsilon$ expresses at all points of $\mathscr{B}$ the volume dilation factor. Building upon formula (2.41) of Chapter 2, where now $\mathbf{L}$ is set equal to $\mathbf{F}_\varepsilon$ and $\mathbf{u}, \mathbf{v}$ are taken as unit vectors tangent to $\partial^* \mathscr{B}$, the reader will easily show that

$$\boldsymbol{\nu}_\varepsilon(f_\varepsilon(x)) = \frac{\mathbf{F}_\varepsilon(x) \boldsymbol{\nu}(x)}{|\mathbf{F}_\varepsilon(x) \boldsymbol{\nu}(x)|} \quad \text{for all } x \in \partial^* \mathscr{B}, \tag{5.24}$$

where $\boldsymbol{\nu}_\varepsilon$ is the outer unit normal to $\mathscr{B}_\varepsilon$ (*cf.* also Gurtin (1981), *loc. cit.*). By (5.20), we finally arrive at the following formulae:

$$|\mathbf{F}_\varepsilon^* \boldsymbol{\nu}| = 1 + \varepsilon \operatorname{div}_s \mathbf{v} + o(\varepsilon), \tag{5.25}$$

$$\boldsymbol{\nu}_\varepsilon = \boldsymbol{\nu} - \varepsilon (\nabla_s \mathbf{v})^T \boldsymbol{\nu} + o(\varepsilon), \tag{5.26}$$

where $\nabla_s \mathbf{v}$ is the surface-gradient of $\mathbf{v}$ and $\operatorname{div}_s \mathbf{v}$ is the trace of $\nabla_s \mathbf{v}$.

In (5.26) the fields $\mathbf{v}$ and $\boldsymbol{\nu}$ are both evaluated at the points of $\partial^*\mathscr{B}$, while $\boldsymbol{\nu}_\varepsilon$ is evaluated at the corresponding points of $\partial^*\mathscr{B}_\varepsilon$.

These preliminary computations are soon to become needed in our development.

For every $\varepsilon \in [-\varepsilon_0, \varepsilon_0]$ let the mapping $\mathbf{n}_\varepsilon : \mathscr{B}_\varepsilon \to \mathbb{S}^2$ be defined by

$$\mathbf{n}_\varepsilon(y) = \mathbf{n}(f_\varepsilon^{-1}(y)) \quad \text{for all } y \in \mathscr{B}_\varepsilon, \tag{5.27}$$

where $\mathbf{n}$ is the orientation field prescribed above in $\mathscr{B}$. It follows from (5.27) that $\mathbf{n}_\varepsilon$ is of class C^2 since both $\mathbf{n}$ and f_ε are so; moreover, also in view of (5.20),

$$\begin{aligned}\nabla\mathbf{n}_\varepsilon(f_\varepsilon(x)) &= \nabla\mathbf{n}(x)\mathbf{F}_\varepsilon^{-1}(x)\\ &= \nabla\mathbf{n}(x) - \varepsilon(\nabla\mathbf{n}(x))(\nabla\mathbf{v}(x)) + o(\varepsilon) \quad \text{for all } x \in \mathscr{B}.\end{aligned} \tag{5.28}$$

As ε spans the interval $[-\varepsilon_0, \varepsilon_0]$, $(\mathscr{B}_\varepsilon, \mathbf{n}_\varepsilon)$ describes a path of configurations in $\mathscr{A}_\mathscr{F}$; furthermore,

$$(\mathscr{B}_\varepsilon, \mathbf{n}_\varepsilon)|_{\varepsilon=0} = (\mathscr{B}, \mathbf{n}).$$

We now introduce another path of configurations, all having the shape $\mathscr{B}$ in common. Let η_0 be given in $\mathbb{R}^+$ and let $\mathbf{u} : \mathscr{B} \to \mathscr{V}$ be a mapping of class C^2. For every $\eta \in [-\eta_0, \eta_0]$ we define the mapping $\mathbf{n}_\eta : \mathscr{B} \to \mathbb{S}^2$ by

$$\mathbf{n}_\eta(x) := \frac{\mathbf{n}(x) + \eta\mathbf{u}(x)}{|\mathbf{n}(x) + \eta\mathbf{u}(x)|} \quad \text{for all } x \in \mathscr{B}. \tag{5.29}$$

The mapping $\mathbf{n}_\eta$ is of class C^2 and such that

$$\mathbf{n}_\eta(x) = \mathbf{n}(x) + \eta\mathbf{P}(\mathbf{n}(x))\mathbf{u}(x) + o(\eta) \quad \text{for all } x \in \mathscr{B}, \tag{5.30}$$

where

$$\mathbf{P}(n) := \mathbf{I} - \mathbf{n} \otimes \mathbf{n}.$$

Clearly, by (5.30), we also have that

$$\nabla\mathbf{n}_\eta = \nabla\mathbf{n} + \eta\nabla(\mathbf{P}(\mathbf{n})\mathbf{u}) + o(\eta). \tag{5.31}$$

Formulae (5.29) and (5.30) are reminiscent of (3.100) and (3.101) of Chapter 3; the equilibrium equations that we are to derive here will indeed include those proved in Section 3.5 of Chapter 3 as a special case.

As η spans the interval $[-\eta_0, \eta_0]$, $(\mathscr{B}, \mathbf{n}_\eta)$ describes a path of admissible configurations in $\mathscr{A}_\mathscr{F}$, and again

$$(\mathscr{B}, \mathbf{n}_\eta)|_{\eta=0} = (\mathscr{B}, \mathbf{n}).$$

Definition 5.1. We give the name **first variations** of $\mathcal{F}^*$ at the configuration $(\mathcal{B},\mathbf{n})$ of $\mathcal{A}_{\mathcal{F}}$ to the linear functionals

$$\delta_1\mathcal{F}^*(\mathcal{B},\mathbf{n})[\mathbf{v}]:=\frac{d}{d\varepsilon}\mathcal{F}^*(\mathcal{B}_\varepsilon,\mathbf{n}_\varepsilon)|_{\varepsilon=0},$$

$$\delta_2\mathcal{F}^*(\mathcal{B},\mathbf{n})[\mathbf{u}]:=\frac{d}{d\eta}\mathcal{F}^*(\mathcal{B},\mathbf{n}_\eta)|_{\eta=0}.$$

Before computing these variations of $\mathcal{F}^*$ we recall for later use some facts that we have already encountered above (*cf.* (2.65) of Chapter 2 and (3.104) of Chapter 3).

Lemma 5.1. If $\mathbf{n}:\mathcal{B}\to\mathbb{S}^2$ is of class C^1, then

$$(\nabla\mathbf{n})^T\mathbf{n}=\mathbf{0}\quad\text{in }\mathcal{B}. \tag{5.32}$$

Furthermore, if the mapping $(\mathbf{n},\nabla\mathbf{n})\mapsto\sigma(\mathbf{n},\nabla\mathbf{n})$ is of class C^1, then

$$\frac{\partial\sigma}{\partial\mathbf{n}}\cdot\mathbf{n}=0\quad\text{and}\quad\left(\frac{\partial\sigma}{\partial\nabla\mathbf{n}}\right)^T\mathbf{n}=\mathbf{0}\quad\text{in }\mathcal{B}. \tag{5.33}$$

Theorem 5.2. If the pair $(\mathcal{B},\mathbf{n})$ of $\mathcal{A}_{\mathcal{F}}$ is an equilibrium configuration of $\mathcal{F}^*$, then the following equations hold:

$$\operatorname{div}\left(\sigma\mathbf{I}-(\nabla\mathbf{n})^T\frac{\partial\sigma}{\partial\nabla\mathbf{n}}\right)=\mathbf{0}\quad\text{in }\mathcal{B}, \tag{5.34}$$

$$\frac{\partial\sigma}{\partial\mathbf{n}}-\operatorname{div}\left(\frac{\partial\sigma}{\partial\nabla\mathbf{n}}\right)=\lambda\mathbf{n}\quad\text{in }\mathcal{B}, \tag{5.35}$$

$$\left((\sigma+\pi_0)\mathbf{I}-(\nabla\mathbf{n})^T\frac{\partial\sigma}{\partial\nabla\mathbf{n}}\right)\boldsymbol{\nu}=\operatorname{div}_s[(\tau\mathbf{I}-\tau'\boldsymbol{\nu}\otimes\mathbf{n})\mathbf{P}(\boldsymbol{\nu})]\quad\text{on }\partial^*\mathcal{B}, \tag{5.36}$$

$$\left(\frac{\partial\sigma}{\partial\nabla\mathbf{n}}+\tau'\mathbf{I}\right)\boldsymbol{\nu}=\lambda\mathbf{n}\quad\text{on }\partial^*\mathcal{B}, \tag{5.37}$$

$$(\tau(\mathbf{n}\cdot\boldsymbol{\nu}_1)\mathbf{I}-\tau'(\mathbf{n}\cdot\boldsymbol{\nu}_1)\boldsymbol{\nu}_1\otimes\mathbf{n})\boldsymbol{\nu}_{1,2}+(\tau(\mathbf{n}\cdot\boldsymbol{\nu}_2)\mathbf{I}-\tau'(\mathbf{n}\cdot\boldsymbol{\nu}_2)\boldsymbol{\nu}_2\otimes\mathbf{n})\boldsymbol{\nu}_{2,1}=\mathbf{0}\quad\text{on }\mathcal{C}, \tag{5.38}$$

where λ is a scalar-valued field on the closure of $\mathcal{B}$, τ' denotes the derivative of τ, and $\mathbf{P}(\boldsymbol{\nu})=\mathbf{I}-\boldsymbol{\nu}\otimes\boldsymbol{\nu}$.

Proof. If $(\mathcal{B},\mathbf{n})$ is an equilibrium configuration for $\mathcal{F}^*$, then both $\delta_1\mathcal{F}^*(\mathcal{B},\mathbf{n})$ and $\delta_2\mathcal{F}^*(\mathcal{B},\mathbf{n})$ must vanish.

We compute $\mathscr{F}^*[\mathscr{B}_\varepsilon, \mathbf{n}_\varepsilon]$, making use of the formulae (5.21)–(5.28) derived above:

$$\mathscr{F}^*[\mathscr{B}_\varepsilon, \mathbf{n}_\varepsilon] = \int_{\mathscr{B}} \sigma(\mathbf{n}, \nabla\mathbf{n} - \varepsilon(\nabla\mathbf{n})(\nabla\mathbf{v}))(1 + \varepsilon \operatorname{div} \mathbf{v})dv$$
$$+ \pi_0\left(v(\mathscr{B}) + \varepsilon \int_{\partial^*\mathscr{B}} \mathbf{v}\cdot\boldsymbol{\nu}\, da\right)$$
$$+ \int_{\partial^*\mathscr{B}} \tau(\mathbf{n}\cdot\boldsymbol{\nu} - \varepsilon\mathbf{n}\cdot(\nabla_s\mathbf{v})^T\boldsymbol{\nu})(1 + \varepsilon \operatorname{div}_s \mathbf{v})da + o(\varepsilon),$$

whence it follows that

$$\delta_1\mathscr{F}^*(\mathscr{B}, \mathbf{n})[\mathbf{v}] = \int_{\mathscr{B}} \left(\sigma \operatorname{div} \mathbf{v} - (\nabla\mathbf{n})^T \frac{\partial\sigma}{\partial\nabla\mathbf{n}}\cdot\nabla\mathbf{v}\right)dv$$
$$+ \int_{\partial^*\mathscr{B}} \{\pi_0\mathbf{v}\cdot\boldsymbol{\nu} + \tau \operatorname{div}_s \mathbf{v} - \tau'\mathbf{n}\cdot(\nabla_s\mathbf{v})^T\boldsymbol{\nu}\}da. \tag{5.39}$$

By the Divergence Theorem the integral over $\mathscr{B}$ in (5.39) can be given the form

$$\int_{\partial^*\mathscr{B}} \left(\sigma\mathbf{I} - (\nabla\mathbf{n})^T \frac{\partial\sigma}{\partial\nabla\mathbf{n}}\right)\mathbf{v}\cdot\boldsymbol{\nu}\, da + \int_{\mathscr{B}} \operatorname{div}\left((\nabla\mathbf{n})^T \frac{\partial\sigma}{\partial\nabla\mathbf{n}} - \sigma\mathbf{I}\right)\cdot\mathbf{v}\, dv,$$

and so

$$\delta_1\mathscr{F}^*(\mathscr{B}, \mathbf{n})[\mathbf{v}] = \int_{\mathscr{B}} \operatorname{div}\left((\nabla\mathbf{n})^T \frac{\partial\sigma}{\partial\nabla\mathbf{n}} - \sigma\mathbf{I}\right)\cdot\mathbf{v}\, dv$$
$$+ \int_{\partial^*\mathscr{B}} \left((\sigma + \pi_0)\mathbf{I} - (\nabla\mathbf{n})^T \frac{\partial\sigma}{\partial\nabla\mathbf{n}}\right)\mathbf{v}\cdot\boldsymbol{\nu}\, da$$
$$+ \int_{\partial^*\mathscr{B}} (\tau \operatorname{div}_s \mathbf{v} - \tau'\mathbf{n}\cdot(\nabla_s\mathbf{v})^T\boldsymbol{\nu})da. \tag{5.40}$$

The last integral in (5.40) can be reduced to a more convenient expression. First, its integrand can be written in the form

$$(\tau\mathbf{I} - \tau'\boldsymbol{\nu}\otimes\mathbf{n})\cdot\nabla_s\mathbf{v},$$

which is equal to

$$\mathbf{A}(\boldsymbol{\nu})\cdot\nabla_s\mathbf{v} \quad \text{with} \quad \mathbf{A}(\boldsymbol{\nu}) := (\tau\mathbf{I} - \tau'\boldsymbol{\nu}\otimes\mathbf{n})\mathbf{P}(\boldsymbol{\nu}), \tag{5.41}$$

as $\nabla_s \mathbf{v}$ is a tangential tensor field on both $\mathcal{S}_1$ and $\mathcal{S}_2$ (*cf.* Section 2.3 of Chapter 2). Second, the following identity holds on $\partial^*\mathcal{B}$.

$$\mathbf{A}(\mathbf{v})\cdot\nabla_s\mathbf{v} = \mathrm{div}_s(\mathbf{A}^T(\mathbf{v})\mathbf{v}) - \mathrm{div}_s\,\mathbf{A}(\mathbf{v})\cdot\mathbf{v}. \tag{5.42}$$

By $(5.41)_2$, $\mathbf{A}^T(\mathbf{v})\mathbf{v}$ is a tangential vector field on $\partial^*\mathcal{B}$, and so, by the Surface-Divergence Theorem,

$$\int_{\partial^*\mathcal{B}} \mathrm{div}_s(\mathbf{A}^T(\mathbf{v})\mathbf{v})da = \int_{\mathscr{C}} (\mathbf{A}(\mathbf{v}_1)\mathbf{v}_{1,2} + \mathbf{A}(\mathbf{v}_2)\mathbf{v}_{2,1})\cdot\mathbf{v}d\ell, \tag{5.43}$$

where ℓ denotes the length-measure on $\mathscr{C}$. Inserting both (5.42) and (5.43) into (5.40), we finally arrive at

$$\begin{aligned}\delta_1\mathcal{F}^*(\mathcal{B},\mathbf{n})[\mathbf{v}] = &\int_{\mathcal{B}} \mathrm{div}\left((\nabla\mathbf{n})^T\frac{\partial\sigma}{\partial\nabla\mathbf{n}} - \sigma\mathbf{I}\right)\cdot\mathbf{v}dv \\ &+ \int_{\partial^*\mathcal{B}}\left\{\left((\sigma+\pi_0)\mathbf{I} - (\nabla\mathbf{n})^T\frac{\partial\sigma}{\partial\nabla\mathbf{n}}\right)\boldsymbol{\nu}\right. \\ &\left. - \mathrm{div}_s((\tau\mathbf{I} - \tau'\boldsymbol{\nu}\otimes\mathbf{n})\mathbf{P}(\boldsymbol{\nu}))\right\}\cdot\mathbf{v}da \\ &+ \int_{\mathscr{C}}\{(\tau(\mathbf{n}\cdot\boldsymbol{\nu}_1)\mathbf{I} - \tau'(\mathbf{n}\cdot\boldsymbol{\nu}_1)\boldsymbol{\nu}_1\otimes\mathbf{n})\mathbf{v}_{1,2} \\ &+ (\tau(\mathbf{n}\cdot\boldsymbol{\nu}_2)\mathbf{I} - \tau'(\mathbf{n}\cdot\boldsymbol{\nu}_2)\boldsymbol{\nu}_2\otimes\mathbf{n})\mathbf{v}_{2,1}\}d\ell. \end{aligned} \tag{5.44}$$

Requiring $\delta_1\mathcal{F}^*(\mathcal{B},\mathbf{n})[\mathbf{v}]$ in (5.44) to vanish for all $\mathbf{v}$ of class C^2 readily leads to (5.34), (5.36) and (5.38).

We now compute $\mathcal{F}^*[\mathcal{B},\mathbf{n}_\eta]$. From (5.30) and (5.31) we get

$$\begin{aligned}\mathcal{F}^*[\mathcal{B},\mathbf{n}_\eta] = &\int_{\mathcal{B}} \sigma(\mathbf{n} + \eta\mathbf{P}(\mathbf{n})\mathbf{u}, \nabla\mathbf{n} + \eta\nabla(\mathbf{P}(\mathbf{n})\mathbf{u}))dv + \pi_0 v(\mathcal{B}) \\ &+ \int_{\partial^*\mathcal{B}} \tau(\mathbf{n}\cdot\boldsymbol{\nu} + \eta\mathbf{P}(\mathbf{n})\mathbf{u}\cdot\boldsymbol{\nu})da + o(\eta),\end{aligned}$$

whence it follows that

$$\begin{aligned}\delta_2\mathcal{F}^*(\mathcal{B},\mathbf{n})[\mathbf{u}] = &\int_{\mathcal{B}}\left(\mathbf{P}(\mathbf{n})\frac{\partial\sigma}{\partial\mathbf{n}}\cdot\mathbf{u} + \frac{\partial\sigma}{\partial\nabla\mathbf{n}}\cdot\nabla(\mathbf{P}(\mathbf{n})\mathbf{u})\right)dv \\ &+ \int_{\partial^*\mathcal{B}} \tau'\mathbf{P}(\mathbf{n})\boldsymbol{\nu}\cdot\mathbf{u}\,da.\end{aligned}$$

By the Divergence Theorem, the right-hand side of this equation also reads

$$\int_{\mathscr{B}} \mathbf{P}(\mathbf{n})\left(\frac{\partial\sigma}{\partial\mathbf{n}} - \operatorname{div}\left(\frac{\partial\sigma}{\partial\nabla\mathbf{n}}\right)\right)\cdot\mathbf{u}\,dv + \int_{\partial^*\mathscr{B}} \mathbf{P}(\mathbf{n})\left(\frac{\partial\sigma}{\partial\nabla\mathbf{n}} + \tau'\mathbf{I}\right)\mathbf{v}\cdot\mathbf{u}\,da.$$

Thus, $\delta_2\mathscr{F}^*(\mathscr{B},\mathbf{n})$ vanishes identically if, and only if, equations (5.35) and (5.37) are satisfied. This concludes the proof of the theorem.

There are a few consequences of the equilibrium equations just established that are worth mentioning.

If τ were constant, (5.37) would reduce to

$$\left(\frac{\partial\sigma}{\partial\nabla\mathbf{n}}\right)\mathbf{v} = \lambda\mathbf{n} \quad \text{on } \partial^*\mathscr{B}. \tag{5.45}$$

By $(5.33)_2$ this equation is satisfied only if λ vanishes on $\partial^*\mathscr{B}$, and so (5.45) is equivalent to equation (3.116) of Chapter 3, which applies to the case of no anchoring, when no surface energy is taken into account.

Furthermore, equations (5.34) and (5.35) are not independent of one another. To be precise, the latter implies the former. A straightforward computation shows that (5.34) can be rewritten as

$$(\nabla\mathbf{n})^T\left(\frac{\partial\sigma}{\partial\mathbf{n}} - \operatorname{div}\left(\frac{\partial\sigma}{\partial\nabla\mathbf{n}}\right)\right) = \mathbf{0},$$

which by (5.32) is a consequence of (5.35).

We shall not discard (5.34) in our development, for it can be given a clearer mechanical interpretation than (5.35). Let $\mathbf{T}$ be the tensor-valued mapping on $\mathscr{B}$ defined by

$$\mathbf{T} := -p\mathbf{I} - (\nabla\mathbf{n})^T\frac{\partial\sigma}{\partial\nabla\mathbf{n}} \quad \text{with} \quad p := -(\sigma + \pi_0). \tag{5.46}$$

Then (5.34) and (5.36) read, respectively,

$$\operatorname{div}\mathbf{T} = \mathbf{0} \quad \text{in } \mathscr{B},$$

$$\mathbf{T}\mathbf{v} = \operatorname{div}_s((\tau\mathbf{I} - \tau'\mathbf{v}\otimes\mathbf{n})\mathbf{P}(\mathbf{v})) \quad \text{on } \partial^*\mathscr{B},$$

and so $\mathbf{T}$ is to be regarded as the stress tensor field in the drop.

We call

$$\mathbf{T}_E := -(\nabla\mathbf{n})^T\frac{\partial\sigma}{\partial\nabla\mathbf{n}} \tag{5.47}$$

Ericksen's stress tensor. As is clear from $(5.46)_1$, it differs from the total stress tensor only by a term proportional to the identity, which by $(5.46)_2$ need not be constant in space: p is to be interpreted as a pressure, but in general only $-\pi_0$ is its hydrostatic component. Furthermore, $\mathbf{T}_E$ may fail to be symmetric, and so does $\mathbf{T}$.

5.2.1 Cusped edges

There are peculiar shapes among those for which (5.17) holds. These are the shapes where $\mathscr{C}$ is a **cusped edge**: they are such that

$$\mathbf{v}_1 = -\mathbf{v}_2 \quad \text{and} \quad \mathbf{v}_{1,2} = \mathbf{v}_{2,1} \quad \text{on } \mathscr{C}. \tag{5.48}$$

Figure 5.2 illustrates the cross-section of such a shape.

Equation (5.38) leads, however, to a contradiction when $\mathscr{C}$ is a cusped edge. Since τ is an even function, τ' is odd and so (5.48) implies that

$$\tau(\mathbf{n}\cdot\mathbf{v}_1) = \tau(\mathbf{n}\cdot\mathbf{v}_2) \quad \text{and} \quad \tau'(\mathbf{n}\cdot\mathbf{v}_1) = -\tau'(\mathbf{n}\cdot\mathbf{v}_2).$$

Thus, (5.38) reduces to

$$(\tau(\mathbf{n}\cdot\mathbf{v}_1)\mathbf{I} - \tau'(\mathbf{n}\cdot\mathbf{v}_1)\mathbf{v}_1 \otimes \mathbf{n})\mathbf{v}_{1,2} = \mathbf{0},$$

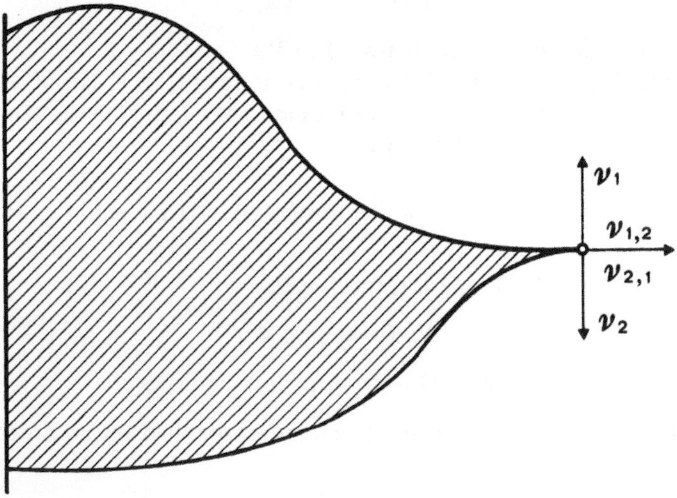

Figure 5.2

whence it follows that

$$\tau(\mathbf{n}\cdot\mathbf{v}_1)\mathbf{v}_{1,2}=\tau'(\mathbf{n}\cdot\mathbf{v}_1)(\mathbf{n}\cdot\mathbf{v}_{1,2})\mathbf{v}_1. \tag{5.49}$$

Since $\mathbf{v}_{1,2}$ and $\mathbf{v}_1$ are orthogonal vectors, equation (5.49) is satisfied if, and only if,

$$\tau(\mathbf{n}\cdot\mathbf{v}_1)=0 \quad \text{and} \quad \tau'(\mathbf{n}\cdot\mathbf{v}_1)(\mathbf{n}\cdot\mathbf{v}_{1,2})=0.$$

As τ is a positive function, we conclude that the equilibrium shapes of a drop cannot bear cusped edges. Furthermore, as (5.38) is an equilibrium equation in local form, we also conclude from (5.49) that the edges of an equilibrium shape cannot be cusped even at a single point.

We close this section by considering special solutions of equation (5.36).

5.2.2 Special solutions

We first learn how to compute in Cartesian co-ordinates the surface-divergence of the tensor $\mathbf{A}(\mathbf{v})$ defined in $(5.41)_2$ above. Let A_{ij} be the components of $\mathbf{A}(\mathbf{v})$ in a given frame. Its surface-divergence on $\partial^*\mathscr{B}$ is the vector whose components are

$$(\operatorname{div}_s \mathbf{A}(\mathbf{v}))_i=\sum_{j=1}^{3} A_{ij;j}:=\sum_{j,h=1}^{3} A_{ij,h}(\delta_{jh}-\nu_j\nu_h), \tag{5.50}$$

where δ_{jh} is the Kronecker symbol and ν_j, for $j=1,2,3$, are the Cartesian components of $\mathbf{v}$.

In (5.50) a new notation has been employed: a *semicolon* between two indices. We already agreed in Section 2.3 of Chapter 2 that a comma represents differentiation with respect to a space variable. Here a semicolon is to represent surface-differentiation. It is worth noting that the same formal rules that apply to commas apply to semicolons as well.

Thus, in particular, the components of the surface-gradient of $\mathbf{v}$ are

$$(\nabla_s\mathbf{v})_{ij}=\nu_{i;j};$$

they satisfy

$$\nu_{i;j}=\nu_{j;i} \quad \text{for all } i,j\in\{1,2,3\}, \tag{5.51}$$

and

$$\sum_{j=1}^{3}\nu_{i;j}\nu_j=0 \quad \text{for } i=1,2,3 \tag{5.52}$$

since $\mathbf{v}$ is a unit vector field. Moreover, the surface-divergence of $\mathbf{v}$ reads

$$\operatorname{div}_s \mathbf{v} = \sum_{i=1}^{3} \nu_{i;i}. \tag{5.53}$$

It follows from $(5.41)_2$ that

$$A_{ij} = \tau\delta_{ij} - \tau\nu_i\nu_j - \tau'\nu_i n_j + \tau'\left(\sum_{h=1}^{3} n_h\nu_h\right)\nu_i\nu_j, \tag{5.54}$$

where the argument of both τ and τ' is

$$\xi = \sum_{h=1}^{3} n_h\nu_h. \tag{5.55}$$

Inserting (5.54) into (5.50) and making use of (5.51), (5.52) and (5.55), we arrive at

$$\begin{aligned}\sum_{j=1}^{3} A_{ij;j} = {} & \tau' \sum_{h=1}^{3} (n_{h;i}\nu_h + n_h\nu_{h;i}) - \tau\left(\sum_{j=1}^{3} \nu_{j;j}\right)\nu_i \\ & - \tau'' \sum_{h,j=1}^{3} (n_{h;j}\nu_h + n_h\nu_{h;j})\nu_i n_j - \tau' \sum_{j=1}^{3} (\nu_{i;j}n_j) \\ & - \tau'\left(\sum_{j=1}^{3} n_{j;j}\right)\nu_i + \tau'\left(\sum_{h=1}^{3} n_h\nu_h\right)\left(\sum_{j=1}^{3} \nu_{j;j}\right)\nu_i,\end{aligned}$$

which in absolute notation reads

$$\begin{aligned}\operatorname{div}_s \mathbf{A}(\mathbf{v}) = {} & \tau'\{(\nabla_s\mathbf{n})^T\mathbf{v} + ((\mathbf{n}\cdot\mathbf{v})\operatorname{div}_s\mathbf{v} - \operatorname{div}_s \mathbf{n})\mathbf{v}\} \\ & - \tau(\operatorname{div}_s \mathbf{v})\mathbf{v} - \tau''(\mathbf{v}\cdot(\nabla_s\mathbf{n})\mathbf{n} + \mathbf{n}\cdot(\nabla_s\mathbf{v})\mathbf{n})\mathbf{v}.\end{aligned} \tag{5.56}$$

It would be an instructive exercise for the reader to prove the following identities:

$$(\mathbf{n}\cdot\mathbf{v})\operatorname{div}_s \mathbf{v} - \operatorname{div}_s \mathbf{n} = -\operatorname{div}_s(\mathbf{P}(\mathbf{v})\mathbf{n}), \tag{5.57}$$

$$\mathbf{v}\cdot(\nabla_s\mathbf{n})\mathbf{n} + \mathbf{n}\cdot(\nabla_s\mathbf{v})\mathbf{n} = \nabla_s(\mathbf{n}\cdot\mathbf{v})\cdot\mathbf{n}. \tag{5.58}$$

By use of both (5.57) and (5.58), (5.56) acquires a shorter form, which is recorded here, though it will not be extensively used in what follows:

$$\begin{aligned}\operatorname{div}_s \mathbf{A}(\mathbf{v}) = {} & \tau'\{(\nabla_s\mathbf{n})^T\mathbf{v} - (\operatorname{div}_s(\mathbf{P}(\mathbf{v})\mathbf{n}))\mathbf{v}\} \\ & - \tau(\operatorname{div}_s \mathbf{v})\mathbf{v} - \tau''(\nabla_s(\mathbf{n}\cdot\mathbf{v})\cdot\mathbf{n})\mathbf{v}.\end{aligned} \tag{5.59}$$

As we learned in Section 2.3 of Chapter 2, $\mathrm{div}_s \mathbf{v}$ may also be written as twice the mean curvature of $\partial^*\mathscr{B}$, that is

$$\mathrm{div}_s \mathbf{v} = \frac{1}{R_1} + \frac{1}{R_2}, \tag{5.60}$$

where R_1 and R_2 are the principal radii of curvature of the reduced boundary, which may also be negative.

Below we apply (5.56) to two special cases; we take $\mathbf{n}$ as constant in both.

Let $\mathscr{B}$ be an infinite cylinder whose cross-section is a plane regular region enclosed by a smooth curve $\mathscr{C}_1$ (see Fig. 5.3). Let $(o, \mathbf{e}_1, \mathbf{e}_2, \mathbf{e}_3)$ be a frame with $\mathbf{e}_3$ parallel to the axis of $\mathscr{B}$. We call $\mathbf{t}$ a unit vector tangent to $\mathscr{C}_1$ and $\mathbf{v}$ the unit outer normal to the lateral surface of $\mathscr{B}$.

Since $\mathbf{v}$ is constant along any straight line parallel to $\mathbf{e}_3$,

$$(\nabla_s \mathbf{v})\mathbf{e}_3 = \mathbf{0};$$

moreover, by (5.52), we also have

$$(\nabla_s \mathbf{v})\mathbf{v} = \mathbf{0}.$$

Thus, since $\nabla_s \mathbf{v}$ is a symmetric tensor, we conclude that

$$\nabla_s \mathbf{v} = (\mathrm{div}_s \mathbf{v})\mathbf{t} \otimes \mathbf{t},$$

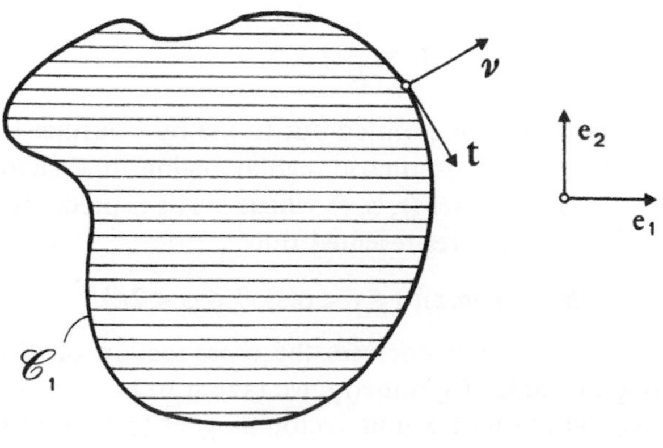

Figure 5.3

which may also be written as

$$\nabla_s \mathbf{v} = -\frac{1}{R}\mathbf{t}\otimes\mathbf{t}; \tag{5.61}$$

here R is the radius of curvature of $\mathscr{C}_1$, which is defined so that

$$\frac{d\mathbf{t}}{ds} = \frac{1}{R}\mathbf{v},$$

where s is the arc-length of $\mathscr{C}_1$.

We now set $\mathbf{n} \equiv \mathbf{e}_1$ and represent $\mathbf{t}$ and $\mathbf{v}$ through the formulae

$$\mathbf{t} = \sin\vartheta\mathbf{e}_1 - \cos\vartheta\mathbf{e}_2, \quad \mathbf{v} = \cos\vartheta\mathbf{e}_1 + \sin\vartheta\mathbf{e}_2 \tag{5.62}$$

so that ϑ is precisely the tilt angle of $\mathbf{n}$ on the lateral boundary of $\mathscr{B}$. Making use of (5.62) and (5.61) in (5.56), we arrive at

$$\operatorname{div}_s \mathbf{A}(\mathbf{v}) = -\frac{1}{R}(\tau'\cos\vartheta - \tau - \tau''\sin^2\vartheta)\mathbf{v}. \tag{5.63}$$

By the change of variables

$$\gamma(\vartheta) := \tau(\cos\vartheta), \tag{5.64}$$

we give (5.63) a neater form:

$$\operatorname{div}_s \mathbf{A}(\mathbf{v}) = \frac{1}{R}(\gamma + \gamma'')\mathbf{v},$$

where a prime now denotes differentiation with respect to ϑ. Since $\sigma \equiv 0$ when $\mathbf{n}$ is constant, we finally see that here (5.36) reads

$$\pi_0 = \frac{1}{R}(\gamma + \gamma''). \tag{5.65}$$

This elegant formula was pointed out to me by Fournier (1992).

Now let $\mathscr{B}$ be an axisymmetric regular region in $\mathscr{E}$ such that in the cylindrical co-ordinates (r, φ, z), whose z-axis is precisely the axis of symmetry for $\mathscr{B}$, it is represented thus:

$$\mathscr{B} = \{(r,\varphi,z) \mid 0 \leqslant r < \rho(z), 0 \leqslant \varphi < 2\pi\}.$$

Let $\mathscr{C}_2$ be the curve that encloses the cross-section of $\mathscr{B}$ in every plane through its axis of symmetry; we take it to be of class C^2 (*vid.* Fig. 5.4). We denote by $\mathbf{t}$ a unit vector tangent to $\mathscr{C}_2$ and by $\mathbf{v}$ the outer unit normal to $\mathscr{B}$; these vectors are represented in the form

$$\mathbf{t} = -\cos\vartheta\mathbf{e}_r + \sin\vartheta\mathbf{e}_z, \quad \mathbf{v} = \sin\vartheta\mathbf{e}_r + \cos\vartheta\mathbf{e}_z, \tag{5.66}$$

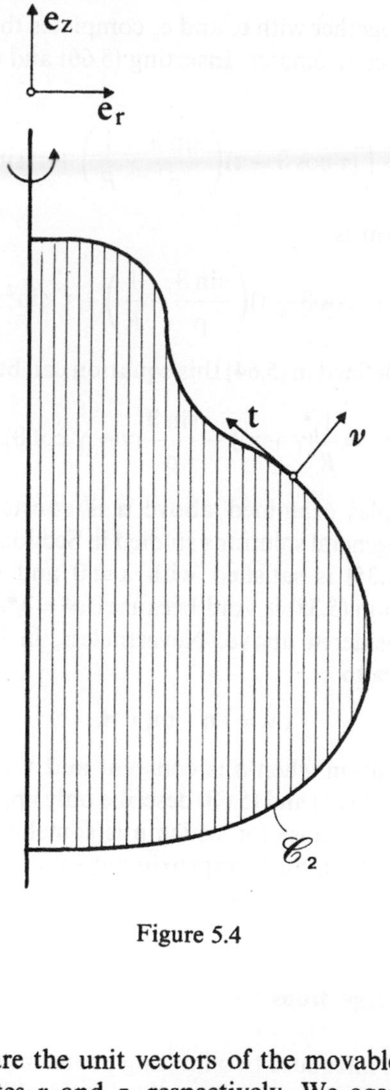

Figure 5.4

where $\mathbf{e}_r$ and $\mathbf{e}_z$ are the unit vectors of the movable frame, relative to the co-ordinates r and z, respectively. We again take $\mathbf{n}$ to be constant in $\mathscr{B}$, and precisely equal to $\mathbf{e}_z$, so that by $(5.66)_2$ ϑ is the tilt angle of $\mathbf{n}$ on $\partial\mathscr{B}$.

It is an instructive exercise to prove that now

$$\nabla_s \mathbf{v} = -\frac{1}{R}\mathbf{t}\otimes\mathbf{t} + \frac{\sin\vartheta}{\rho}\mathbf{e}_\varphi\otimes\mathbf{e}_\varphi, \tag{5.67}$$

where R now denotes the radius of curvature of $\mathscr{C}_2$ and $\mathbf{e}_\varphi$ is the unit vector that together with $\mathbf{e}_r$ and $\mathbf{e}_z$ completes the movable frame of the cylindrical co-ordinates. Inserting (5.66) and (5.67) into (5.56), we obtain

$$\operatorname{div}_s \mathbf{A}(\mathbf{v}) = \left((\tau' \cos\vartheta - \tau)\left(\frac{\sin\vartheta}{\rho} - \frac{1}{R} \right) + \tau'' \sin^2\vartheta \frac{1}{R} \right)\mathbf{v},$$

and so (5.36) becomes

$$\pi_0 = (\tau' \cos\vartheta - \tau)\left(\frac{\sin\vartheta}{\rho} - \frac{1}{R} \right) + \tau'' \sin^2\vartheta \frac{1}{R}.$$

In the variable γ defined in (5.64) this equation can be given the form

$$\pi_0 = \frac{1}{R}(\gamma + \gamma'') - \frac{\sin\vartheta}{\rho}(\gamma + \gamma' \cot\vartheta). \tag{5.68}$$

In both examples examined above $\mathbf{n}$ is constant in $\mathscr{B}$. This is indeed one of the general solutions studied in Section 3.6 of Chapter 3; thus, equation (5.35) is satisfied with $\lambda \equiv 0$ and so also is (5.34). Moreover, equation (5.38) is void because $\partial\mathscr{B} = \partial^*\mathscr{B}$ for the equilibrium shapes considered above. Nevertheless, in both cases equation (5.37) reduces to

$$\tau' \mathbf{v} = \lambda \dot{\mathbf{n}} \quad \text{on } \partial\mathscr{B},$$

which has no solution when $\mathbf{n}$ is constant and τ' does not vanish.

Thus, equations (5.65) and (5.68) describe only approximate equilibrium configurations: those for which $\mathbf{n}$ is somehow held fixed, and so suffers no variation. Such approximate solutions will come into play again in Section 5.4 below.

5.3 Small and large drops

In this section we introduce two limiting problems which are slightly easier to solve than the general equilibrium problem posed above.

Let ℓ_0 be the unity of length in $\mathscr{E}$. We set $\beta := (\ell_0\alpha)^3$ and define in $\mathscr{E}$ a change of variable $x \mapsto y(x)$ as follows

$$y(x) := \frac{x}{\alpha}. \tag{5.69}$$

Let a fit region $\mathscr{B}$ of volume β be given in $\mathscr{E}$. The region $\mathscr{B}_1 := y(\mathscr{B})$

is again a fit region and has unit volume. For every C^2-mapping $\mathbf{n}: \mathscr{B} \to \mathbb{S}^2$ we define the mapping $\mathbf{n}_1: \mathscr{B}_1 \to \mathbb{S}^2$ by

$$\mathbf{n}_1(y) := \mathbf{n}(\alpha y) \quad \text{for all } y \in \mathscr{B}_1; \tag{5.70}$$

$\mathbf{n}_1$ is again of class C^2 and

$$\nabla \mathbf{n}_1(y) = \alpha \nabla \mathbf{n}(\alpha y) \quad \text{for all } y \in \mathscr{B}_1. \tag{5.71}$$

If $\sigma: (\mathbf{n}, \nabla \mathbf{n}) \mapsto \sigma(\mathbf{n}, \nabla \mathbf{n})$ satisfies (5.13), it is easily seen that the ratio of the bulk energy to the surface energy of the drop which occupies $\mathscr{B}$ is given by

$$\frac{\mathscr{F}_B[\mathscr{B}, \mathbf{n}]}{\mathscr{F}_S[\mathscr{B}, \mathbf{n}]} = \alpha^{(1-\nu)} \frac{\mathscr{F}_B[\mathscr{B}_1, \mathbf{n}_1]}{\mathscr{F}_S[\mathscr{B}_1, \mathbf{n}_1]}. \tag{5.72}$$

In the limit as α tends to 0, that is when the drop is **small**, the bulk energy prevails over the surface energy, and so the total energy is minimum only if the bulk energy is. Since by (5.14) the bulk energy is always positive, except for $\mathbf{n} \equiv \mathbf{n}_0$ when it vanishes, minimizing $\mathscr{F}$ in (5.16) amounts to solving the following variational problem.

Problem 1. Find a fit region $\mathscr{B}$ such that the functional

$$\mathscr{F}_1[\mathscr{B}] := \int_{\partial^* \mathscr{B}} \tau(\mathbf{n}_0 \cdot \boldsymbol{\nu}) da \tag{5.73}$$

is minimized subject to

$$v(\mathscr{B}) = \beta. \tag{5.74}$$

This problem is subsumed under another, first considered by Wulff (1901), which will be examined in some detail in the next section.

Conversely, in the limit as $\alpha \to \infty$, that is when the drop is **large**, the surface energy prevails over the bulk energy of the drop, and so the total energy is minimum only if the surface energy is. Thus, assumption (5.6) on τ leads us to a variational problem which is the counterpart of Problem 1 above.

Problem 2. Find a fit region $\mathscr{B}$ and a differentiable mapping $\mathbf{n}: \mathscr{B} \to \mathbb{S}^2$ such that the functional

$$\mathscr{F}_2[\mathscr{B}, \mathbf{n}] := \int_{\mathscr{B}} \sigma(\mathbf{n}, \nabla \mathbf{n}) dv + \tau(\xi_0) a(\partial^* \mathscr{B}) \tag{5.75}$$

is minimized subject to (5.74) and to

$$(\mathbf{n}\cdot\mathbf{v})|_{\partial^*\mathscr{B}} = \xi_0, \tag{5.76}$$

where ξ_0 is as in (5.6).

This problem was first proposed by Ericksen (1988).

For completeness, another variational problem must be stated here. It is the main problem of this chapter, the one from which both Problems 1 and 2 above derive.

Problem 3. Find a fit region $\mathscr{B}$ and a differentiable mapping $\mathbf{n}: \mathscr{B} \to \mathbb{S}^2$ such that the functional

$$\mathscr{F}[\mathscr{B},\mathbf{n}] = \int_{\mathscr{B}} \sigma(\mathbf{n}, \nabla\mathbf{n})dv + \int_{\partial^*\mathscr{B}} \tau(\mathbf{n}\cdot\mathbf{v})da \tag{5.77}$$

is minimized subject to (5.74).

Lin and Poon (1992) have recently studied in some detail both Problems 2 and 3, reaching quite neat conclusions. They adopted the one-constant approximation to Frank's formula for the bulk free energy density (*cf.* (3.58) of Chapter 3), and so

$$\sigma = \kappa|\nabla\mathbf{n}|^2$$

in both (5.75) and (5.77). They also took τ as in (5.10) above, and so in both (5.75) and (5.76) either

$$\xi_0 = 1 \quad \text{or} \quad \xi_0 = 0.$$

Correspondingly, they considered two versions of Problem 2, namely Problems A and B below.

Problem A. Find a fit region $\mathscr{B}$ and a differentiable mapping $\mathbf{n}: \mathscr{B} \to \mathbb{S}^2$ such that the functional

$$\mathscr{F}_A[\mathscr{B},\mathbf{n}] := \kappa\int_{\mathscr{B}} |\nabla\mathbf{n}|^2 dv + \tau_A a(\partial^*\mathscr{B}) \tag{5.78}$$

is minimized subject to (5.74) and to

$$\mathbf{n}|_{\partial^*\mathscr{B}} = \mathbf{v}, \tag{5.79}$$

τ_A being a given positive constant.

Problem B. Find a fit region $\mathscr{B}$ and a differentiable mapping $\mathbf{n}: \mathscr{B} \to \mathbb{S}^2$ such that the functional

$$\mathscr{F}_B[\mathscr{B}, \mathbf{n}] := \kappa \int_{\mathscr{B}} |\nabla \mathbf{n}|^2 dv + \tau_B a(\partial^* \mathscr{B}) \tag{5.80}$$

is minimized subject to (5.74) and to

$$(\mathbf{n} \cdot \boldsymbol{\nu})|_{\partial^* \mathscr{B}} = 0, \tag{5.81}$$

τ_B being a given positive constant.

We also rephrase Problem 3 above under the same assumptions posited by Lin and Poon.

Problem C. Find a fit region $\mathscr{B}$ and a differentiable mapping $\mathbf{n}: \mathscr{B} \to \mathbb{S}^2$ such that the functional

$$\mathscr{F}_C[\mathscr{B}, \mathbf{n}] := \kappa \int_{\mathscr{B}} |\nabla \mathbf{n}|^2 dv + \tau_0 \int_{\partial^* \mathscr{B}} (1 + \omega(\mathbf{n} \cdot \boldsymbol{\nu})^2) da \tag{5.82}$$

is minimized subject to (5.74), τ_0 and ω being given, respectively, in $]0, \infty[$ and $]-1, \infty[$.

I briefly state below four theorems proved by Lin and Poon. For simplicity, I shall confine attention to a special class of admissible configurations for the drop:

$$\mathscr{A}_C := \{(\mathscr{B}, \mathbf{n}) | \mathscr{B} \text{ is a convex fit region and } \mathbf{n}: \mathscr{B} \to \mathbb{S}^2 \text{ is differentiable}\}.$$

Theorem 5.3. There are solutions in $\mathscr{A}_C$ to Problems A, B and C.

The following theorem concerns the regularity of the configurations that solve the above problems in $\mathscr{A}_C$.

Theorem 5.4. If the pair $(\mathscr{B}, \mathbf{n})$ of $\mathscr{A}_C$ solves either Problem A or Problem B, then $\mathbf{n}$ is smooth in $\mathscr{B}$, possibly away from isolated points. If $(\mathscr{B}, \mathbf{n})$ solves Problem C in $\mathscr{A}_C$, then $\partial\mathscr{B}$ differs from $\partial^*\mathscr{B}$ at most by a set of Hausdorff dimension 1 and $\mathbf{n}$ is everywhere smooth in $\mathscr{B}$.

Problem A has essentially a unique solution in $\mathscr{A}_C$.

Theorem 5.5. If the pair $(\mathscr{B}, \mathbf{n})$ solves Problem A in $\mathscr{A}_C$, then $\mathscr{B} = \mathscr{B}_o$, where $\mathscr{B}_o$ is a ball of volume β, and **n** is the radial field

$$\mathbf{n}_o(x) = \frac{x - o}{|x - o|},$$

where o is the centre of $\mathscr{B}_o$.

The configuration $(\mathscr{B}_o, \mathbf{n}_o)$ plays a peculiar role also within classes of configurations whose shapes are not convex.

Theorem 5.6. The configuration $(\mathscr{B}_o, \mathbf{n}_o)$ is always locally stable for $\mathscr{F}_A$, and also for $\mathscr{F}_C$ if $\omega \leqslant 0$.

The heuristic argument which has led us to state both Problems 1 and 2 above gives rise to the following question:

Do small *and* large *refer only to the geometric size of the drop*?

The answer is: *No*. The drop cannot be either so small as to cause the continuum theory to fail or so large as not to be a drop any more. The correct interpretation of the argument above relies on an accurate estimate of the ratio

$$\iota := \frac{\mathscr{F}_B[\mathscr{B}_1, \mathbf{n}_1]}{\mathscr{F}_S[\mathscr{B}_1, \mathbf{n}_1]} \tag{5.83}$$

occurring on the right-hand side of (5.72). If, for example, we employ formula (5.12) for σ and formula (5.10) for τ, then

$$\iota \approx \frac{k}{\tau_0} \ell_0, \tag{5.84}$$

where k is a characteristic value of Frank's constants. Using experimental data of Haller (1972) and Langevin (1972) for MBBA near 25°C, we get $\iota \approx 10^{-8}$ when $\ell_0 = 1$ cm. Thus, all possible drops of MBBA at room temperature are *large*, regardless of their actual size.

We shall see in the following section that there are also nematic liquid crystals, perhaps slightly exotic, which form *small* drops.

5.4 Wulff's construction

Wulff (1901) studied the growth of solid crystals. He assumed that when a lump of a crystalline solid grows, it tends to acquire a shape that minimizes the functional

$$\mathscr{F}_W[\mathscr{B}] := \int_{\partial^*\mathscr{B}} \chi(\mathbf{v})\,da, \tag{5.85}$$

without changing its volume. In (5.85) $\mathscr{B}$ is to be regarded as a fit region and $\partial^*\mathscr{B}$ is its reduced boundary; $\chi:\mathbb{S}^2 \to \mathbb{R}^+$ is a continuous function which satisfies the invariance requirements that reflect the symmetries of the crystal.

Wulff conjectured that there is essentially only one minimizer of $\mathscr{F}_W$ subject to (5.1): namely, the region of volume β similar to

$$\mathscr{W}(\chi) := \{p \in \mathscr{E} \mid (p-o)\cdot\mathbf{v} \leqslant \chi(\mathbf{v}) \text{ for all } \mathbf{v} \in \mathbb{S}^2\}, \tag{5.86}$$

where o is a given point of $\mathscr{E}$. All other minimizers of $\mathscr{F}_W$ would differ from the set in (5.86) by a translation.

What Wulff conjectured is indeed a theorem. The major contributions to its proof have been given by Dinghas (1944), Herring (1951) and Taylor (1974 and 1978). Various classes of shapes have been employed in these papers to prove Wulff's conjecture. An elegant and short proof can also be found in Chandrasekhar (1966). More recently, Fonseca (1991) has given a proof within a class of shapes more germane to our development (see also Fonseca and Müller (1990)).

The set $\mathscr{W}(\chi)$ in (5.86) results from **Wulff's construction**, whose steps are now described.

Select a point o in $\mathscr{E}$ and take a vector $p(\mathbf{v}) - o$ of length $\chi(\mathbf{v})$ along the direction of every unit vector $\mathbf{v} \in \mathbb{S}^2$. Through every point $p(\mathbf{v})$ trace the plane $\pi(\mathbf{v})$ orthogonal to $\mathbf{v}$ and call $\Sigma(\mathbf{v})$ the half-space bounded by $\pi(\mathbf{v})$ that includes o. $\mathscr{W}(\chi)$ is then the intersection of all $\Sigma(\mathbf{v})$ as $\mathbf{v}$ spans $\mathbb{S}^2$. Clearly, Wulff's construction gives a convex set.

Since the functional $\mathscr{F}_W$ in (5.85) reduces to the functional $\mathscr{F}_1$ defined in (5.73) above when $\chi(\mathbf{v}) = \tau(\mathbf{n}_0 \cdot \mathbf{v})$ for all $\mathbf{v} \in \mathbb{S}^2$ and a given $\mathbf{n}_0 \in \mathbb{S}^2$, Problem 1 of Section 5.3 can be solved by use of Wulff's construction. The corresponding sets $\mathscr{W}(\tau)$ thus obtained are axially symmetric about $\mathbf{n}_0$ and symmetric with respect to the plane orthogonal to $\mathbf{n}_0$ and passing through o.

Figure 5.5 illustrates one step of Wulff's construction on a plane containing $\mathbf{n}_0$: the line $l(\mathbf{v})$ is the trace of $\pi(\mathbf{v})$ and the half-plane $S(\mathbf{v})$ is the cross-section of $\Sigma(\mathbf{v})$. The length of $p(\mathbf{v}) - o$ is $\tau(\cos\alpha)$, where α is the angle between $\mathbf{v}$ and $\mathbf{n}_0$. The co-ordinates of the points where

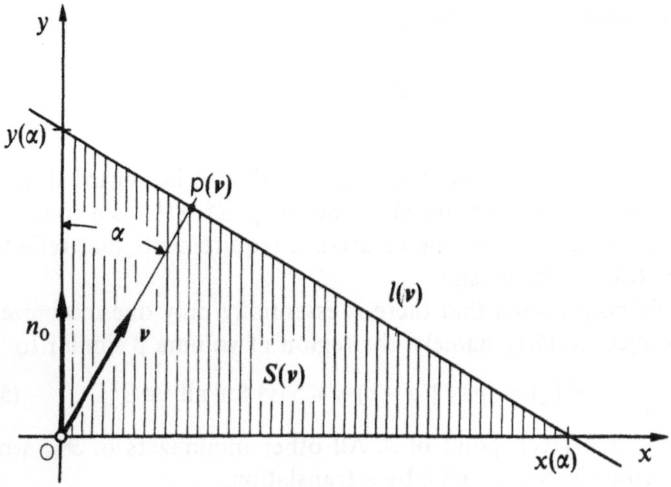

Figure 5.5

$l(\mathbf{v})$ intersects the lines through o that are, respectively, parallel to $\mathbf{n}_0$ and orthogonal to it are $(0, y(\alpha))$ and $(x(\alpha), 0)$ (see Fig. 5.5).

An easy computation shows that

$$y(\alpha) = \frac{\tau(\cos\alpha)}{\cos\alpha} \quad \text{for all } \alpha \in [0, \tfrac{1}{2}\pi[, \tag{5.87}$$

$$x(\alpha) = \frac{\tau(\cos\alpha)}{\sin\alpha} \quad \text{for all } \alpha \in]0, \tfrac{1}{2}\pi]. \tag{5.88}$$

Closely examining Wulff's construction, the reader will realize that if the function y defined by (5.87) is decreasing in a neighbourhood of $\alpha = 0$, then the outer normal field to $\mathscr{W}(\tau)$ is singular at the poles: $\mathscr{W}(\tau)$ looks like a sharply pointed American football and it is called a **tactoid**. If, on the other hand, y is not decreasing near $\alpha = 0$, then the poles of $\mathscr{W}(\tau)$ are regular points. Likewise, if the function x defined by (5.88) is increasing in a neighbourhood of $\alpha = \pi/2$ then the outer normal field of $\mathscr{W}(\tau)$ is singular along the equator: $\mathscr{W}(\tau)$ is **lens-shaped**. If x is not increasing near $\alpha = \pi/2$ then the equator of $\mathscr{W}(\tau)$ is not an edge of its boundary.

This is a criterion for the regularity of $\mathscr{W}(\tau)$, which leads to the following theorem:

Theorem 5.7. Let the function τ in (5.73) be of class C^2. If

$\tau(1)-\tau'(1)<0$, then $\mathscr{W}(\tau)$ is a tactoid. If $\tau'(0)<0$ or $\tau'(0)=0$ and $\tau(0)+\tau''(0)<0$, then $\mathscr{W}(\tau)$ is lens-shaped. If the reverse inequalities hold, then $\mathscr{W}(\tau)$ is regular at poles or along the equator, respectively.

When τ is the function given in (5.10) the theorem above implies that $\mathscr{W}(\tau)$ is lens-shaped for $-1<\omega<-\frac{1}{2}$, it is regular at the poles and along the equator for $-\frac{1}{2}<\omega<1$, and it is a tactoid for $\omega>1$. Figure 5.6 illustrates the cross-section of $\mathscr{W}(\tau)$ for four values of ω. The corresponding solutions to Problem 1 of Section 5.3 are obtained by revolving these shapes about $\mathbf{n}_0$ and rescaling them so that they will be of the desired volume.

Tactoids have indeed been observed by Zocher (1925) in colloidal suspensions and by Bernal and Fankuchen (1941) in aqueous solutions of plant virus, which behave like nematic liquid crystals (*cf.* Section 1.1 of Chapter 1). All the shapes observed by Zocher and by Bernal and Frankuchen are much like that illustrated in Fig. 5.6d.

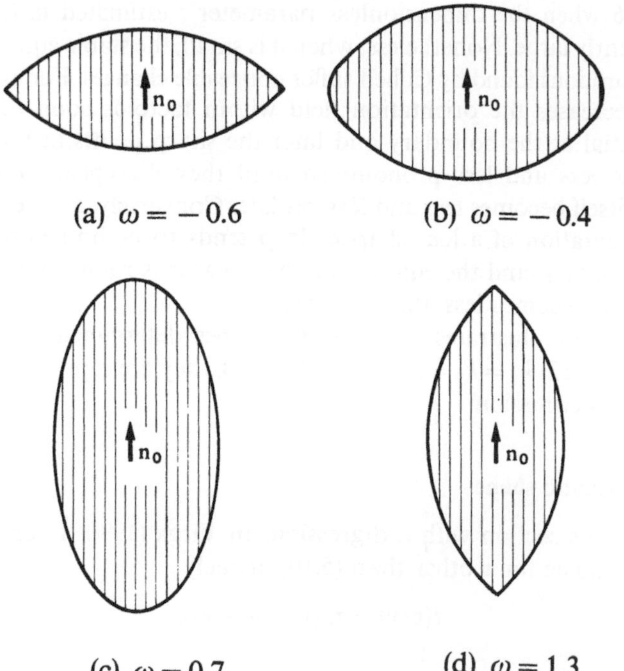

(a) $\omega=-0.6$ (b) $\omega=-0.4$

(c) $\omega=0.7$ (d) $\omega=1.3$

Figure 5.6

This confirms once more that formula (5.10) is fit to serve as anisotropic surface tension for nematic liquid crystals. The other shapes predicted here and illustrated in Fig. 5.6a–c seem not to have been observed so far. As ω is expected to depend on the temperature, these shapes will indeed be observed only if the nematic phase persists in the range of temperature where $-1 < \omega < 1$.

Bernal and Fankuchen described in detail the orientation of the optic axis inside the drops that they had observed. They found $\mathbf{n}$ to be parallel to the axis of the drops in their core, but tangential to their boundary. Thus, while Fig. 5.6d illustrates quite well the shape of a drop, it fails to represent the orientation field within it.

Cohen (1992) has recently studied in great generality Problem 3 of Section 5.3. His analysis is mainly numerical: he developed a scheme based on successive approximations which led him to determine the stable equilibrium configurations of $\mathscr{F}$ in (5.77) for appropriate choices of the functions σ and τ. Taking the former as in (5.12) and the latter as in (5.10), he retraced all the shapes illustrated in Fig. 5.6 when the dimensionless parameter ι estimated in (5.84) is sufficiently large. Nonetheless, when it is small, the stable equilibrium configurations found by Cohen suffer noticeable changes. For example, as ι decreases the orientation field within tactoids soon becomes tangential at the boundary and later the sharp points at the poles become less and less pronounced until they disappear, while the shape itself becomes less and less prolate. Conversely, as ι decreases, the orientation of a lens-shaped drop tends to be homeotropic at the boundary, and the edge along the equator is rounded off, while its shape becomes less and less oblate.

Figure 5.7 illustrates some stable shapes determined by Cohen when in (5.12) $k_1 = k_2 = k_3 = 1$ and $k_4 = 0$; they are reproduced here with his permission.

5.4.1 Exotic shapes

I close this section with a digression. In Virga (1990) I considered two formulae for τ other than (5.10), namely

$$\tau(\mathbf{n}\cdot\mathbf{v}) = \tau_0(1 + \omega|\mathbf{n}\cdot\mathbf{v}|), \tag{5.89}$$

and

$$\tau(\mathbf{n}\cdot\mathbf{v}) = \tau_0 \max\{1, \bar{\omega}|\mathbf{n}\cdot\mathbf{v}|\}, \tag{5.90}$$

where $\tau_0 > 0$, $\omega > -1$, and $\bar{\omega} > 1$.

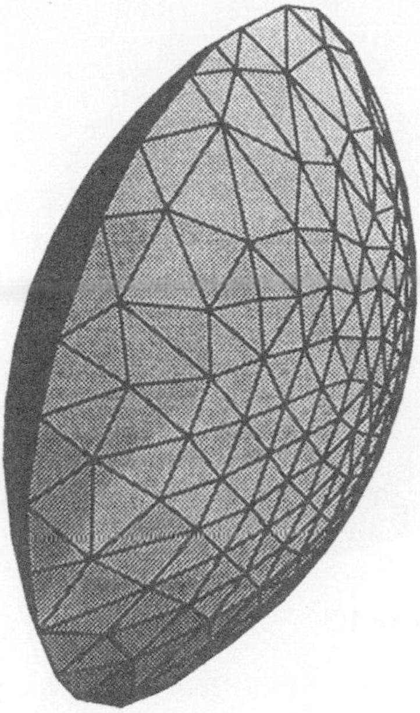

(a) $\omega = -0.6$ $\tau_0 = 1$

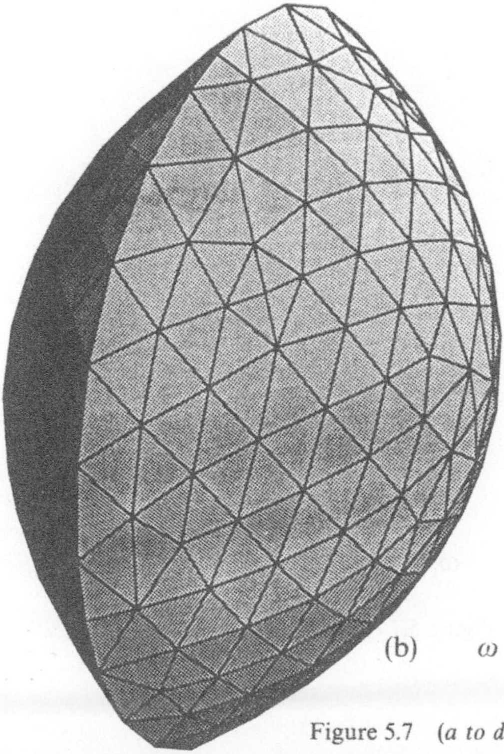

(b) $\omega = -0.6$ $\tau_0 = 10$

Figure 5.7 (*a to d*)

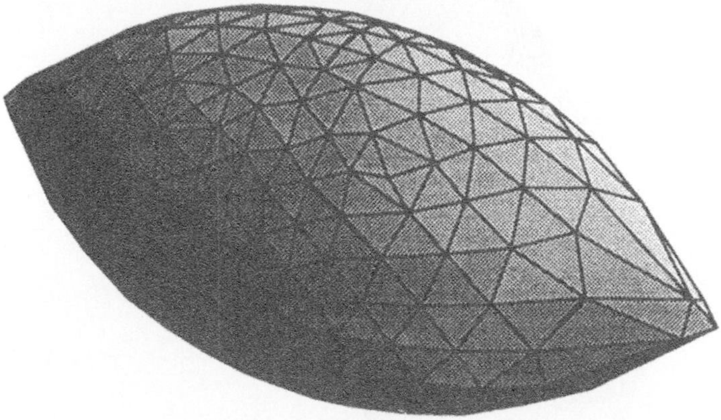

(c) $\omega = 1.3$ $\tau_0 = 1$

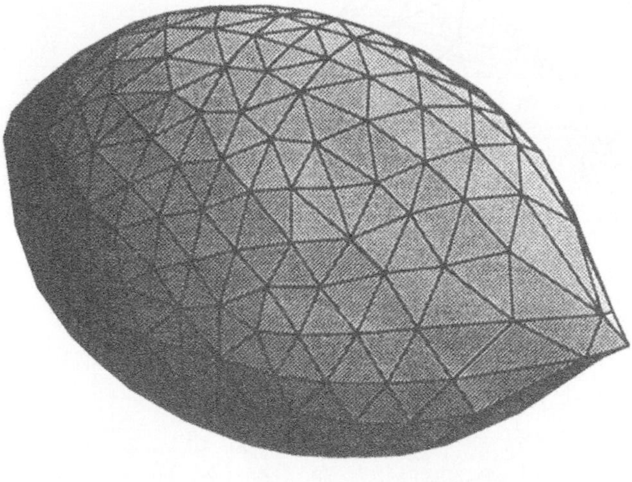

(d) $\omega = 1.3$ $\tau_0 = 10$

Figure 5.7 (*cont'd*)

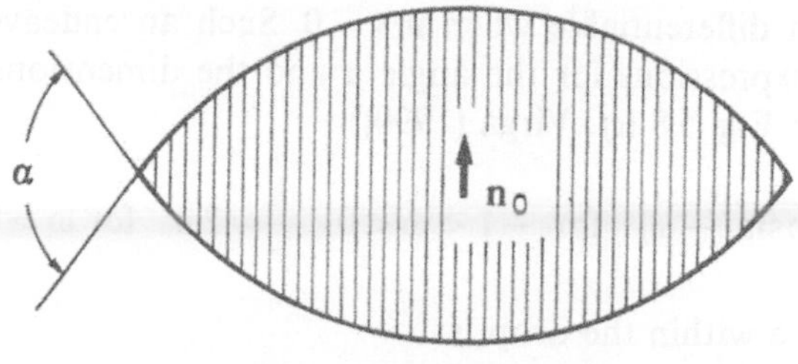

(a) $-1 < \omega < 0$

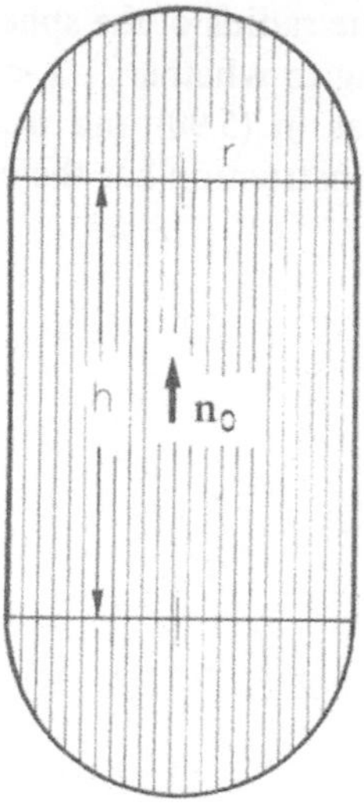

(b) $\omega > 0$

Figure 5.8

Figure 5.8 represents the cross-section of $\mathscr{W}(\tau)$ when τ is as in (5.89) with $\mathbf{n} \equiv \mathbf{n}_0$, and ω takes values in different intervals. For $-1 < \omega < 0$, $\mathscr{W}(\tau)$ is the region between two equal spherical caps, while for $\omega > 0$, $\mathscr{W}(\tau)$ is a cylinder completed by two half-balls matching the circles that bound it.

It is possible to show that the shapes illustrated in Fig. 5.8 satisfy the equilibrium equations for $\mathscr{F}$. Here, however, the analysis carried out in Section 5.2 is to be slightly recast, because the function in

(5.89) is not differentiable when $\mathbf{n}\cdot\mathbf{v}=0$. Such an endeavour gives analytical expressions for the angle α and the dimensions h and r indicated in Fig. 5.8 (*cf.* Virga (1990)):

$$\alpha = 2\arcsin(-\omega) \quad \text{for } -1<\omega<0, \quad \frac{h}{r}=2\omega \quad \text{for } \omega>0. \tag{5.91}$$

The pressure within the drop is

$$\pi_0 = -\frac{2\tau_0}{r}, \tag{5.92}$$

where r denotes either the radius of the spherical caps or the radius of the cylinder, according to whether $-1<\omega<0$ or $\omega>0$.

Likewise, when τ is as in (5.90) and $\mathbf{n}\equiv\mathbf{n}_0$, the domain $\mathscr{W}(\tau)$ resulting from Wulff's construction has the cross-section depicted in Fig. 5.9.

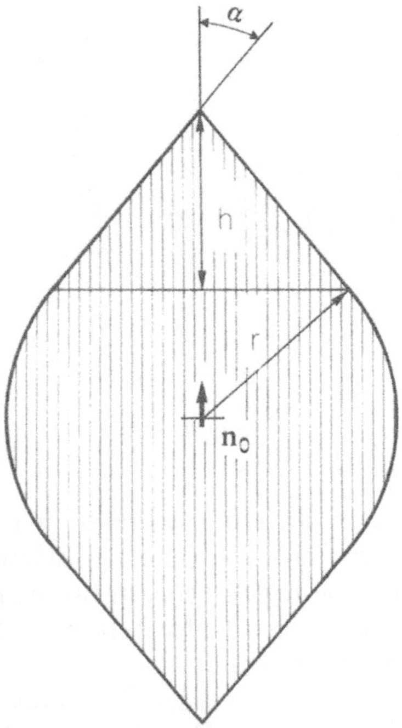

Figure 5.9

It is the portion of a ball bounded by two parallel planes and completed by two cones which fit it so that the boundary of $\mathscr{W}(\tau)$ possesses no edge. Equation (5.91) is now replaced by

$$\alpha = \arcsin\left(\frac{1}{\bar{\omega}}\right) \quad \text{and} \quad \frac{h}{r} = \bar{\omega} - \frac{1}{\bar{\omega}} \quad \text{for all } \bar{\omega} > 1; \tag{5.93}$$

equation (5.92) still gives the pressure in the drop, though r now has a different geometric meaning.

5.5 Floating drops

In this section, following a paper by Roccato and Virga (1990), I propose an approximate solution to a fascinating problem, namely to the problem of finding the stable equilibrium shape of a drop free to float on a dense substrate. The opportunity to study this problem came from a picture, apparently shot by P. Piéranski, which is reproduced in Kléman (1983) (*cf.* Plate III of p. xvi). It shows drops of nematic liquid crystal floating on gelatine; they appear as balls bearing a crater underneath the line of buoyancy. We see here under which circumstances this is indeed a stable shape for a drop partially immersed in another fluid.

More precisely, consider a drop of nematic liquid crystal floating on a dense isotropic liquid (say, gelatine) and in contact with a gas above it (say, air). Suppose that it is a large drop according to the terminology of Section 5.3 and give the names $\mathscr{S}^+$ and $\mathscr{S}^-$, respectively, to the emerged and the submerged boundary of the drop. The former is the interface between the liquid crystal and the gas, while the latter is the interface between the liquid crystal and the liquid.

We take $\mathscr{S}^+$ and $\mathscr{S}^-$ as disjoint surfaces, and we denote their closures by $\bar{\mathscr{S}}^+$ and $\bar{\mathscr{S}}^-$. Let $\mathscr{B}$ be the shape of the drop; thus,

$$\partial\mathscr{B} = \bar{\mathscr{S}}^+ \cup \bar{\mathscr{S}}^-. \tag{5.94}$$

Equation (5.94) is very different from (5.17), though it looks exactly the same: $\mathscr{S}_1$ and $\mathscr{S}_2$ may overlap, whereas $\mathscr{S}^+$ and $\mathscr{S}^-$ may not; furthermore, both $\mathscr{S}_1$ and $\mathscr{S}_2$ are regular orientable surfaces, whereas $\mathscr{S}^+$ and $\mathscr{S}^-$ need not be. Thus the boundary of the drop, besides having an edge along the border of both $\mathscr{S}^+$ and $\mathscr{S}^-$ if their normals do not match, may also have another edge either on $\mathscr{S}^+$ or on $\mathscr{S}^-$.

The volume of $\mathscr{B}$ is subject to (5.1) above:

$$v(\mathscr{B}) = \beta. \tag{5.95}$$

Another constraint comes from the law of buoyancy: the volume of the submerged part of the drop is also prescribed, and so

$$v(\mathscr{B}^-) = \beta^-, \tag{5.96}$$

where $\mathscr{B}^- \subset \mathscr{B}$ is the region under the line of buoyancy and β^- is a given positive constant such that

$$\beta^- < \beta.$$

We assume that the energy density at the interface between the drop and the gas attains its minimum when the tilt angle ϑ vanishes (*cf.* Section 5.1). There is indeed enough evidence that for many nematic liquid crystals the homeotropic boundary condition holds at the interface with air (*cf. e.g.* p. 57 of Kléman (1983)).

If β^- is much smaller than β, one may take the emerged boundary of the drop to be much larger than the submerged boundary, and one may also think of the former as playing just the same role as the whole boundary of the drop in the argument leading to the Problem 2 of Section 5.3. Theorems 5.5 and 5.6 of that section then justify the assumption that the emerged boundary of the drop is part of the boundary of a ball $\mathscr{B}_o$, and that the orientation field within the drop is given by

$$\mathbf{n}_o(x) := \frac{x - o}{|x - o|}, \tag{5.97}$$

where o is the centre of $\mathscr{B}_o$.

Thus, only the submerged boundary $\mathscr{S}^-$ remains undetermined. Under these assumptions, the problem of minimizing the total free energy of a floating drop reduces to:

Find the surface $\mathscr{S}^-$ that minimizes the functional

$$\mathscr{F}[\mathscr{S}^-, \vartheta] := \tau_0 \int_{\mathscr{S}^-} (1 + \omega \cos^2 \vartheta)\, da, \tag{5.98}$$

where ϑ is the tilt angle of $\mathbf{n}$ on $\mathscr{S}^-$.

In (5.98) the energy density on $\mathscr{S}^-$ is expressed as in (5.10) above; $\tau_0 > 0$ and $\omega > -1$.

We now describe the special class of surfaces in which we look for a solution of this problem. We take $\mathscr{S}^-$ to be symmetric about

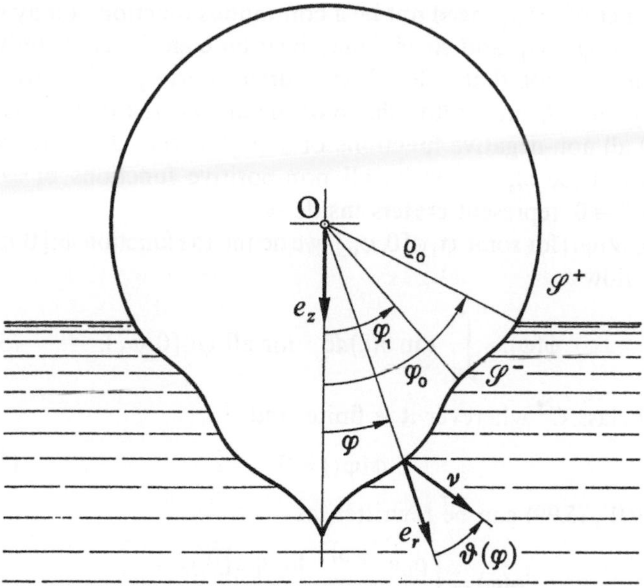

Figure 5.10

e_z, the vertical axis passing through o. We assume that the trace of $\mathscr{S}^-$ on each plane through e_z may be represented by a function $\varphi \mapsto \rho(\varphi)$, where (ρ, φ) are the polar co-ordinates illustrated in Fig. 5.10, which represents a cross-section of the drop.

Here both ρ_0, the radius of $\mathscr{B}_o$, and φ_0, the angle at which $\bar{\mathscr{S}}^-$ and $\bar{\mathscr{S}}^+$ touch each other, are taken as given. They play the role of parameters to be chosen so as to fulfil both (5.95) and (5.96).

The function $\varphi \mapsto \rho(\varphi)$ can easily be expressed in terms of the tilt angle ϑ:

$$\rho(\varphi) = \rho_0 \exp\left(-\int_{\varphi_0}^{\varphi} \tan \vartheta(\xi) d\xi \right). \tag{5.99}$$

As is shown in Fig. 5.10, we allow the surface $\mathscr{S}^-$ to adhere to $\partial\mathscr{B}_o$ on a zone underneath the line of buoyancy. Such a surface is formally described as follows. Let $\varphi_1 \in [0, \varphi_0]$ be given; we define the class of functions

$$\mathscr{A}(\varphi_1) := \{\vartheta : [0, \varphi_0] \to [-\tfrac{1}{2}\pi, \tfrac{1}{2}\pi] \mid \vartheta|_{[\varphi_1, \varphi_0]} = 0,\ \vartheta|_{[0, \varphi_1[} \in C^1\}. \tag{5.100}$$

A member of $\mathscr{A}(\varphi_1)$ need not be a continuous function: it may suffer a jump at $\varphi = \varphi_1$, and so $\mathscr{S}^-$ may bear an edge. There is only one function in $\mathscr{A}(0)$, that is $\vartheta \equiv 0$; the surface corresponding to it lies entirely on $\partial \mathscr{B}_o$, and so the shape of the drop is the ball $\mathscr{B}_o$ itself. If $\varphi_1 > 0$ all non-negative functions of $\mathscr{A}(\varphi_1)$, except $\vartheta \equiv 0$, represent **bumps** outside $\mathscr{B}_o$; similarly, all non-positive functions of $\mathscr{A}(\varphi_1)$, except $\vartheta \equiv 0$, represent **craters** inside $\mathscr{B}_o$.

If $\vartheta \in \mathscr{A}(\varphi_1)$ for some $\varphi_1 \in [0, \varphi_0]$, we define the function $\phi : [0, \varphi_1] \to \mathbb{R}$ as follows:

$$\phi(\varphi) := \int_{\varphi_1}^{\varphi} \tan \vartheta(\xi) d\xi \quad \text{for all } \varphi \in [0, \varphi_1]; \tag{5.101}$$

ϕ is of class C^2 wherever it is finite, and

$$\phi(\varphi_1) = 0. \tag{5.102}$$

By (5.101), (5.99) can be rewritten as

$$\rho(\varphi) = \begin{cases} \rho_0 e^{-\phi(\varphi)} & \text{for } \varphi \in [0, \varphi_1], \\ \rho_0 & \text{for } \varphi \in [\varphi_1, \varphi_0]. \end{cases} \tag{5.103}$$

By the use of (5.103) the functional $\mathscr{F}$ in (5.98) becomes a functional of ϕ which depends on φ_1:

$$F[\varphi_1, \phi] := 2\pi\rho_0^2\tau_0 \left(\int_0^{\varphi_1} e^{-2\phi(\varphi)} \frac{1 + \omega + \phi'(\varphi)^2}{\sqrt{1 + \phi'(\varphi)^2}} \sin\varphi \, d\varphi \right.$$
$$\left. + (1 + \omega)(\cos\varphi_1 - \cos\varphi_0) \right). \tag{5.104}$$

The admissible shapes of the drop are described here by the pairs (φ_1, ϕ). An equilibrium shape corresponds to a pair where F is stationary. A stable shape corresponds to a pair that minimizes F.

Thus, the variational problem that we face here can be stated as follows:

Find $\varphi_1 \in [0, \varphi_0]$ *and* ϕ *of class* C^2 *in* $]0, \varphi_1[$ *that minimize* F *subject to* (5.102).

We now derive the equilibrium equations for F.

5.5.1 Equilibrium equations

If the pair (φ_1, ϕ) represents an equilibrium shape of the drop, both the variation of F relative to φ_1 and that relative to ϕ must vanish.

The former is nothing but the derivative of F with respect to φ_1 while the latter is the usual variation of a functional in integral form, depending on functions free at one end-point (*cf. e.g.* p. 208 of Courant and Hilbert (1989), Vol. 1). Requiring these variations to vanish, we obtain the following equations:

$$\frac{1+\omega+\phi'(\varphi_1)^2}{\sqrt{1+\phi'(\varphi_1)^2}}-(1+\omega)=0, \tag{5.105}$$

$$\lim_{\varphi\to 0^+}\left(e^{-2\phi(\varphi)}\frac{1-\omega+\phi'(\varphi)^2}{\sqrt{(1+\phi'(\varphi)^2)^3}}\phi'(\varphi)\sin\varphi\right)=0, \tag{5.106}$$

$$\frac{d}{d\varphi}\left(e^{-2\phi(\varphi)}\frac{1-\omega+\phi'(\varphi)^2}{\sqrt{(1+\phi'(\varphi)^2)^3}}\phi'(\varphi)\sin\varphi\right)$$
$$+2e^{-2\phi(\varphi)}\frac{1+\omega+\phi'(\varphi)^2}{\sqrt{1+\phi'(\varphi)^2}}\sin\varphi=0 \quad \text{in }]0,\varphi_1[. \tag{5.107}$$

These are the equilibrium equations of F. For ϕ of class C^2 both ϕ and ϕ' are bounded in $[0,\varphi_1]$, and so equation (5.106) is always satisfied.

By use of the change of variables

$$\tau := \tan\varphi, \quad t := \tan\vartheta = \phi', \tag{5.108}$$

equations (5.105) and (5.107) become

$$\frac{1+\omega+t^2(\tau_1)}{\sqrt{1+t^2(\tau_1)}}-(1+\omega)=0, \tag{5.109}$$

$$t' = -\frac{1+t^2}{1+\tau^2}\,\frac{2\tau(1+\omega+(2\omega+1)t^2)+t(1-\omega+t^2)}{\tau(1-\omega+(2\omega+1)t^2)} \quad \text{in }]0,\tau_1[, \tag{5.110}$$

where $\tau_1 := \tan\varphi_1$ and the function $t:[0,\tau_1]\to\mathbb{R}$ is of class C^1. Henceforth in this section a prime will denote differentiation with respect to τ.

Equation (5.109) assigns the value of t at τ_1. Let h be the function defined by

$$h(t) := \frac{1+\omega+t^2}{\sqrt{1+t^2}}-(1+\omega). \tag{5.111}$$

The graph of h is plotted in Fig. 5.11. When $-1<\omega\leqslant 1$, h vanishes

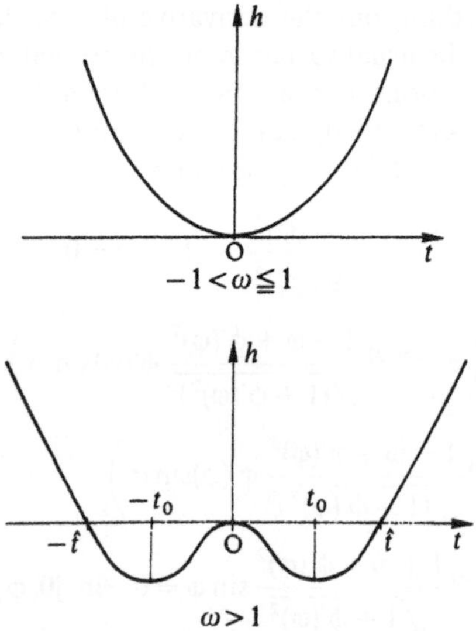

Figure 5.11

only at 0. When $\omega > 1$, h vanishes at $-\hat{t}, 0, \hat{t}$ and it attains its minimum at $-t_0$ and t_0, where

$$\hat{t} := \sqrt{\omega^2 - 1}, \quad t_0 := \sqrt{\omega - 1}. \tag{5.112}$$

Thus, $t(\tau_1)$ is uniquely determined when $-1 < \omega \leqslant 1$ while it may take three different values when $\omega > 1$. The role played by t_0 will soon become clear.

Now, for every $\omega > -1$, we seek $\tau_1 \geqslant 0$ and a solution of (5.110) such that $t(\tau_1)$ satisfies (5.109). Of course, $\tau_1 = 0$ and $t(\tau_1) = 0$ always represent a solution in the class that we employ. The shape of the drop corresponding to such a solution is the whole ball $\mathscr{B}_o$.

If $-1 < \omega < -\frac{1}{2}$ the half-plane $\{(\tau, t) \in \mathbb{R}^2 \mid \tau \geqslant 0\}$ splits into several regions where the sign of t' is the same, as is shown in Fig. 5.12.

Along the lines defined by the equations $\tau = 0, t = \tilde{t}$ and $t = -\tilde{t}$, where

$$\tilde{t} := \sqrt{\frac{\omega - 1}{2\omega + 1}}, \tag{5.113}$$

t' becomes unbounded. It is easily seen that the graph of any solution of (5.110) can touch the axis $\tau = 0$ only at the origin. On close

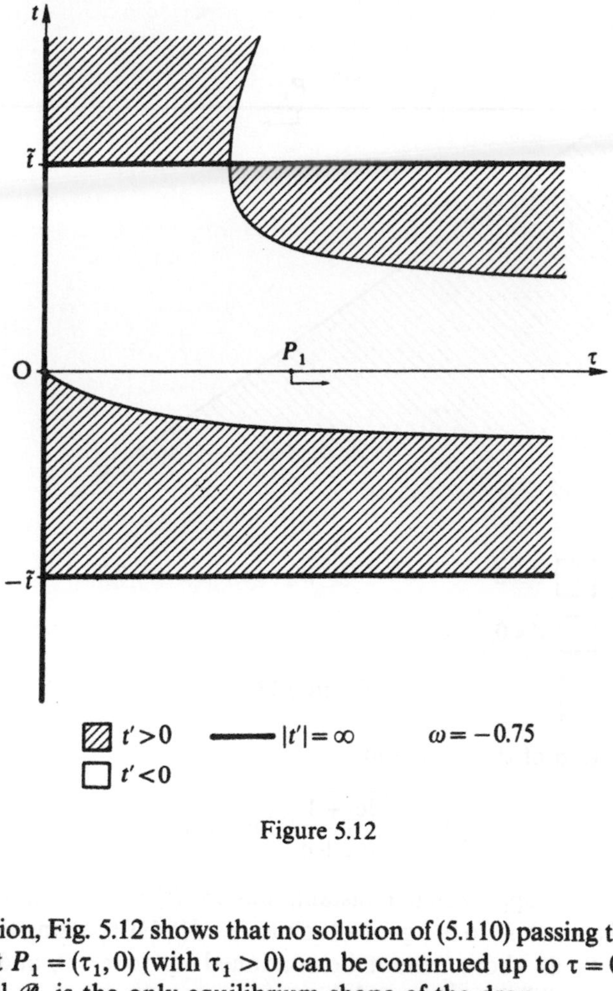

Figure 5.12

inspection, Fig. 5.12 shows that no solution of (5.110) passing through a point $P_1 = (\tau_1, 0)$ (with $\tau_1 > 0$) can be continued up to $\tau = 0$. Thus the ball $\mathscr{B}_o$ is the only equilibrium shape of the drop.

If $-\frac{1}{2} \leqslant \omega \leqslant 1$, there is only one region where t' is positive and only one where it is negative, as is shown in Fig. 5.13. The analysis above still applies, but a new case arises: the solution of (5.110) passing through a point P_1 in Fig. 5.13 might approach $t = \infty$ as $\tau \to 0$. This case, however, does not actually occur. The approximate form of (5.110) when $t \to \infty$ is the following:

$$t' = -\frac{1}{2\omega + 1} \frac{t^3}{\tau}. \tag{5.114}$$

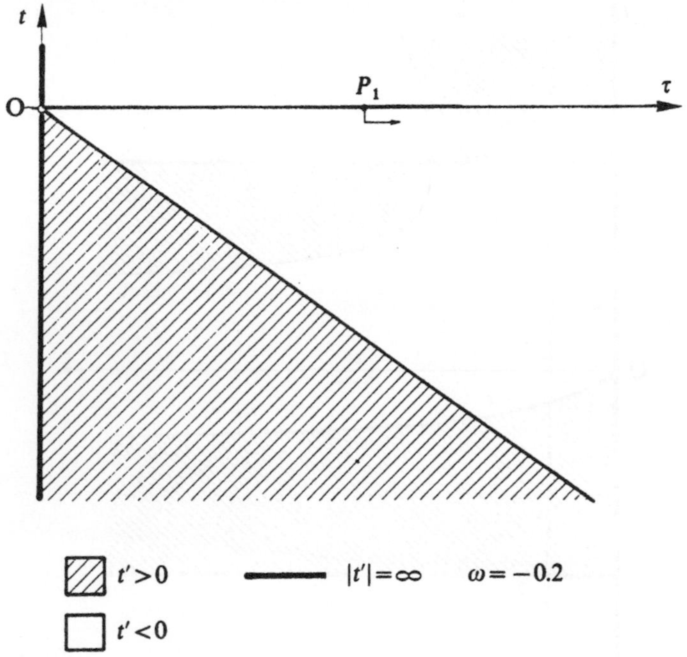

Figure 5.13

The solution of this equation is

$$t(\tau)=\sqrt{\frac{2\omega+1}{\ln\tau+c}}\quad \text{for } \tau>e^{-c}, \tag{5.115}$$

where c is an appropriate constant, and so $t(\tau)$ is not even defined when τ is too close to 0. Again all solutions of (5.110) passing through a point P_1 as above cannot be continued up to $\tau=0$, and $\mathscr{B}_o$ is the only equilibrium shape of the drop.

If $\omega>1$, the analysis of (5.110) is more involved. Figure 5.14 illustrates the regions with the same sign of t'. Here the solutions of (5.110) can touch the axis $\tau=0$ either at $t=0$ or at $t=\pm t_0$. The points P_1, P_1', and P_1'' in Fig. 5.14 correspond to the three possible values of $t(\tau_1)$. Following again the same lines of thought as outlined above, we show that no solution of (5.110) passing through P_1' or P_2'' can be continued up to $\tau=0$.

It remains to consider those through P_1: only one solution can be continued up to $\tau=0$, that is the one that reaches the point $(0,-t_0)$.

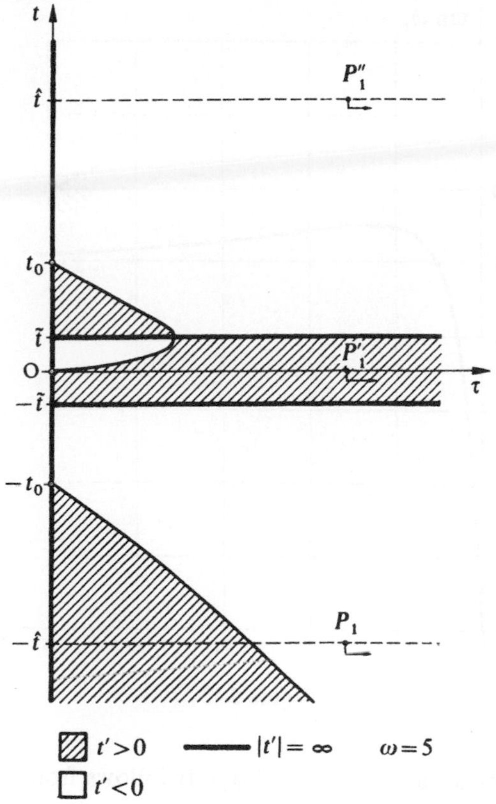

Figure 5.14

Thus we integrate (5.110) numerically starting from this point and we denote by $\hat{P}_1$ the point where this solution intersects the line $t = -\hat{t}$. If $\hat{\tau}_1$ denotes the abscissa of $\hat{P}_1$, $\hat{\varphi}_1 := \arctan \hat{\tau}_1$ gives the only value of φ_1 where F is stationary. The corresponding solution of (5.110) represents an equilibrium shape $\hat{\mathscr{B}}$ that bears a crater underneath the line of buoyancy.

Figure 5.15 shows the graph of $\tan \hat{\varphi}_1$ as a function of ω. It is proved in Section 5 of Roccato and Virga (1990) that the line $\tan \hat{\varphi}_1 = \frac{1}{4}$ is indeed an asymptote to the graph in Fig. 5.15.

We conclude that when $-1 < \omega \leqslant 1$ the only equilibrium shape of the drop is the ball $\mathscr{B}_o$, with φ_0 and ρ_0 chosen so as to satisfy (5.95) and (5.96). When $\omega > 1$ there are two equilibrium shapes of the drop, namely $\mathscr{B}_o$ and $\hat{\mathscr{B}}$. Through $(5.108)_2$, the latter can be

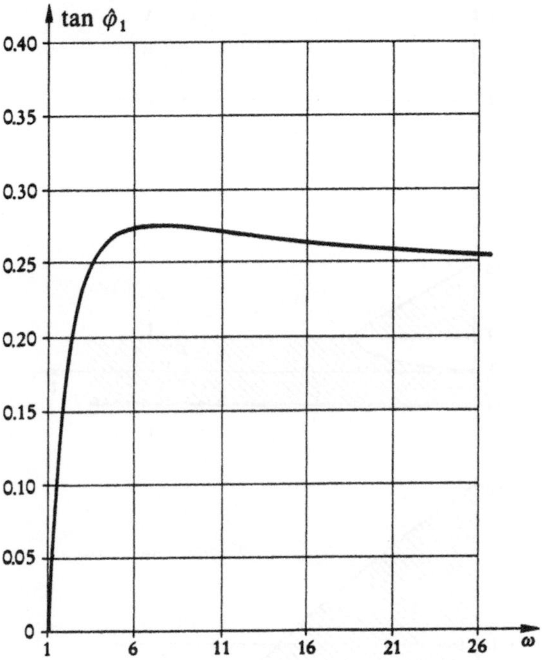

Figure 5.15

represented by a function $\hat{\vartheta}$ of $\mathscr{A}(\phi_1)$. It follows from $(5.112)_2$ that $\hat{\vartheta}$ satisfies

$$\hat{\vartheta}(0) = -\arctan\sqrt{\omega - 1}. \tag{5.116}$$

We explore below the stability of both $\mathscr{B}_o$ and $\hat{\mathscr{B}}$.

5.5.2 Stable shape

We see from Fig. 5.15 that $\hat{\varphi}_1$ approaches 0 as ω approaches 1. Thus $\mathscr{B}_o$ and $\hat{\mathscr{B}}$ represent, in a sense, the branches of a bifurcation occurring at $\omega = 1$. To figure out which of the two shapes is stable, we estimate the energy of both. We denote by F_0 and $\hat{F}$ the energy of $\mathscr{B}_o$ and $\hat{\mathscr{B}}$, respectively. Since $\hat{\varphi}_1$ is sufficiently small for all values of ω (*cf.* Fig. 5.15), to estimate $\hat{F}$ we may replace $\hat{\vartheta}$ by the function

$$\vartheta_0(\varphi) := \begin{cases} \vartheta(0) & \text{for } 0 \leqslant \varphi \leqslant \hat{\varphi}_1, \\ 0 & \text{otherwise.} \end{cases} \tag{5.117}$$

This amounts to replacing the crater of $\hat{\mathscr{B}}$ by another whose profile is an arc of a logarithmic spiral. An easy computation shows that

$$F_0 - \hat{F} = 2\pi\rho_0^2\tau_0\sqrt{\omega}(g(\omega) - 1)\hat{\phi}_1^2 + o(\hat{\phi}_1^2), \tag{5.118}$$

where the function g is defined by

$$g(\omega) := \frac{\omega + 1}{2\sqrt{\omega}}. \tag{5.119}$$

The graph of g is plotted in Fig. 5.16.

Since $g(\omega) > 1$ for all $\omega > 1$, we conclude that

$$\hat{F} < F_0.$$

Thus $\hat{\mathscr{B}}$ is the stable shape of the drop when $\omega > 1$.

At first glance the crater of $\hat{\mathscr{B}}$ looks like a cone. Figures 5.17 and 5.18 illustrate this shape for $\omega = 2$ and $\omega = 10$. The angle γ' is the semi-angle of the cone and the segment QP is a generator.

On closer inspection, the crater of $\hat{\mathscr{B}}$ differs slightly from a cone, as is shown in Fig. 5.19, where the length scale of the vertical axis considerably exceeds that of the horizontal axis.

It follows from (5.116) that the semi-angle of the cone tangent in Q to the crater is given by

$$\gamma' + \gamma'' = \arcsin\frac{1}{\sqrt{\omega}},$$

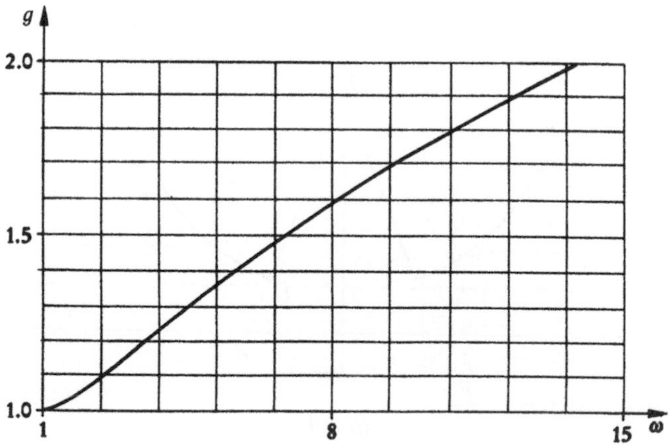

Figure 5.16

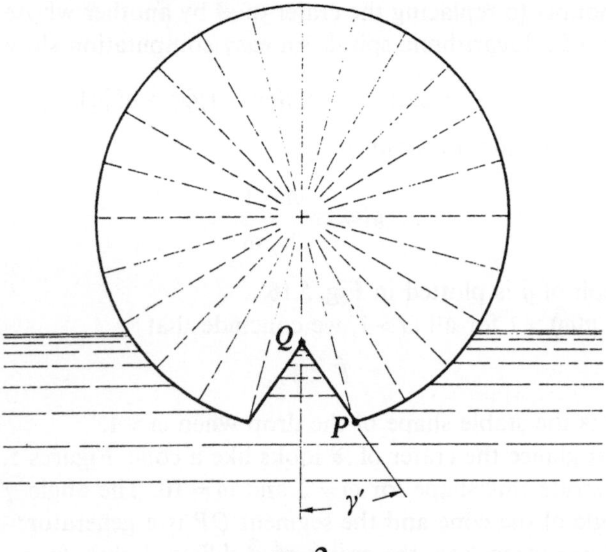

Figure 5.17

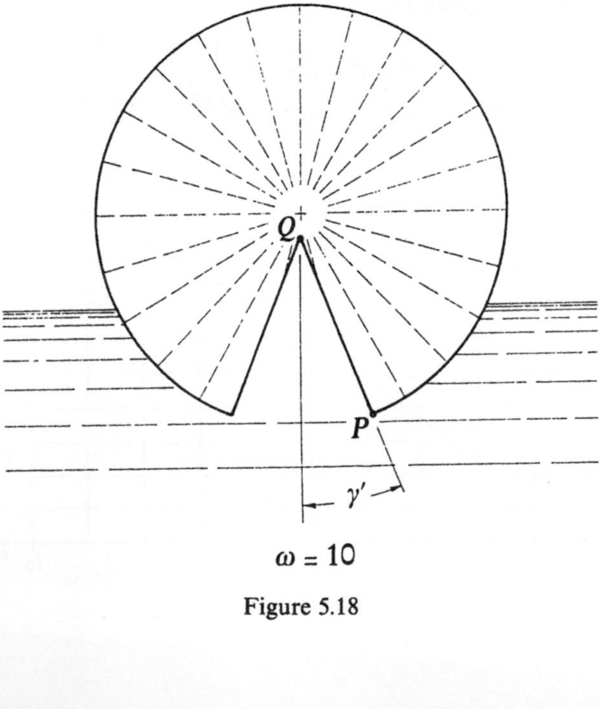

Figure 5.18

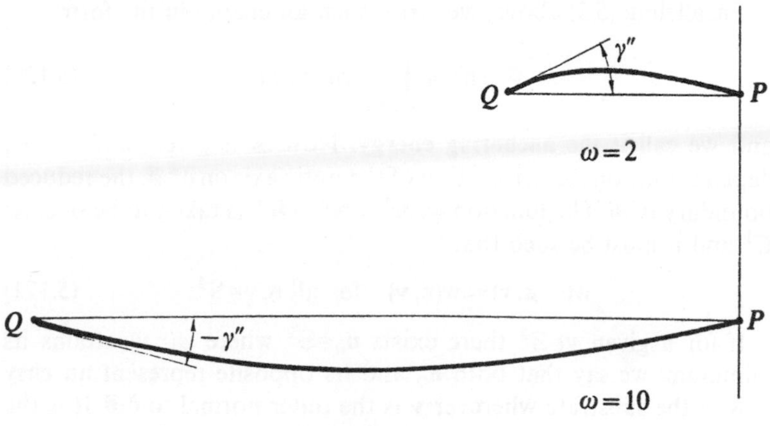

Figure 5.19

where the angle γ'' measures how the tangent at Q to the actual profile of the crater departs from the segment QP.

5.6 Weak anchoring

The content of this section is not strictly germane to the subject of this chapter in that it does not concern drops; nevertheless, the role played here by surface energies makes this the most appropriate place in our development to treat a fairly common way of imposing boundary conditions on liquid crystals.

So far we have considered either strong anchoring conditions on the boundary of a prescribed domain or free-boundary problems where the region occupied by the liquid crystal is *a priori* unknown and its boundary bears an energy on its own. The former models are fit to describe vessels confining liquid crystals, while the latter apply to drops in contact with isotropic environments.

Let $\mathscr{B}$ be a given regular region occupied by a liquid crystal. In practice, a certain orientation of the optic axis can be induced on $\partial\mathscr{B}$ by an appropriate treatment of the bounding surface, which often consists in evaporating SiO at an angle against the side in contact with the liquid crystal. In doing so, however, a tiny interface is created between the material and its container, and, as explained in Section 5.1, a surface energy also should come into play wherever $\partial\mathscr{B}$ has been so treated.

Paralleling (5.2) above, we write such an energy in the form

$$\mathscr{F}_A[\mathbf{n}] := \int_{\partial^*\mathscr{B}} w(\mathbf{n}, \mathbf{v})\, da, \tag{5.120}$$

and we call it the **anchoring energy**. Now $\mathscr{B}$ is given, and so $\mathscr{F}_A$ depends only on the orientation of the optic axis on $\partial^*\mathscr{B}$, the reduced boundary of $\mathscr{B}$. The function $w\colon \mathbb{S}^2 \times \mathbb{S}^2 \to \mathbb{R}^+$ is taken to be of class C^1 and it must be such that

$$w(-\mathbf{n}, \mathbf{v}) = w(\mathbf{n}, \mathbf{v}) \quad \text{for all } \mathbf{n}, \mathbf{v} \in \mathbb{S}^2. \tag{5.121}$$

If for a given $\mathbf{v} \in \mathbb{S}^2$ there exists $\mathbf{n}_0 \in \mathbb{S}^2$ where $w(\cdot, \mathbf{v})$ attains its minimum, we say that both $\mathbf{n}_0$ and its opposite represent an **easy axis** of the substrate wherever $\mathbf{v}$ is the outer normal to $\partial\mathscr{B}$. It is the orientation that the boundary would induce by itself if the surface energy $\mathscr{F}_A$ did not compete against the bulk energy $\mathscr{F}_B$. Both the strong and the conical anchoring conditions considered in Section 3.5 of Chapter 3 are to be regarded as limiting cases, akin to that of a large drop, where the functional $\mathscr{F}_A$ prevails over $\mathscr{F}_B$. Thus, the strong anchoring results from a surface energy which possesses only one easy axis, while the conical anchoring results from a surface energy which possesses a whole cone of easy axes around the normal.

When $\mathscr{F}_A$ does not prevail over $\mathscr{F}_B$ so as not to be merely translated into a boundary condition for the optic axis, we must face the problem of minimizing the total free energy, namely

$$\mathscr{F} := \mathscr{F}_B + \mathscr{F}_A,$$

subject to *no* boundary condition for $\mathbf{n}$.

This is to say that $\mathbf{n}$ is subject to a **weak anchoring**. I treat it here, close to the free-boundary problems for drops in contact with isotropic fluids, to emphasize that the variational problems where the orientation of the optic axis is somehow prescribed on the boundary, problems which have received much attention in the literature of the past decades, stand on the same footing as the limiting cases for large and small drops introduced above.

Furthermore, the function w is generally required to reflect the material symmetries of the substrate in contact with the liquid crystal. Since at each point of $\partial^*\mathscr{B}$ these properties pertain, in a sense, to the tangent plane, we express these requirements through the equation

$$w(\mathbf{Q}\mathbf{n}, \mathbf{v}) = w(\mathbf{n}, \mathbf{v}) \quad \text{for all } \mathbf{Q} \in G(\mathbf{v}), \tag{5.122}$$

where $G(\mathbf{v})$ is an appropriate subgroup of $SO(\mathbf{v})$.

Here, to simplify matters, we take w as φ in Section 5.1, that is in the form

$$w(\mathbf{n},\mathbf{v}) = \tau(\mathbf{n}\cdot\mathbf{v}), \tag{5.123}$$

where τ is an even function of class C^1 of $[-1,1]$ into $\mathbb{R}^+$. In particular, we shall employ below formula (5.10) for τ:

$$\tau(\mathbf{n}\cdot\mathbf{v}) = \tau_0(1+\omega(\mathbf{n}\cdot\mathbf{v})^2), \tag{5.124}$$

with $\tau_0 > 0$ and $\omega > -1$. Thus, for $-1 < \omega < 0$ the easy axis is parallel to the normal, while for $\omega > 1$ all orientations tangent to the boundary are easy axes. It is to be noted that the function in (5.124) satisfies (5.122) for all $\mathbf{Q}\in SO(\mathbf{v})$.

5.6.1 Weak Freedericks's transition

Making use of an anchoring energy, we now treat a variational problem already studied above as a classical boundary-value problem with strong anchoring, that is Freedericks's transition. We shall see that the weak anchoring affects the critical magnetic field at which the transition takes place, also when it is not too weak.

Below we follow in part a paper by Rapini and Papoular (1969) extending their analysis slightly.

Let $\mathscr{B}$ be the cell considered in Section 4.2 of Chapter 4, which in the frame $(o,\mathbf{e}_1,\mathbf{e}_2,\mathbf{e}_3)$ is represented as follows:

$$\mathscr{B} = \{p = o + x\mathbf{e}_1 + y\mathbf{e}_2 + z\mathbf{e}_3 \,|\, x, y\in]0,\ell[,\ z\in]-d,d[\}.$$

The plates $\mathscr{S}^-$ and $\mathscr{S}^+$ that bound it are represented by the equations $z = -d$ and $z = +d$, and so the outer normal to $\mathscr{B}$ is given there by

$$\mathbf{v}|_{\mathscr{S}^-} = -\mathbf{e}_3 \quad \text{and} \quad \mathbf{v}|_{\mathscr{S}^+} = \mathbf{e}_3. \tag{5.125}$$

Here we prescribe $\mathbf{n}$ neither on $\mathscr{S}^-$ nor on $\mathscr{S}^+$, but we rather write the anchoring energy as

$$\mathscr{F}_A[\mathbf{n}] = \int_{\mathscr{S}^-\cup\mathscr{S}^+} \tau_0(1+\omega(\mathbf{n}\cdot\mathbf{v})^2)da. \tag{5.126}$$

In Section 4.2 of Chapter 4 $\mathbf{n}$ was taken to be $\mathbf{e}_1$ on both $\mathscr{S}^-$ and $\mathscr{S}^+$. To mimic this strong boundary condition we take ω to be positive so that $\mathscr{F}_A$ attains its minimum when $\mathbf{n}$ is tangent to both $\mathscr{S}^-$ and $\mathscr{S}^+$. Clearly, the larger ω is, the closer is this weak anchoring to being a strong one.

We further take $\mathscr{F}_B$ as in (4.7) of Chapter 4 with $\sigma_e = 0$ and

$$\sigma_m = -\tfrac{1}{2}(\chi_\perp \mathbf{h}\cdot\mathbf{h} + \chi_a(\mathbf{h}\cdot\mathbf{n})^2), \tag{5.127}$$

where the diamagnetic anisotropy χ_a is positive and the magnetic field $\mathbf{h}$ is parallel to $\mathbf{e}_3$:

$$\mathbf{h} = H\mathbf{e}_3. \tag{5.128}$$

The total free energy is then

$$\mathscr{F}[\mathbf{n}] := \mathscr{F}_B[\mathbf{n}] + \mathscr{F}_A[\mathbf{n}]. \tag{5.129}$$

We seek the minimizers of $\mathscr{F}$ within the class of orientation fields represented by

$$\mathbf{n} = \cos\vartheta(z)\mathbf{e}_1 + \sin\vartheta(z)\mathbf{e}_3, \tag{5.130}$$

where ϑ is a function of $[-d, d]$ into $[-\pi/2, \pi/2]$.

Making use of (5.130), (5.127) and (5.126) in (5.129), with the aid of computations already performed in Section 4.2 of Chapter 4, we arrive at the following formula for $\mathscr{F}$:

$$\mathscr{F}[\mathbf{n}] = \ell^2(F[\vartheta] - \chi_\perp H^2 d + 2\tau_0),$$

where F is defined by

$$F[\vartheta] := \int_{-d}^{+d} \{k(\vartheta)\vartheta'^2 + h(\vartheta)\}\,dz + \tau_0\omega\{\sin^2\vartheta(-d) + \sin^2\vartheta(d)\} \tag{5.131}$$

and k and h are as in (4.41) and (4.42) of Chapter 4, namely

$$k(\vartheta) := k_1\cos^2\vartheta + k_3\sin^2\vartheta, \tag{5.132}$$

$$h(\vartheta) := -\tfrac{1}{2}\chi_a H^2\sin^2\vartheta. \tag{5.133}$$

When $\mathbf{n}$ is represented as in (5.130), $\mathscr{F}$ attains its minimum whenever F does. Guided by the conclusions reached in Chapter 4, where we first studied Freedericks's transition, we seek the minimizers of F within the class

$$\mathscr{A}_F := \{\vartheta : [-d, d] \to [-\tfrac{1}{2}\pi, \tfrac{1}{2}\pi] \mid \vartheta \in C^2 \text{ and } \vartheta(-z) = \vartheta(z) \text{ for all } z \in [-d, d]\}.$$

Nevertheless, to obtain the equilibrium equations for F in this class, the analysis unfolded in Section 4.2 of Chapter 4 is to be slightly amended because here the functions of $\mathscr{A}_F$ are no longer prescribed at the end-points of the interval $[-d, d]$. In computing the first variation of F we must now take account also of the variations of ϑ at these points.

As explained in any elementary book on calculus of variations (*cf. e.g.* p. 209 of Courant and Hilbert (1989), Vol. 1) the equilibrium equations for F in $\mathscr{A}_F$ are

$$2k(\vartheta)\vartheta'' + \frac{dk}{d\vartheta}\vartheta'^2 + \chi_a H^2 \sin\vartheta\cos\vartheta = 0 \quad \text{in }]-d, d[\qquad (5.134)$$

and

$$\{k(\vartheta(z))\vartheta'(z) + \tau_0\omega\sin\vartheta(z)\cos\vartheta(z)\} = 0 \quad \text{for } z\in\{-d, d\}. \quad (5.135)$$

Equation (5.134) is precisely (4.45) of Chapter 4, but here we do not look for solutions that take prescribed values for $z\in\{-d, d\}$, rather for those that satisfy (5.135).

Two solutions of both (5.134) and (5.135) are easily found; they are

$$\vartheta \equiv 0 \quad \text{and} \quad \vartheta \equiv \frac{\pi}{2},$$

and will play here the role of trivial solutions. We explore below how other solutions may depart from these.

5.6.2 Nonlinear analysis

We first set up the general nonlinear analysis of equations (5.134) and (5.135), though here we cannot easily pursue it as far as we did in Chapter 4 above for equation (4.45).

It is easily seen that the solutions of (5.134) and (5.135) come in pairs, in each of which one function is the opposite of the other. Hereafter we restrict attention to *positive* solutions of (5.134) and (5.135) and leave it to the reader to show that the minimizers of F in $\mathscr{A}_F$ do not change sign in $[-d, d]$ and are decreasing on $[0, d]$, like those of the functional studied in Section 4.2 of Chapter 4 (*cf.* Lemma 4.5).

Since the members of $\mathscr{A}_F$ are even functions, ϑ' vanishes at $z = 0$ and ϑ there attains its maximum. It is thus sufficient to study both equations (5.134) and (5.135) in the interval $[0, d]$. Letting

$$\vartheta(0) = \vartheta_m, \qquad (5.136)$$

and reasoning as in Section 4.2 of Chapter 4, from (5.134) we arrive at the following equation:

$$k(\vartheta)\vartheta'^2 = \tfrac{1}{2}\chi_a H^2(\sin^2\vartheta_m - \sin^2\vartheta). \tag{5.137}$$

Furthermore, by setting

$$\vartheta_0 := \vartheta(d) \quad \text{and} \quad \rho := \frac{1}{\tau_0\omega}, \tag{5.138}$$

equation (5.135) becomes

$$k(\vartheta_0)\vartheta'(d) = -\frac{1}{\rho}\sin\vartheta_0\cos\vartheta_0. \tag{5.139}$$

Making use of (5.137) in (5.139), we may remove $\vartheta'(d)$ from the latter equation and convert it into a relation between ϑ_0 and ϑ_m:

$$\sin^2\vartheta_m = \sin^2\vartheta_0\left(1 + \alpha\frac{\cos^2\vartheta_0}{k(\vartheta_0)}\right), \tag{5.140}$$

where

$$\alpha := \frac{2}{\chi_a\rho^2H^2}. \tag{5.141}$$

Since both ϑ_m and ϑ_0 belong to $[0, \pi/2]$, it follows from (5.140) that $\vartheta_m > \vartheta_0$, as it must be.

Theorem 5.8. For any given $\vartheta_m \in [0, \pi/2]$ there is precisely one ϑ_0 that satisfies (5.140); it can be expressed as a function of class C^1 of ϑ_m and α,

$$\vartheta_0 = \varphi(\vartheta_m, \alpha), \tag{5.142}$$

such that

$$\frac{\partial\varphi}{\partial\vartheta_m} > 0 \quad \text{and} \quad \frac{\partial\varphi}{\partial\alpha} < 0, \tag{5.143}$$

$$\lim_{\alpha\to 0^+}\varphi(\vartheta_m, \alpha) = \vartheta_m \quad \text{and} \quad \lim_{\alpha\to\infty}\varphi(\vartheta_m, \alpha) = 0. \tag{5.144}$$

Furthermore, if $\alpha < k_3$ then $\varphi(\cdot, \alpha)$ maps $[0, \pi/2]$ into itself, whereas if $\alpha > k_3$ its image does not cover $[0, \pi/2]$.

Proof. Let f be the function defined by

$$f(\vartheta) := \sin^2\vartheta\left(1 + \alpha\frac{\cos^2\vartheta}{k(\vartheta)}\right),$$

so that (5.140) reads

$$\sin^2 \vartheta_m = f(\vartheta_0).$$

A straightforward computation yields

$$\frac{df}{d\vartheta} = \frac{2\tan\vartheta(k_3(k_3 - \alpha)\tan^4\vartheta + 2k_1k_3\tan^2\vartheta + k_1(k_1 + \alpha))}{(1 + \tan^2\vartheta)(k_1 + k_3\tan^2\vartheta)^2}.$$

Thus, the function f is increasing on $[0, \pi/2]$ if $\alpha < k_3$, whereas if $\alpha > k_3$ it possesses an isolated maximum and there is an interval $[0, \hat{\vartheta}] \subset [0, \pi/2]$ that is monotonically mapped onto $[0, 1]$. Figure 5.20 illustrates two typical graphs of f, one for $\alpha < k_3$ and the other for $\alpha > k_3$.

These properties of f and the use of Dini's Theorem on implicit functions will now allow the reader to prove both (5.143) and (5.144).

Integrating (5.137) by separation of variables, we see that any equilibrium configuration for F in $\mathscr{A}_F$ other than a constant is

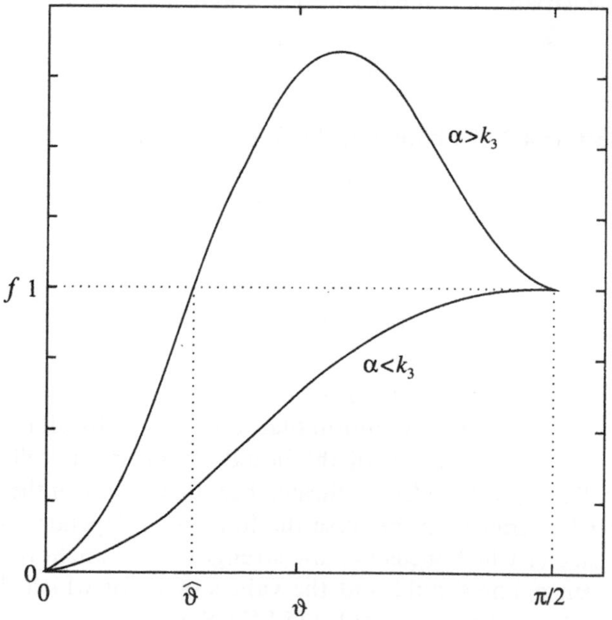

Figure 5.20

represented in $[0, d]$ by a function ϑ^* whose inverse is

$$z = d - \frac{1}{H}\sqrt{\frac{2}{\chi_a}}\int_{\varphi(\vartheta_m,\alpha)}^{\vartheta^*}\sqrt{\frac{k(\vartheta)}{\sin^2\vartheta_m - \sin^2\vartheta}}\,d\vartheta, \tag{5.145}$$

where φ is the function defined in (5.142). Thus, for given H and ρ, equation (5.145) delivers a non-trivial solution of the equilibrium equations for F, provided that ϑ^* satisfies (5.135), that is

$$Hd\sqrt{\frac{\chi_a}{2}} = \int_{\varphi(\vartheta_m,\alpha)}^{\vartheta_m}\sqrt{\frac{k(\vartheta)}{\sin^2\vartheta_m - \sin^2\vartheta}}\,d\vartheta. \tag{5.146}$$

Inserting (5.141) into (5.146), we arrive at the equation

$$Hd\sqrt{\frac{\chi_a}{2}} = g(\vartheta_m, \rho, H), \tag{5.147}$$

where

$$g(\vartheta_m, \rho, H) := \int_{\lambda_0(\vartheta_m,\rho,H)}^{\pi/2} G(\lambda, \vartheta_m)\,d\lambda.$$

Here the variable λ is related to ϑ through

$$\sin\lambda = \frac{\sin\vartheta}{\sin\vartheta_m},$$

as in Section 4.2 of Chapter 4; the function G is defined by

$$G(\lambda, \vartheta_m) := \sqrt{\frac{k\arcsin(\sin\vartheta_m\sin\lambda)}{1 - (\sin\vartheta_m\sin\lambda)^2}},$$

and

$$\sin\lambda_0(\vartheta_m, \rho, H) := \frac{\sin\varphi(\vartheta_m, \alpha)}{\sin\vartheta_m},$$

where φ is as in (5.142) and α as in (5.141).

Equation (5.147) is a condition that must be met by all non-trivial equilibrium configurations of the liquid crystal in the cell, if there are any. The scene is more complicated here than it was for the classical Freedericks's transition, because the function g depends also on ρ, the parameter which measures the strength of the anchoring. Thus, for a given ρ, one should find the values of H for which there are values of ϑ_m in $[0, \pi/2]$ that solve (5.147). Such an analysis, however, is rather involved in the fully nonlinear case. If the analysis carried out in Chapter 4 for the classical Freedericks's transition may suggest anything relevant also to this case, we would expect that there are two critical values of H, say H'_c and H''_c, such that for $H \in [H'_c, H''_c]$ there is precisely one ϑ_m in $[0, \pi/2]$ that solves (5.147); it should correspond to a non-trivial solution with less energy than the trivial ones.

This amounts to saying that a weak Freedericks's transition is similar to the classical one, unless the magnetic field becomes so strong as to make the minimum energy shift towards the configuration where **n** and **h** are everywhere parallel in the cell.

We shall not pursue our analysis as far as to prove this conjecture in detail; a skilful numerical study of (5.147) might help the reader to decide whether or not it is true. Here we take it as an assumption and we compute both H'_c and H''_c resorting to the linear analysis first described in Section 4.3 of Chapter 4 for the classical Freedericks's transition.

5.6.3 Critical fields

Let us consider an orientation in the form (5.130) with ϑ very small:

$$\vartheta = \varepsilon u, \tag{5.148}$$

where ε is a given parameter and u is a function of class C^2 on $[-d, d]$, which is taken to be even to make ϑ belong to $\mathscr{A}_F$.

Making use of (5.148) in (5.131), we arrive at

$$F[\vartheta] = \varepsilon^2 F'_2[u] + o(\varepsilon^2),$$

where

$$F'_2[u] := \int_{-d}^{+d} \{k_1 u'^2 - \tfrac{1}{2}\chi_a H^2 u^2\}\, dz + \frac{1}{\rho}\{u^2(-d) + u^2(d)\}. \tag{5.149}$$

The equilibrium equations for F'_2 are

$$k_1 u'' = -\tfrac{1}{2}\chi_a H^2 u \quad \text{in }]-d, d[\tag{5.150}$$

and

$$k_1 u'(z) + \frac{1}{\rho} u(z) = 0 \quad \text{for } z \in \{-d, d\}. \tag{5.151}$$

The even solutions of the former equation read as

$$u(z) = u_0 \cos \lambda z, \tag{5.152}$$

where now

$$\lambda := \sqrt{\frac{\chi_a H^2}{2k_1}} \tag{5.153}$$

and u_0 is an arbitrary constant. Inserting (5.152) into (5.151) we obtain

$$\cot \lambda d = k_1 \rho \lambda. \tag{5.154}$$

Thus, the values of λ that solve (5.154) define through (5.153) the values of H for which there is an equilibrium configuration of F close to $\vartheta \equiv 0$. The least of these latter, which we call H_c', marks the onset of the transition; it is a function of ρ, whose inverse $\rho_1(H_c')$ can easily be derived from (5.154). Let H_1 be the critical value of H at which the classical Freedericks's transition takes place; by (4.43) of Chapter 4 it is given by

$$H_1 := \frac{\pi}{2d}\sqrt{\frac{2k_1}{\chi_a}}. \tag{5.155}$$

It follows from both (5.153) and (5.155) that

$$\rho_1(H_c') = \frac{2d}{\pi k_1}\frac{H_1}{H_c'}\cot\left(\frac{\pi}{2}\frac{H_c'}{H_1}\right). \tag{5.156}$$

An easy computation, which is left to the reader, shows that when $H = H_c'$, F_2' vanishes on the functions in (5.152), and so the energy does not suffer jumps when this new branch of solutions bifurcates from $\vartheta \equiv 0$.

Similarly, we perturb the other trivial solution, that is $\vartheta \equiv \pi/2$. We write ϑ in the form

$$\vartheta = \frac{\pi}{2} + \varepsilon v, \tag{5.157}$$

where ε is the same parameter as above and v is an even function of class C^2 on $[-d, d]$. Inserting (5.157) into (5.131) we get

$$F[\vartheta] = 2\left(\frac{1}{\rho} - \frac{1}{2} d\chi_a H^2\right) + \varepsilon^2 F_2''[v] + o(\varepsilon^2),$$

where

$$F_2''[\vartheta] := \int_{-d}^{+d} \{k_3 v'^2 + \tfrac{1}{2}\chi_a H^2 v^2\}\,dz - \frac{1}{\rho}\{v^2(-d) + v^2(d)\}.$$

The equilibrium equations for F_2'' are the following:

$$k_3 v'' = \tfrac{1}{2}\chi_a H^2 v \quad \text{in }]-d, d[\tag{5.158}$$

and

$$k_3 v'(z) - \frac{1}{\rho} v(z) = 0 \quad \text{for } z \in \{-d, d\}. \tag{5.159}$$

The even solutions of the former equation are

$$v(z) = v_0 \cosh \mu z,$$

where v_0 is an arbitrary constant and

$$\mu := \sqrt{\frac{\chi_a H^2}{2k_3}}.$$

These functions obey (5.159) whenever

$$\coth \mu d = k_3 \rho \mu.$$

The reader will easily see that there is only one value of μ that solves this equation; the corresponding value of H, which we call H_c'', will be seen below to mark the point above which the trivial solution $\vartheta \equiv \pi/2$ becomes stable.

H_c'' is a function of ρ, whose inverse is given by

$$\rho_3(H_c'') := \frac{2d}{\pi k_3} \frac{H_3}{H_c''} \coth\left(\frac{\pi}{2} \frac{H_c''}{H_3}\right), \tag{5.160}$$

where

$$H_3 := \frac{\pi}{2d} \sqrt{\frac{2k_3}{\chi_a}}.$$

Figure 5.21 illustrates the graphs of both H_c' and H_c'' as functions of ρ.

Three features of these functions are to be noted. First, $H_c'' > H_c'$ for all values of ρ, because $\rho_3(H) > \rho_1(H)$ for all $H > 0$. Second, H_c' converges to H_1 when ρ tends to 0, while H_c'' diverges, and so there is only one critical value of H when the anchoring energy is infinitely strong, in complete agreement with our analysis of the classical Freedericks's transition. Third, H_c' and H_c'' have the same asymptotic behaviour as ρ tends to ∞: they both decrease to 0, and so the two critical values of H collapse to zero when the anchoring energy is infinitely weak. More precisely, a glance at both (5.156) and (5.160)

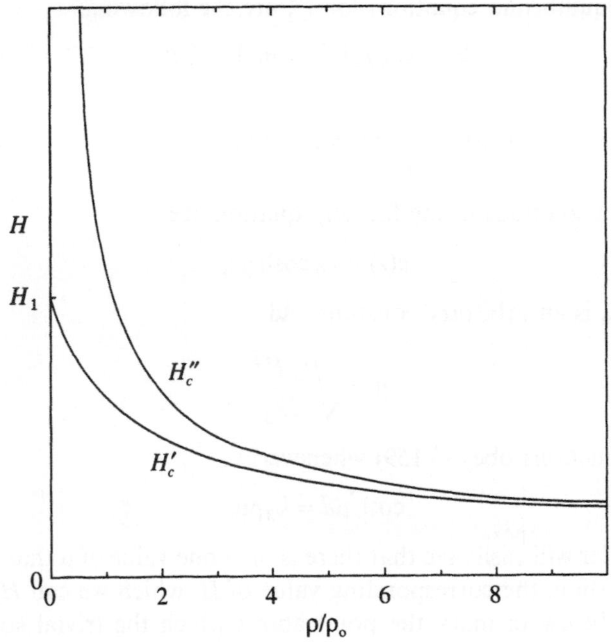

Figure 5.21

suffices to show that

$$\rho_1(H) = \rho_3(H) \approx \left(\frac{2}{d\chi_a}\right)\frac{1}{H^2} \quad \text{as } H \to 0.$$

If H is so close to H_1 that we can write it as

$$H = (1 - \eta)H_1$$

up to terms of order 2 in the parameter η, equation (5.156) readily leads to

$$\rho = \frac{d}{k_1}\eta = \frac{d}{k_1}\frac{H - H_1}{H_1},$$

and so when the anchoring is very strong, but not infinitely so, the critical field of Freedericks's transition decreases according to the formula

$$\frac{\Delta H_c}{H_c} = -\frac{k_1 \rho}{d}. \tag{5.161}$$

Rapini and Papoular (1969) estimated ρ to be approximately $1\,\text{dyn}^{-1}\,\text{cm}$ and so for a cell of thickness $2d \approx 10\,\mu\text{m}$ and $k_1 \approx 10^{-6}$ dyne (5.161) leads to

$$\frac{\Delta H_c}{H_c} \approx -2 \times 10^{-3}.$$

Comparing this figure and that computed at the end of Section 4.2 in Chapter 4, we conclude that a weak anchoring at the plates of the cell is in general responsible for an uncertainty on H_c much smaller than that due to a misalignment of the field.

5.6.4 Stability

We close this section by exploring the stability of the trivial equilibrium configurations described by $\vartheta \equiv 0$ and $\vartheta \equiv \pi/2$.

Theorem 5.9. If $H < H_c'$ then $\vartheta \equiv 0$ is an equilibrium configuration locally stable for F, while it is unstable if $H > H_c'$. Likewise, if $H > H_c''$ then $\vartheta \equiv \pi/2$ is an equilibrium configuration locally stable for F, while it is unstable if $H < H_c''$.

Proof. Deciding the stability of F in the vicinity of both trivial equilibrium solutions presumes the analysis of its second variation at both configurations. On close inspection, the functionals F_2' and F_2'' introduced above are seen to be the second variation of F at $\vartheta \equiv 0$ and $\vartheta \equiv \pi/2$, respectively.

Let ρ be given. If $H < H_c'$, by (5.149) we have that

$$F_2'[u] > \int_{-d}^{+d} \{k_1 u'^2 - \tfrac{1}{2}\chi_a H_c'^2 u^2\}\,dz + \frac{1}{\rho}\{u^2(-d) + u^2(d)\} =: F_c'[u].$$

The minimizers of F_c' are as in (5.152) with

$$\lambda = \lambda_c' := \sqrt{\frac{\chi_a H_c'^2}{2k_1}},$$

and its minimum is zero. Thus, $F_2'[u]$ is strictly positive for all functions u in $\mathscr{A}_F$. This shows that $\vartheta \equiv 0$ is locally stable. On the other hand, if $H > H_c'$ we obtain that

$$F_2'[u] < F_c'[u] \quad \text{for all } u \text{ in } \mathscr{A}_F.$$

Since $F_c'[u]$ vanishes when u is as in (5.152) and $\lambda = \lambda_c'$, we see that

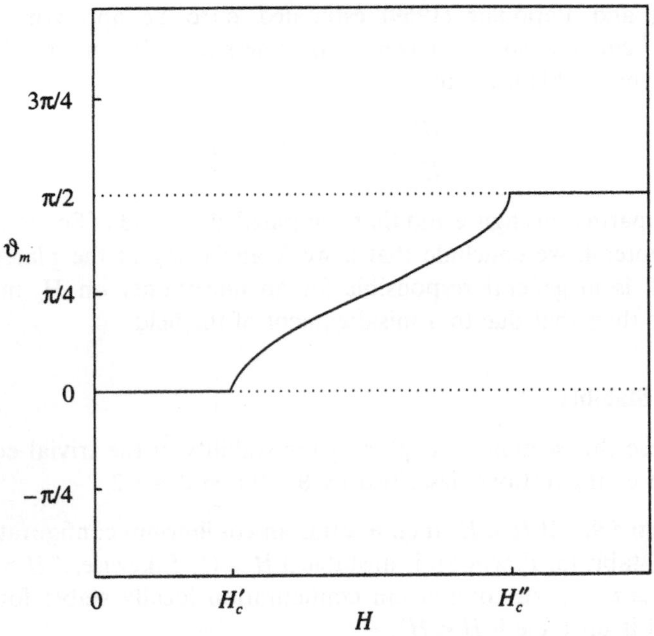

Figure 5.22

F_2' is negative for some functions in $\mathscr{A}_F$, and this is enough for us to conclude that $\vartheta \equiv 0$ is unstable.

To complete the proof of the theorem the reader should apply this argument again to the other trivial solution and note that when $H = H_c''$ the minimum of the functional F_2'' is also zero.

Figure 5.22 illustrates a bifurcation diagram compatible with the analysis just completed. Solid lines refer to the stable equilibrium configurations, while broken lines refer to the unstable ones. It must be noted, however, that for $H \in [H_c', H_c'']$ the graph shown in Fig. 5.22 is a mere figment of the imagination because our non-linear analysis above has not been pursued far enough to support it.

5.7 References

Ambrosio, L. (1989a) A compactness theorem for a new class of functions of bounded variation, *Boll. Un. Mat. Ital.*, **3B**, 857–881.

Ambrosio, L. (1989b) Variational problems in SBV and image segmentation, *Acta Appl. Math.*, **17**, 1–40.

Ambrosio, L. (1990) Existence theory for a new class of variational problems, *Arch. Rational Mech. Anal.*, **111**, 291–322.

Bernal, J.D. and Fankuchen, I. (1941) X-ray and crystallographic studies of plant virus preparations, *J. Gen. Physiol.*, **25**, 111–165.

Chandrasekhar, S. (1966) Surface tension of liquid crystals, in: *Liquid crystals*, G.H. Brown, G.J. Dienes and M.M. Labes eds., Gordon & Breach, New York, 331–340.

Cohen, R. (1992) A numerical algorithm and computational results for liquid crystal droplet problems, Lecture given at the ALCOM/IMA Symposium on Computational Problems in Liquid Crystals held at Kent State University, Kent, Ohio, 13–14 November 1992.

Courant, R. and Hilbert, D. (1989) *Methods of mathematical physics*, Volumes 1 & 2, J. Wiley & Sons, New York. Republication in the Wiley Classics Library Series of the first English edition (1953), Interscience Publishers.

De Giorgi, E. and Ambrosio, L. (1988) Un nuovo tipo di funzionale del calcolo delle variazioni, *Atti Acc. Lincei Rend. fis.*, **82**, 199–210.

Dinghas, A. (1944) Über einen geometrischen Satz von Wulff für die Gleichgewichtsform von Kristallen, *Z. Kristallographie*, **105**, 304–314.

Ericksen, J.L. (1988) Static theory of point defects in nematic liquid crystals, Private communication.

Fonseca, I. (1991) The Wulff theorem revisited, *Proc. Roy. Soc. Lond.*, **A432**, 125–145.

Fonseca, I. and Müller, S. (1990) A uniqueness proof for the Wulff problem, *Proc. Roy. Soc. Edin.*, **A119**, 125–136.

Fournier, J.B. (1992) Germination of drops of liquid crystals from their melt, Private communication.

Friedel, G. (1922) Les états mésomorphes de la matière, *Ann. Phys. (Paris)*, **18**, 273–474.

Gurtin, M.E. (1981) *An introduction to continuum mechanics*, Academic Press, New York.

Haller, I. (1972) Elastic constants of nematic liquid crystalline phase of *p*-methoxybenzylidene-*p*-*n*-butylaniline (MBBA), *J. Chem. Phys.*, **57**, 1400–1405.

Herring, C. (1951) Some theorems on the free energies of crystal surfaces, *Phys. Rev.*, **82**, 87–93.

Kléman, M. (1983) *Points, lines and walls*, J. Wiley & Sons, Chichester.

Langevin, D. (1972) Analyse spectrale de la lumière diffusée par la surface libre d'un cristal liquide nématique. Mesure de la tension superficielle et des coefficients de viscosité, *J. Phys.* (*Paris*), **33**, 249–256.

Lin, F.-H. and Poon, C.-C. (1992) On nematic liquid crystal droplets, Lecture given by F.-H. Lin at the ALCOM/IMA Symposium on Computational Problems in Liquid Crystals held at Kent State University, Kent, Ohio, 13–14 November 1992.

Naemura, S. (1978) Measurements of anisotropic interfacial interactions between a nematic liquid crystal and various substrates, *Appl. Phys. Lett.*, **33**, 1–3.

Naemura, S. (1979) Anisotropic interactions between MBBA and surface-treated substrates, *J. Phys.* (*Paris*), **40**, **C3**, 514–518.

Noll, W. and Virga, E.G. (1990) On edge interactions and surface tension, *Arch. Rational Mech. Anal.*, **111**, 1–31.

Oseen, C.W. (1931) Probleme für die Theorie der anisotropen Flüssigkeiten, *Z. Kristallographie*, **79**, 173–185.

Oseen, C.W. (1933) The theory of liquid crystals, *Trans. Faraday Soc.*, **29**, 883–899.

Rapini, A. and Papoular, M. (1969) Distorsion d'une lamelle nématique sous champ magnétique. Conditions d'ancrage aux parois, *J. Phys.* (*Paris*), *Colloque* **C4**, 54–56.

Rivière, D., Lévy, G. and Guyon, C. (1979) Determination of anchoring energies from surface tilt angle measurements in a nematic liquid crystal, *J. Phys. Lett.* (*Paris*), **40**, L215–L218.

Roccato, D. and Virga, E.G. (1990) Drops of nematic liquid crystals floating on a liquid, *Riv. Mat. Univ. Parma*, **16**, 47–61.

Sluckin, T.J. and Poniewierski, A. (1986) Orientational wetting transitions and related phenomena in nematics, in: *Fluid interfacial phenomena*, C.A. Croxton ed., J. Wiley & Sons, Chichester, 215–253.

Taylor, J.E. (1974) Existence and structure of solutions to a class of nonelliptic variational problems, *Symposia Mathematica*, **14**, 499–508.

Taylor, J.E. (1978) Crystalline variational problems, *Bull. Amer. Math. Soc.*, **84**, 568–588.

Virga, E.G. (1989) Drops of nematic liquid crystals, *Arch. Rational Mech. Anal.*, **107**, 371–390.

Virga, E.G. (1990) Sulle forme di equilibrio di una goccia di cristallo liquido, *Atti Sem. Mat. Fis. Univ. Modena*, **38**, 29–38.

Wulff, G. (1901) Zur Frage der Geschwindigkeit des Wachsthums und der Auflösung der Kristallflächen, *Z. Kristallographie und Mineralogie*, **34**, 449–550.
Zocher, H. (1925) Über freiwillige Strukturbildung in Solen, *Z. Anorg. Chem.*, **147**, 91–110.

5.8 Further reading

Surface free energy

Barbero, G., Gabbasova, Z. and Kosevich, Yu.A. (1991) On the generalization of the Rapini-Papoular expression of the surface energy for nematic liquid crystals, *J. Phys. II France*, **1**, 1505–1513.
Chiarelli, P., Faetti, S. and Fronzoni, L. (1983a) Experimental measurement of the director orientation at the free surface of a nematic liquid crystal, *Lett. Nuovo Cimento*, **36**, 60–64.
Chiarelli, P., Faetti, S. and Fronzoni, L. (1983b) Determination of the molecular orientation at the free surface of liquid crystals from Brewster angle measurements, *Optics Comm.*, **46**, 9–13.
Chiarelli, P., Faetti, S. and Fronzoni, L. (1983c) Structural transition at the free surface of the nematic liquid crystals MBBA and EBBA, *J. Phys. (Paris)*, **44**, 1061–1067.
Goossens, W.J.A. (1985) Bulk, interfacial and anchoring energies of liquid crystals, *Mol. Cryst. Liq. Cryst.*, **124**, 305–331.
Jenkins, J.T. and Barratt, P.J. (1974) Interfacial effects in the static theory of nematic liquid crystals, *Quart. J. Mech. Appl. Math.*, **27**, 111–127.
Kralj, S., Žumer, S. and Allender, D.W. (1991) Nematic–isotropic phase transition in a liquid-crystal droplet, *Phys. Rev. A*, **43**, 2943–2952.
Langevin, D. and Bouchiat, M.A. (1973) Molecular order and surface tension for the nematic–isotropic interface of MBBA, deduced from light reflectivity and light scattering measurements, *Mol. Cryst. Liq. Cryst.*, **22**, 317–331.
Lavrentovich, O.D. and Tarakhan, L.N. (1990) Temperature dependence of the surface tension on the liquid crystal–isotropic liquid interface, *Poverkhnost*, 39–44 (Russian).
Smith, G.W., Vaz, N.A. and Vansteenkiste, T.H. (1989) The interfacial free energy of nematogen droplets in an isotropic matrix: deter-

mination of its temperature dependence from coalescence kinetics, *Mol. Cryst. Liq. Cryst.*, **174**, 49–64.
Williams, R. (1967) Interfaces in nematic liquids, in: Ordered fluids and liquid crystals, *Advances in Chemistry Series, Vol. 63*, R.F. Gould ed., American Chemical Society Publications, Washington.
Žumer, S. and Kralj, S. (1992) Influence of K_{24} on the structure of nematic liquid crystal droplets, *Liq. Crystals*, **12**, 613–624.

Small and large drops

Bezic, J. and Žumer, S. (1992) Structures of the cholesteric liquid crystal droplets with parallel surface anchoring, *Liq. Crystals*, **11**, 593–619.
Candau, S., Le Roy, P. and Debeauvais, F. (1973) Magnetic field effects in nematic and cholesteric droplets suspended in an isotropic liquid, *Mol. Cryst. Liq. Cryst.*, **23**, 283–297.
Drzaic, P.S. (1988) A new director alignment for droplets of nematic liquid crystal with low bend-to-splay ratio, *Mol. Cryst. Liq. Cryst.*, **154**, 289–306.
Drzaic, P.S. and Muller, A. (1989) Droplet shape and reorientation fields in nematic droplet: polymer films, *Liq. Crystals*, **5**, 1467–1475.
Erdmann, J.H., Žumer, S. and Doane, J.W. (1990) Configuration transition in a nematic liquid crystal confined to a small spherical cavity, *Phys. Rev. Lett.*, **64**, 1907–1910.
Golovataya, N.M., Kurik, M.V. and Lavrentovich, O.D. (1990) Self-organization of polymer dispersed nematic droplets, *Liq. Crystals*, **7**, 287–291.
Kilian, A. and Hess, S. (1989) Derivation and application of an algorithm for the numerical calculation of the local orientation of nematic liquid crystals, *Z. Naturforsch.*, **44a**, 693–703.
Kilian, A. and Hess, S. (1990) On the simulation of the director field of a nematic liquid crystal, *Liq. Crystals*, **8**, 465–472.
Lavrentovich, O.D. (1988) Flexoelectricity of droplets of a nematic liquid crystal, *Sov. Tech. Phys. Lett.*, **14**, 73–76.
Lavrentovich, O.D. and Nastishin, Yu.A. (1985) Division of drops of a liquid crystal in the case of a cholesteric–smectic-A phase transition, *JEPT Lett.*, **40**, 1015–1019.
Lavrentovich, O.D. and Nastishin, Yu.A. (1990) Parity-breaking phase transition in tangentially anchored nematic drops, *Nuovo Cimento*, **12D**, 1219–1222.

Le Roy, P., Debeauvais, F. and Candau, S. (1972) Effets de champ magnétique sur la structure de gouttelettes de cristaux liquides nématiques en émulsion dans un liquide isotrope, *C.R. Acad. Sc. Paris*, **274(B)**, 419–422.

Leslie, F.M. (1968) Some thermal effects in cholesteric liquid crystals, *Proc. Roy. Soc. Lond.* **307A**, 358–372.

Press, M.J. and Artrott, A.S. (1974) Theory and experiments on configurations with cylindrical symmetry in liquid crystals droplets, *Phys. Rev. Lett.*, **33**, 403–406.

Volovik, G.E. and Lavrentovich, O.D. (1983) Topological dynamics of defects: boojums in nematic drops, *Sov. Phys. JEPT*, **58**, 1159–1166.

Wulff's construction

Hildebrandt, S. and Tromba, A. (1985) *Mathematics and optimal forms*, Scientific American Books, New York.

Weak anchoring

Eidner, K., Mayer, G. and Schuster, R. (1988) Determination of the surface anchoring energy of a nematic liquid crystal by polarization azimuth measurement, *Mol. Cryst. Liq. Cryst.*, **159**, 27–36.

Faetti, S. (1987) Azimuthal anchoring energy of a nematic liquid crystal at a grooved interface, *Phys. Rev. A*, **36**, 408–410.

Faetti, S., Gatti, M. and Palleschi, V. (1986) Measurements of surface elastic torques in liquid crystals: a method to measure elastic constants and anchoring energies, *Revue Phys. Appl.*, **21**, 451–461.

Faetti, S., Gatti, M., Palleschi, V. and Sluckin, T.J. (1985) Almost critical behavior of the anchoring energy at the interface between a nematic liquid crystal and a SiO substrate, *Phys. Rev. Lett.*, **55**, 1681–1684.

Faetti, S. and Palleschi, V. (1987) The twist elastic constant and anchoring energy of the nematic liquid crystal 4-*n*-octyl-4′-cyano-biphenyl, *Liq. Crystals*, **2**, 261–268.

Faetti, S., Palleschi, V. and Schirone, A. (1988) A reflectometric method to measure the azimuthal anchoring energy of a nematic liquid crystal, *Nuovo Cimento*, **10D**, 1313–1324.

Gleeson, J.T. and Palffy-Muhoray, P. (1989) Determination of the surface anchoring potential of a nematic in contact with a substrate, *Liq. Crystals*, **5**, 663–671.

Haas, G., Fritsch, M., Wöhler, H. and Mlynski, D.A. (1989) Polar anchoring energy and order parameter at a nematic liquid crystal–wall interface, *Liq. Crystals*, **5**, 673–681.

Jérôme, B. and Piéranski, P. (1989) Statics and dynamic of wetting: its contribution to the anchoring of nematics, *Liq. Crystals*, **5**, 683–691.

Kralj, S. and Zumer, S. (1992) Fréedericks transitions in supra-μm nematic droplets, *Phys. Rev. A*, **45**, 2461–2470.

Nobili, M., Lazzari, C., Schirone, A. and Faetti, S. (1992) Azimuthal anchoring energy at a SiO_x–nematic interface, *Mol. Cryst. Liq. Cryst.*, **212**, 97–106.

Strigazzi, A. (1988) Surface elasticity and Fréedericks threshold in a nematic cell weakly anchored, *Nuovo Cimento*, **10D**, 1335–1344.

Yokoyama, H. (1988) Surface anchoring of nematic liquid crystals, *Mol. Cryst. Liq. Cryst.*, **165**, 265–316.

6

The new theory

This closing chapter is mainly devoted to a new equilibrium theory of liquid crystals, which has been recently formulated by Ericksen (1991) in a rigorous way, though it had been anticipated in some earlier works of De Gennes (1969 and 1971), Fan (1971) and Fan and Stephen (1970). This theory is primarily meant to describe through the same mathematical model all defects occurring in liquid crystals, those for which the classical theory is successful as well as those for which it fails.

6.1 Classical defects

The literature on defects studied within the classical theory of liquid crystals is huge, as is shown by the list of further reading appended to this chapter. Here we confine attention to special issues such as the space dimension of defects in the minimizes of Frank's energy functional.

We shall see that only point defects are explained by the theory we have developed so far, whereas both plane and line defects are outside its scope. We shall also see that when the one-constant approximation to Frank's energy applies there is indeed only one point defect which is essentially the prototype of all others, in that it generates all singularities admissible in a minimizer of the energy functional.

6.1.1 Existence of minimizers

Throughout this section we assume that the region $\mathscr{B}$ occupied by a nematic liquid crystal has a smooth boundary. More precisely, in the language of Section 2.3 of Chapter 2, we assume that $\mathscr{B}$ is a regular region whose boundary $\partial\mathscr{B}$ has no edge and no vertex. This assumption is made for simplicity; to establish most of the theorems reviewed below it would suffice to assume that $\partial\mathscr{B}$ can be represented locally as the graph of a Lipschitz continuous function (*cf.* Definition 6.2 in Section 6.3 below).

For the convenience of the reader I recall from Sections 3.1 and 3.2 of Chapter 3 the free energy functional of the classical theory, which is here denoted by $\mathscr{F}_F$ to indicate that it is the same as the functional in (3.2) of Chapter 3 with $\sigma = \sigma_F$, as in formula (3.23) of that chapter:

$$\mathscr{F}_F[\mathbf{n}] := \int_{\mathscr{B}} \{k_1(\operatorname{div}\mathbf{n})^2 + k_2(\mathbf{n}\cdot\operatorname{curl}\mathbf{n})^2 + k_3|\mathbf{n}\wedge\operatorname{curl}\mathbf{n}|^2 + (k_2+k_4)(\operatorname{tr}(\nabla\mathbf{n})^2 - \operatorname{div}\mathbf{n})^2\}\,dv, \tag{6.1}$$

where the natural twist τ has been set equal to zero since the liquid crystal is nematic.

Let $W^{1,2}(\mathscr{B},\mathscr{V})$ be the space[1] of all L^2-mappings of $\mathscr{B}$ into the translation space $\mathscr{V}$ with first weak derivatives in L^2. In this space we consider the closed subset

$$W^{1,2}(\mathscr{B},\mathbb{S}^2) := \{\mathbf{n}\in W^{1,2}(\mathscr{B},\mathscr{V}) \mid \ |\mathbf{n}| = 1 \text{ almost everywhere in } \mathscr{B}\}. \tag{6.2}$$

It is a simple exercise to show that the field $\mathbf{n}_{x_0}:\mathscr{B}\to\mathbb{S}^2$ defined by

$$\mathbf{n}_{x_0}(x) := \frac{x-x_0}{|x-x_0|} \tag{6.3}$$

belongs to $W^{1,2}(\mathscr{B},\mathbb{S}^2)$ for all $x_0\in\mathscr{B}$. When $\mathbf{n}$ is subject to strong anchoring conditions on the whole of $\partial\mathscr{B}$ the class of admissible mappings for $\mathscr{F}_F$ is[2]

$$\mathscr{A}(\mathbf{n}_0) := \{\mathbf{n}\in W^{1,2}(\mathscr{B},\mathbb{S}^2) \mid \mathbf{n}_0 \text{ is the trace of } \mathbf{n} \text{ on } \partial\mathscr{B}\}, \tag{6.4}$$

where $\mathbf{n}_0$ is a given mapping of $\partial\mathscr{B}$ into $\mathbb{S}^2$.

Hardt, Kinderlehrer and Lin (1986) proved that the class $\mathscr{A}(\mathbf{n}_0)$ is not empty for any admissible choice of $\mathbf{n}_0$, and then they established the following theorem.

Theorem 6.1. For any given $\mathbf{n}_0:\partial\mathscr{B}\to\mathbb{S}^2$ there is a mapping $\mathbf{n}^*\in\mathscr{A}(\mathbf{n}_0)$ where $\mathscr{F}_F$ attains its minimum:

$$\mathscr{F}_F[\mathbf{n}^*] = \inf_{\mathbf{n}\in\mathscr{A}(\mathbf{n}_0)} \mathscr{F}_F[\mathbf{n}]. \tag{6.5}$$

---

1 The reader is referred to Chapters 1 and 2 of Ziemer (1989) for both Lebesgue and Sobolev spaces such as L^2 and $W^{1,2}$.

2 The *trace* of a Sobolev function is defined, for example, on p. 190 of Ziemer (1989).

It is central to the proof of this theorem to remark that

$$\sigma_L(\mathbf{n}, \nabla\mathbf{n}) := \operatorname{tr}(\nabla\mathbf{n})^2 - (\operatorname{div}\mathbf{n})^2 \tag{6.6}$$

is a null Lagrangian and that for all $\mathbf{n}\in\mathscr{A}(\mathbf{n}_0)$ its integral over $\mathscr{B}$ depends only on $\mathbf{n}_0$ (*cf.* Section 3.8 of Chapter 3).

6.1.2 Regularity of minimizers

Theorem 6.1 assures us that the problem of finding the stable equilibrium configurations of a nematic liquid crystal which occupies $\mathscr{B}$ is well posed when the optic axis is prescribed on the whole of $\partial\mathscr{B}$, but it does not say anything about the defects that an energy minimizer can bear. Hardt, Kinderlehrer and Lin (1986, 1988) have determined the degree of regularity of all minimizers of $\mathscr{F}_F$.

Before getting into further details we should ask why the minimizers of $\mathscr{F}_F$ cannot be expected to be smooth, or even continuous. Hardt and Kinderlehrer (1987) give two reasons to this effect: one is topological and the other is analytical; we discuss both below.

For any continuous mapping $\mathbf{u}$ of $\mathbb{S}^2$ onto itself, there is an integer $\deg(\mathbf{u})$ associated with it, which we call the **degree** of $\mathbf{u}$. Intuitively, $|\deg(\mathbf{u})|$ tells us how many times $\mathbf{u}(x)$ covers $\mathbb{S}^2$ as x ranges in $\mathbb{S}^2$, while its sign indicates whether it preserves the orientation of $\mathbb{S}^2$ or not. A more formal definition of degree can be found *e.g.* in Section 6.14 of Hocking and Young (1961). If $\mathscr{S}$ is any oriented surface and $\mathbf{u}$ is a mapping of $\mathscr{S}$ onto $\mathbb{S}^2$, the degree of $\mathbf{u}$ can be defined likewise. It can be computed through the following formula (see *e.g.* Volovik and Lavrentovich (1983)):

$$\deg(\mathbf{u}) = \frac{1}{4\pi}\int_{\mathscr{S}} (\nabla_s\mathbf{u})^*\boldsymbol{\nu}\cdot\mathbf{u}\,da, \tag{6.7}$$

where $(\nabla_s\mathbf{u})^*$ is the adjugate of the surface gradient of $\mathbf{u}$ on $\mathscr{S}$ (*cf.* Section 2.2 of Chapter 2), $\boldsymbol{\nu}$ is one normal field of $\mathscr{S}$, and a denotes the area-measure. This formula shows that the degree of $\mathbf{u}$ on $\mathscr{S}$ changes sign if the orientation of $\mathscr{S}$ is reversed.

As an example, consider the mapping $\mathbf{n}_f : \mathbb{S}^2 \to \mathbb{S}^2$ defined by

$$\mathbf{n}_f := \pi^{-1}\circ f\circ\pi, \tag{6.8}$$

where $\pi : \mathbb{S}^2\setminus\{N\}\to\mathbb{C}$ is the stereographic projection onto the complex plane $\mathbb{C}$ from the north pole $N\in\mathbb{S}^2$, and f is a continuous mapping of $\mathbb{C}$ into itself. If $f(z) = z^m$ with $m\in\mathbb{N}$ then $\deg(\mathbf{n}_f) = |m|$.

If $\mathbf{n}$ is a mapping of $\mathscr{B}$ into $\mathbb{S}^2$ which is continuous near $x_0\in\mathscr{B}$, but not at x_0, then the degree of $\mathbf{n}$ at x_0 is the integer defined by

$$\deg(\mathbf{n}, x_0) := \deg(\mathbf{n}|_{\partial B_r(x_0)}), \tag{6.9}$$

where $B_r(x_0)$ is any ball centred at x_0 with a radius r sufficiently small. That $\deg(\mathbf{n}|_{\partial B_r(x_0)})$ is independent of r follows from classical theorems on homotopy (*cf. e.g.* pp. 266–267 of Hocking and Young (1961)). The same theorems lead to the following assertion.

Lemma 6.2. If the degree of the mapping $\mathbf{n}_0 : \partial\mathscr{B} \to \mathbb{S}^2$ is zero then there is a continuous mapping $\mathbf{n}$ of the closure of $\mathscr{B}$ into $\mathbb{S}^2$ such that

$$\mathbf{n}|_{\partial\mathscr{B}} = \mathbf{n}_0. \tag{6.10}$$

If the degree of $\mathbf{n}_0$ is not zero, there is no continuous mapping $\mathbf{n}$ that satisfies (6.10).

Thus, if the degree of $\mathbf{n}_0$ is not zero the minimizers of $\mathscr{F}_F$ in $\mathscr{A}(\mathbf{n}_0)$ cannot be continuous due to a topological obstruction.

The simplest example of a mapping $\mathbf{n}_0$ with nonzero degree is each normal field of $\partial\mathscr{B}$: its degree is $+1$ or -1, according to whether it is the orientation of $\partial\mathscr{B}$ chosen in formula (6.7) or its oppsite. To be more specific, let $\mathscr{B}$ be the ball $B_r(x_0)$. Its outward normal field is given by

$$\mathbf{v}(x) = \mathbf{n}_{x_0}(x) \quad \text{for all } x \in \partial B_r(x_0), \tag{6.11}$$

where $\mathbf{n}_{x_0}$ is as in (6.3). An easy computation shows that

$$\nabla_s \mathbf{v} = \frac{1}{r}(\mathbf{I} - \mathbf{v} \otimes \mathbf{v}), \tag{6.12}$$

hence

$$(\nabla_s \mathbf{v})^* = \frac{1}{r^2} \mathbf{v} \otimes \mathbf{v}. \tag{6.13}$$

By (6.7), we conclude that

$$\deg(\mathbf{v}) = 1. \tag{6.14}$$

Thus, if $\mathbf{n}$ is prescribed on $\partial B_r(x_0)$ as the outer normal field $\mathbf{v}$, the minimizers of $\mathscr{F}_F$ are not smooth in $B_r(x_0)$. Furthermore, one would expect that the radial field $\mathbf{n}_{x_0}$ in (6.3), which exhibits a point defect at x_0, be the only minimizer of $\mathscr{F}_F$. As is shown below, this is indeed the case when Frank's constants k_1, k_2 and k_3 are all equal, but it is not when they differ too much. Following Volovik and Lavrentovich (1983), we give the name **hedgehogs** to all orientation fields with point defects such as that in x_0 for $\mathbf{n}_{x_0}$. We shall see at the end of this section when a hedgehog is in general a minimizer of $\mathscr{F}_F$.

Even when $\mathbf{n}_0$ has degree zero, and so there is a continuous extension of it to the whole of $\mathscr{B}$, the minimizers of $\mathscr{F}_F$ may still exhibit singularities. This was first shown by Hardt and Lin (1986) through an explicit construction in which $\mathscr{B}$ was the unit ball. Hardt and Kinderlehrer (1987) applied the same idea in a dumbbell-shaped region consisting of *two* unit balls connected by a thin tube. These constructions prove the following theorem.

Theorem 6.3. There are regions $\mathscr{B}$ and mappings $\mathbf{n}_0:\partial\mathscr{B}\to\mathbb{S}^2$ of degree zero such that every minimizer of $\mathscr{F}_F$ in $\mathscr{A}(\mathbf{n}_0)$ is not continuous.

To characterize the set where a minimizer of $\mathscr{F}_F$ fails to be regular, we introduce the following definition.

Definition 6.1. Let $\mathbf{n}\in\mathscr{A}(\mathbf{n}_0)$ be given; for all $x\in\mathscr{B}$ define

$$S_{\mathbf{n}}(x):=\limsup_{r\to 0}\frac{1}{8\pi r}\int_{B_r(x)}|\nabla\mathbf{n}|^2\,dv, \tag{6.15}$$

and define as $\mathscr{S}(\mathbf{n})$ the following subset of $\mathscr{B}$:

$$\mathscr{S}(\mathbf{n}):=\{x\in\mathscr{B}\,|\,S_{\mathbf{n}}(x)>0\}. \tag{6.16}$$

It has been proved in Hardt, Kinderlehrer and Lin (1986) that any minimizer $\mathbf{n}$ of $\mathscr{F}_F$ in $\mathscr{A}(\mathbf{n}_0)$ is smooth away from $\mathscr{S}(\mathbf{n})$. More precisely:

Theorem 6.4. If $\mathbf{n}\in\mathscr{A}(\mathbf{n}_0)$ is a minimizer of $\mathscr{F}_F$, then $\mathbf{n}$ is analytic in $\mathscr{B}\setminus\mathscr{S}(\mathbf{n})$, $\mathscr{S}(\mathbf{n})$ is relatively closed in $\mathscr{B}$ and has one-dimensional Hausdorff measure zero.

Thus, $\mathscr{S}(\mathbf{n})$ is the **singular set** of the minimizer $\mathbf{n}$. From Hardt, Kinderlehrer and Lin (1988) we learn more about this set. Their main conclusion is stated in a corollary, which stems from the two theorems below.

Theorem 6.5 Let $\mathbf{n}$ be a minimizer of $\mathscr{F}_F$ in $\mathscr{A}(\mathbf{n}_0)$. For any compact subset $\mathscr{K}$ of $\mathscr{B}$ there is a constant C, which depends on $\mathscr{K}$ and k_1, k_2 and k_3, such that

$$\int_{\mathscr{K}}|\nabla\mathbf{n}|^2\,dv\leqslant C. \tag{6.17}$$

Theorem 6.6. For every $\mathbf{n}$ that minimizes $\mathscr{F}_F$ in $\mathscr{A}(\mathbf{n}_0)$ there is a

$q > 2$ depending on k_1, k_2 and k_3 such that

$$\int_{\mathscr{K}} |\nabla \mathbf{n}|^q dv < \infty \tag{6.18}$$

for any compact set $\mathscr{K} \subset \mathscr{B}$.

Corollary 6.7. The singular set $\mathscr{S}(\mathbf{n})$ of a minimizer of $\mathscr{F}_F$ in $\mathscr{A}(\mathbf{n}_0)$ is such that

$$H^{3-q}(\mathscr{S}(\mathbf{n})) = 0, \tag{6.19}$$

where H denotes the Hausdorff measure and q is as in Theorem 6.6.

In particular, it follows from (6.19) that the Hausdorff dimension of the singular set is *strictly less* than 1 for all minimizers of $\mathscr{F}_F$.

Theorems similar to these hold also when electric or magnetic fields are applied to the material (*cf.* Hardt, Kinderlehrer and Lin (1986, 1988)).

We have just learned that the space dimension of the singular set of an energy minimizer is bounded by a number that depends only on the material constants k_1, k_2 and k_3. Suppose now that $x_0 \in \mathscr{B}$ is a point defect for a minimizer $\mathbf{n}$ of $\mathscr{F}_F$ in $\mathscr{A}(\mathbf{n}_0)$. We might wonder whether we can estimate the degree of $\mathbf{n}$ at x_0. Hardt, Kinderlehrer and Lin (1986) proved that there is a finite scalar-valued function M of k_1, k_2 and k_3 such that

$$S_{\mathbf{n}}(x_0) \leqslant M(k_1, k_2, k_3).$$

This inequality allows us to show that the degree of a point defect for a minimizer is bounded by a number that depends on the material constants alone; in particular, it does not depend on the boundary condition $\mathbf{n}_0$. Thus, stable point defects cannot be arbitrarily complicated (*cf.* also Hardt and Kinderlehrer (1987) and Kinderlehrer (1991)).

6.1.3 Harmonic mappings

We have seen in Section 3.3 of Chapter 3 that if for a nematic liquid crystal

$$k_1 = k_2 = k_3 = \kappa \quad \text{and} \quad k_4 = 0, \tag{6.20}$$

then the elastic free energy takes the form

$$\mathscr{F}_F[\mathbf{n}] = \kappa \int_{\mathscr{B}} |\nabla \mathbf{n}|^2 dv. \tag{6.21}$$

It follows from (3.111) and (3.115) of Chapter 3 that the equilibrium equation for the functional in (6.21) is

$$\Delta \mathbf{n} + |\nabla \mathbf{n}|^2 = 0, \tag{6.22}$$

which holds wherever $\mathbf{n}$ is regular in $\mathscr{B}$. Any solution of (6.22) is a **harmonic mapping**. Thus, all minimizers of the one-constant approximation to the energy functional are harmonic mappings, though not all harmonic mappings are energy minimizers.

A great deal is known about the singular set of minimizing harmonic mappings. We collect from various sources the theorems relevant to our development; the conclusions reached here are neater than those obtained for the minimizers of the general functional $\mathscr{F}_F$.

Here we follow in part the expository paper by Almgren and Lieb (1990). The general method of Schoen and Uhlenbeck (1982, 1983) leads us to the following theorem.

Theorem 6.8. If $\mathbf{n}$ is a minimizing harmonic mapping of $\mathscr{A}(\mathbf{n}_0)$ then it is really analytic except at finitely many points of discontinuity.

The proof of Theorem 6.8 is based on an estimate of monotonicity which has no analogue when (6.20) are not satisfied. Let $\mathbf{n}$ be a minimizing harmonic mapping of $\mathscr{B}$ into $\mathbb{S}^2$, and let x be any point in $\mathscr{B}$. If $0 < r_1 < r_2 < r$, where r is such that the ball $B_r(x)$ lies inside $\mathscr{B}$, then

$$\frac{1}{r_1}\int_{B_{r_1}(x)} |\nabla \mathbf{n}|^2 \, dv \leqslant \frac{1}{r_2}\int_{B_{r_2}(x)} |\nabla \mathbf{n}|^2 \, dv. \tag{6.23}$$

Inequality (6.23) allows us to show that for all $x \in \mathscr{B}$ there is a **tangential approximation** to $\mathbf{n}$ near x (see Almgren and Lieb (1988)). A deeper result of Simon (1983) then ensures that for a minimizing harmonic mapping such an approximation is unique, though it is not the same near all points of $\mathscr{B}$. More precisely, let $\mathbf{n}$ be a harmonic mapping of $\mathscr{B}$ into $\mathbb{S}^2$ that minimizes the energy functional (6.21) in $\mathscr{A}(\mathbf{n}_0)$, for a given $\mathbf{n}_0$. If x_0 is a regular point of $\mathbf{n}$ then in the vicinity of x_0 $\mathbf{n}$ can be approximated by the constant mapping that takes the value $\mathbf{n}(x_0)$. If, on the other hand, x_0 is a singular point for $\mathbf{n}$, there is precisely *one* mapping $\mathbf{f}_{x_0}: \mathbb{S}^2 \to \mathbb{S}^2$ such that

$$\mathbf{n}(x) \approx \mathbf{f}_{x_0}\left(\frac{x - x_0}{|x - x_0|}\right) \tag{6.24}$$

for all x sufficiently close to x_0. In general, the approximating mappings are different at different singular points, but they all share a common feature: they are *harmonic*.

If $\mathbf{f}$ is a mapping of $\mathbb{S}^2$ into itself, we extend it to the whole space by defining the mapping $\mathbf{f}^*:\mathscr{E}\to\mathbb{S}^2$ as

$$\mathbf{f}^*(x):=\mathbf{f}\left(\frac{x-x_0}{|x-x_0|}\right), \tag{6.25}$$

where x_0 is any given point of $\mathscr{E}$. It is to be noted that the degree of $\mathbf{f}^*$ at x_0 is precisely the degree of $\mathbf{f}$, whereas the degree of $\mathbf{f}^*$ at all points of $\mathscr{E}$ other than x_0 is zero. We say that the mapping $\mathbf{f}$ is harmonic whenever $\mathbf{f}^*$ is.

All harmonic mappings of $\mathbb{S}^2$ into itself have been completely classified (*cf.* Lemaire (1978)). Every mapping $\mathbf{f}:\mathbb{S}^2\to\mathbb{S}^2$ can be represented by a mapping $f:\mathbb{C}\to\mathbb{C}$ through the stereographic projection of $\mathbb{S}^2$ onto $\mathbb{C}$ (*cf.* (6.8) above):

$$\mathbf{f}=\pi^{-1}\circ f\circ\pi. \tag{6.26}$$

The mapping $\mathbf{f}$ in (6.26) is harmonic if, and only if, the corresponding mapping f can be given one of the following forms:

$$f(z)=\frac{P(z)}{Q(z)}\quad\text{or}\quad f(z)=\frac{P(\bar{z})}{Q(\bar{z})}, \tag{6.27}$$

where P and Q are relatively prime complex polynomials and $\bar{z}$ denotes the complex conjugate of z. When f is as in (6.27), the degree of the mapping $\mathbf{f}$ that corresponds to f through (6.26) is given by

$$\deg(\mathbf{f})=\begin{cases}\max\{\deg(P),\deg(Q)\} & \text{for } f \text{ as in } (6.27)_1,\\ -\max\{\deg(P),\deg(Q)\} & \text{for } f \text{ as in } (6.27)_2.\end{cases} \tag{6.28}$$

Furthermore, when $\mathbf{f}$ is harmonic the energy of its extension $\mathbf{f}^*$ to a ball can be computed in terms of its degree:

$$\int_{B_r(x_0)}|\nabla\mathbf{f}^*|^2\,dv=8\pi r|\deg(\mathbf{f})|. \tag{6.29}$$

A major theorem of Brezis, Coron and Lieb (1986) shows that every minimizing harmonic mapping $\mathbf{n}$ has either degree $+1$ or degree -1 and it tells us which is the tangential approximation to $\mathbf{n}$ near its singular points.

Theorem 6.9. If $\mathbf{n}$ is a minimizing harmonic mapping of $\mathscr{A}(\mathbf{n}_0)$, then each singular point x_0 of $\mathbf{n}$ has degree $+1$ or -1. Moreover, near a singular point x_0 of $\mathbf{n}$

$$\mathbf{n}(x) \approx \mathbf{R}_{x_0}\left(\frac{x - x_0}{|x - x_0|}\right), \tag{6.30}$$

where $\mathbf{R}_{x_0}$ is an orthogonal tensor depending on x_0.

Employing the notation introduced in (6.24), we can also write (6.30) in the form

$$\mathbf{n}(x) \approx \mathbf{R}_{x_0}\mathbf{n}_{x_0} \quad \text{for all } x \text{ near } x_0, \tag{6.31}$$

whence it follows that

$$\deg(\mathbf{n}, x_0) = \det \mathbf{R}_{x_0}. \tag{6.32}$$

Remark. For any energy minimizer given in $\mathscr{A}(\mathbf{n}_0)$, elementary theorems of topology require the algebraic sum of the degrees of all singularities to be equal to the degree of $\mathbf{n}_0$.

The graphs in Fig. 6.1 illustrate the fields $\mathbf{n}_{x_0}$ and $\mathbf{R}_{x_0}\mathbf{n}_{x_0}$ in the vicinity of x_0 when $\mathbf{R}_{x_0}$ is a rotation of angle $\pi/4$.

It is to be noted that the class of mappings with tangential approximations as in (6.31) does not even include all harmonic mappings of degree ± 1.

The proof of Theorem 6.9 proceeds by comparison. If $\mathbf{n}$ is a harmonic mapping of $\mathscr{A}(\mathbf{n}_0)$ such that at a singular point x_0 the absolute value of its degree is greater than 1, then one can decrease its energy by splitting the singularity at x_0 into two nearby singularities of lower degree. If the degree of $\mathbf{n}$ at x_0 is $+1$ or -1, but the

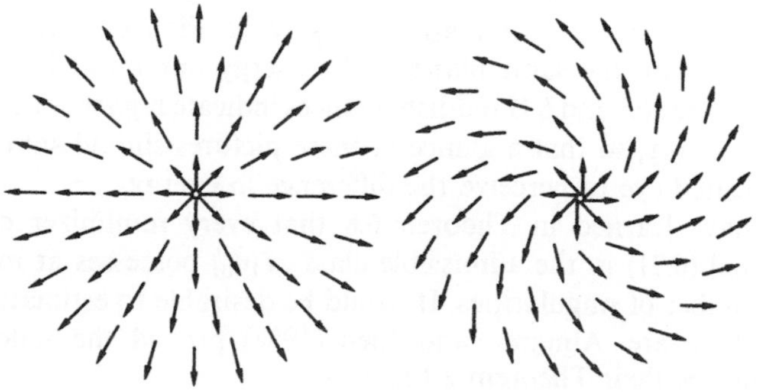

(a) (b)

Figure 6.1

tangential approximation to $\mathbf{n}$ at x_0 is not in the form (6.31), one can decrease the energy of $\mathbf{n}$ by slightly displacing x_0 (see Section 7 of Brezis, Coron and Lieb (1986)).

The instability of defects with degree other than $+1$ and -1 was first observed numerically by Cohen, Hardt, Kinderlehrer, Lin and Luskin (1987) (see also Cohen, Lin and Luskin (1989) for further details on the numerical scheme). They established by example that defects of degree greater than 1 are indeed saddle points of the energy functional in (6.21). They studied harmonic mappings in a unit cube $\mathscr{Q}$, which in Cartesian co-ordinates is described by

$$\mathscr{Q} = \{(x_1, x_2, x_3) \mid x_1, x_2, x_3 \in]0, 1[\}.$$

The field $\mathbf{n}_0$ was prescribed on $\partial\mathscr{Q}$ so that there is an extension $\mathbf{n}_0^*$ of it to the whole closure of $\mathscr{Q}$ with a singularity at the point $x_0 = (\frac{1}{2}, \frac{1}{2}, \frac{1}{2})$ with degree greater than 1. An iterative numerical algorithm starting from $\mathbf{n}_0^*$ and devised so as to decrease the energy without affecting the boundary data, then led to a mapping with only singularities of degree 1, apart from one another.

Plates 6.I and 6.II, which are reproduced here by courtesy of M. Luskin, illustrate the first and the last step of this method for a singularity of degree 2.

In this example $\mathbf{n}_0^*$ is given by the formula

$$\mathbf{n}_0^*(x) = \pi^{-1}\left(\left(\pi\left(\frac{x - x_0}{|x - x_0|}\right)\right)^2\right), \quad \text{for all } x \in \mathscr{Q}, \tag{6.33}$$

which amounts to saying that in (6.8) $f(z) = z^2$ for all $z \in \mathbb{C}$. The centre of $\mathscr{Q}$ is a singularity of degree 2 for $\mathbf{n}_0^*$; Plate 6.I shows the projection of $\mathbf{n}_0^*$ onto the plane $x_3 = \frac{1}{2}$, while Plate 6.II shows the projection onto the same plane of the energy minimizer in $\mathscr{A}(\mathbf{n}_0)$. In both Plates 6.I and 6.II reddish nuances indicate higher concentrations of energy, so that a glance at these pictures should suffice for the educated eye to perceive the difference in energy.

We have learned in Theorem 6.8 that every minimizer of the functional (6.21) in the admissible class $\mathscr{A}(\mathbf{n}_0)$ possesses at most a finite number of singularities. It would be desirable to estimate how many there are. Almgren and Lieb (1988) proved the following theorem (see their Theorem 2.12).

Theorem 6.10. Let $\mathbf{n}$ be a minimizing harmonic mapping in the class $\mathscr{A}(\mathbf{n}_0)$ and denote by $\#\mathscr{S}(\mathbf{n})$ the cardinality of its singular set.

There is a constant $C_{\mathscr{B}}$, possibly depending on the domain $\mathscr{B}$, such that

$$\#\mathscr{S}(\mathbf{n}) \leqslant C_{\mathscr{B}} \int_{\partial\mathscr{B}} |\nabla_s \mathbf{n}_0|^2 \, da, \tag{6.34}$$

where $\nabla_s \mathbf{n}_0$ is the surface gradient of $\mathbf{n}_0$ on $\partial\mathscr{B}$ and a denotes the area-measure.

Though there is no indication about the value of $C_{\mathscr{B}}$ and about how sensitive it is to the domain $\mathscr{B}$, to my knowledge (6.34) is the best estimate of $\#\mathscr{S}(\mathbf{n})$ available in the literature.

We close this digression on harmonic mappings with a theorem of Almgren and Lieb (1988) which tells us that the singularities of a minimizer cannot be too close, if they are well inside the domain $\mathscr{B}$.

Theorem 6.10. Suppose that $\mathbf{n}$ is a minimizing harmonic mapping having a singular point at $x_0 \in \mathscr{B}$. Let d_{x_0} be the distance of x_0 from $\partial\mathscr{B}$. There is a universal constant c, independent of anything, such that there is no other singularity of $\mathbf{n}$ within distance $c d_{x_0}$ of x_0.

The proof of this theorem is by contradiction; no estimate of the constant c is available yet.

6.1.4 Symmetry breaking

One might be inclined to believe that if both the domain $\mathscr{B}$ and the boundary data $\mathbf{n}_0$ enjoy a certain symmetry, so also the minimizers of $\mathscr{F}_F$ in $\mathscr{A}(\mathbf{n}_0)$ should enjoy the same symmetry. There are examples by Almgren and Lieb (1988) and Hardt, Kinderlehrer and Lin (1990) showing that this may not be the case. In both the examples below the energy functional is taken to be as in (6.21).

Almgren and Lieb (1988) consider harmonic mappings from a unit ball B_1 into $\mathbb{S}^2$, that have $\mathbf{n}_0$ as trace on ∂B_1: this is a distortion of the identity that in two small caps around the poles cover, respectively, the northen and southern hemisphere of $\mathbb{S}^2$, so that each of these mappings is the mirror image of the other. Away from these caps, $\mathbf{n}_0$ takes values in the equator of $\mathbb{S}^2$ so as to be symmetric under reflections about the equatorial plane (see Fig. 6.2).

Almgren and Lieb (1988) show that the caps around the poles can be chosen so that the minimizing harmonic mappings subject to the above boundary conditions all have singularities near the poles. This is sufficient to conclude by contradiction that the minimizers do not

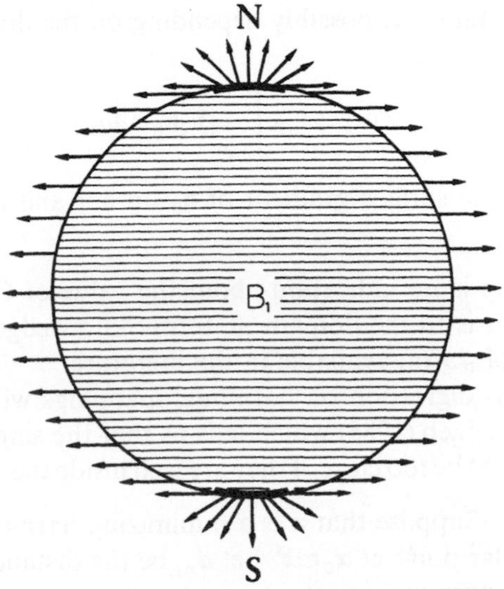

Figure 6.2

inherit the mirror symmetry of $\mathbf{n}_0$. In fact, all minimizers have degree $+1$ on ∂B_1, and so they must display an odd number of singularities in B_1. If they were symmetric, they would have at least one singularity in the equatorial plane, which is a contradiction.

Hardt, Kinderlehrer and Lin (1990) take as domain the cylinder

$$\mathscr{C} = \{(r, \vartheta, z) | r \in [0, 1[, \vartheta \in [0, 2\pi[, z \in]-1, 1[\},$$

where (r, ϑ, z) denote cylindrical co-ordinates. They describe as **axisymmetric** each mapping $\mathbf{n}: \mathscr{C} \to \mathbb{S}^2$ that can be represented in the form

$$\mathbf{n} = \cos\vartheta \sin\varphi \mathbf{e}_r + \sin\vartheta \sin\varphi \mathbf{e}_\vartheta + \cos\varphi \mathbf{e}_z$$

for some real-valued function φ on $\mathscr{C}$ which does not depend on the azimuth ϑ. Let $\mathbf{n}_0$ be the trace of an axisymmetric mapping. We write it in the form

$$\mathbf{n}_0 = \cos\vartheta \sin\varphi_0 \mathbf{e}_r + \sin\vartheta \sin\varphi_0 \mathbf{e}_\vartheta + \cos\varphi_0 \mathbf{e}_z, \tag{6.35}$$

where φ_0 is a function defined on $\partial\mathscr{C}$ so that

$$\varphi_0(r, 1) = 0, \ \varphi_0(r, -1) = k\pi \ \textit{for all} \ r \in [0, 1] \ \textit{and some} \ k \in \mathbb{N},$$

$$\textit{and} \quad \varphi_0(1, \cdot) \ \textit{is monotonic on} \ [-1, 1].$$

Hardt, Kinderlehrer and Lin proved that if k is sufficiently large, every minimizing harmonic mapping that agrees with (6.35) on $\partial\mathscr{C}$ is *not* axisymmetric because it possesses at least one singularity away from the axis of $\mathscr{C}$.

Let $\mathbf{n}$ be any such minimizer. Acting on it through an appropriate family of rotations about the axis of $\mathscr{C}$ we get a continuum of fields with the same trace on $\partial\mathscr{C}$ as $\mathbf{n}$ and the same energy. Thus, when the symmetry is broken the minimizer fails to be unique. Kinderlehrer (1991) presents an argument due to Ericksen which shows that such a lack of uniqueness is to be expected also for the general functional $\mathscr{F}_F$ in (6.1).

6.1.5 Stable hedgehogs

A remarkable case of uniqueness arises when Frank's constants are as in (6.20) and $\mathscr{B}$ is the ball $B_r(x_0)$.

Theorem 6.11. If $\mathbf{v}_r : \partial B_r(x_0) \to \mathbb{S}^2$ is the outer normal field to $B_r(x_0)$ then the hedgehog $\mathbf{n}_{x_0}$ in (6.3) is the unique minimizer of the functional (6.21) within the class $\mathscr{A}(\mathbf{v}_r)$.

This theorem is proved in Section 7 of Brezis, Coron and Lieb (1986); it has the following consequence.

Corollary 6.12. Let the mapping $\mathbf{n}_\mathbf{R} : \partial B_r(x_0) \to \mathbb{S}^2$ be defined by

$$\mathbf{n}_\mathbf{R}(x) := \mathbf{R}\left(\frac{x - x_0}{r}\right), \tag{6.36}$$

where $\mathbf{R}$ is given in $O(\mathscr{V})$. The extension of $\mathbf{n}_\mathbf{R}$ to the whole of $B_r(x_0)$ defined by

$$\mathbf{n}^*_\mathbf{R}(x) := \mathbf{R}\left(\frac{x - x_0}{|x - x_0|}\right) \tag{6.37}$$

is the unique minimizing harmonic mapping in $\mathscr{A}(\mathbf{n}_\mathbf{R})$.

We have learned in Section 3.6 of Chapter 3 that the hedgehog $\mathbf{n}_{x_0}$ is a solution to the equilibrium equation of $\mathscr{F}_F$ for all values of Frank's constants. We now wonder whether it is also a minimizer of $\mathscr{F}_F$ when equations (6.20) do not apply. Moreover, in the light of Corollary 6.12, we shall ask the same question also for the field $\mathbf{n}^*_\mathbf{R}$ in (6.37).

The first answer came from a paper of Hélein (1987) and it was negative.

Theorem 6.13. If Frank's constants satisfy the inequality

$$8(k_2 - k_1) + k_3 < 0, \tag{6.38}$$

then the hedgehog $\mathbf{n}_{x_0}$ is not an energy minimizer in $\mathscr{A}(\mathbf{v})$.

Hélein's proof is constructive: he exhibits a field with an energy which becomes smaller than that of the hedgehog when (6.38) is satisfied. It is remarkable that the theorem proved later by Cohen and Taylor (1990) is the exact counterpart of Hélein's.

Theorem 6.14. If

$$8(k_2 - k_1) + k_3 \geqslant 0, \tag{6.39}$$

then the second variation of $\mathscr{F}_F$ at $\mathbf{n}_{x_0}$ is positive in an appropriate class of perturbations, and so this hedgehog is locally stable.

This theorem is built upon the following inequalities, which are hard to prove. Let B_r denote for brevity the ball $B_r(x_0)$. For all bounded mappings $\mathbf{u}$ that belong to $W^{1,2}(B_r, \mathscr{V})$ and are such that $\mathbf{n}_{x_0} \cdot \mathbf{u} = 0$ almost everywhere in B_r, we have that

$$\int_{B_r} \{(\operatorname{div} \mathbf{u})^2 + (\mathbf{n}_{x_0} \cdot \operatorname{curl} \mathbf{u})^2\} \, dv \geqslant \int_{B_r} \frac{2}{|x - x_0|} |\mathbf{u}|^2 \, dv, \tag{6.40}$$

$$\int_{B_r} |\mathbf{n}_{x_0} \wedge \operatorname{curl} \mathbf{u}|^2 \, dv \geqslant \frac{1}{4} \int_{B_r} \frac{1}{|x - x_0|^2} |\mathbf{u}|^2 \, dv. \tag{6.41}$$

The latter inequality is strict if $\mathbf{u}$ does not vanish identically in B_r. The proofs of (6.40) and (6.41) given by Cohen and Taylor (1990) are rather involved. Much simpler proofs can be found in Kinderlehrer and Ou (1992), in which a central role is played by the null Lagrangian σ_L in (6.6).

We know from Corollary 6.12 that the feld $\mathbf{n}_{\mathbf{R}}^*$ in (6.37) is a harmonic mapping for all $\mathbf{R}$ in $O(\mathscr{V})$. We now ask whether there is any $\mathbf{R}$ such that $\mathbf{n}_{\mathbf{R}}^*$ is also an equilibrium configuration for $\mathscr{F}_F$ when equations (6.20) do not apply, and, if so, for which values of Frank's constants $\mathbf{n}_{\mathbf{R}}^*$ is an energy minimizer. This question was addressed by Kinderlehrer, Ou and Walkington (1993).

If $\mathbf{R}$ is either the identity or its opposite, we already know whether $\mathbf{n}_{\mathbf{R}}^*$ is an energy minimizer or not. Here we restrict attention to the case where $\mathbf{R} \in O(\mathscr{V}) \setminus \{\mathbf{I}, -\mathbf{I}\}$.

Theorem 6.15. If $\mathbf{n}_{\mathbf{R}}^*$ solves the equilibrium equation of $\mathscr{F}_F$ for some

$\mathbf{R}$ in $O(\mathscr{V})\backslash\{\mathbf{I}, -\mathbf{I}\}$ then we must have

$$k_1 = k_3 \quad \text{and} \quad \mathbf{R} = -(\mathbf{I} - 2\mathbf{e}\otimes\mathbf{e}) \quad \text{for some } \mathbf{e}\in\mathbb{S}^2. \tag{6.42}$$

Furthermore, when both these conditions are satisfied, $\mathbf{n}_{\mathbf{R}}^*$ minimizes the energy in B_r among all fields that share its trace on ∂B_r if

$$k_2 \geqslant k_1, \tag{6.43}$$

whereas it fails to do so if

$$k_2 < \tfrac{149}{184}k_1. \tag{6.44}$$

Note that the tensor $\mathbf{R}$ in $(6.42)_2$ represents a rotation of angle π about $\mathbf{e}$. If conditions (6.42) are satisfied, but $(149/184)k_1 \leqslant k_2 < k_1$, there is as yet no indication as to whether $\mathbf{n}_{\mathbf{R}}^*$ is stable or unstable.

In these theorems about the stability of hedgehogs we have taken the domain to be a ball centred at the defect. When hedgehogs arise in domains other than balls, this analysis enables us to explore their stability against perturbations concentrated in small balls embracing the defect. This notion of local stability for defects was introduced by Ericksen (1988). It was Kinderlehrer (1991)[3] who first realized its relevance.

Most of the issues raised in this section have also been addressed from a different perspective in a series of papers by Giaquinta, Modica and Souček (1989a and b, 1990, 1991). They make use of Cartesian currents and other techniques of geometric measure theory. A readily accessible account of their methods and conclusions can be found in the expository paper of Giaquinta (1992).

6.2 Variable degree of orientation

So far we have taken the degree of orientation s as constant throughout the region occupied by the liquid crystal. In Section 1.3 of Chapter 1 we learned how both s and $\mathbf{n}$ are related to the order tensor $\mathbf{Q}$ for uniaxial nematics:

$$\mathbf{Q} = s(\mathbf{n}\otimes\mathbf{n} - \tfrac{1}{3}\mathbf{I}), \tag{6.45}$$

where $\mathbf{I}$ is the identity tensor. In this book the microscopic states of a material element of nematic liquid crystal are described through

[3] This paper reproduces a lecture given in October 1988 at the *Colloque* in honour of J.-L. Lions.

s and $\mathbf{n}$ separately, rather than through $\mathbf{Q}$. Clearly, as long as s is constant in space and different from zero, as is the case in the classical theory employed so far, choosing $\mathbf{n}$ instead of $\mathbf{Q}$ to represent the microstate of an element may introduce fictitious defects. In most problems studied in this book fictitious defects are easily singled out, and so the traditional view, which regards the optic axis as an element of $\mathbb{S}^2$, is generally well grounded. On the other hand, when s also is allowed to vary in space, the picture becomes more complicated, and describing the microstate through the pair $(s, \mathbf{n})$ may differ considerably from describing it through $\mathbf{Q}$.

Here I assume that the free energy functional depends on the fields s and $\mathbf{n}$ defined over the region $\mathscr{B}$ occupied by the liquid crystal. For completeness, following Ericksen (1991), I also discuss below the conditions to be imposed on a specific form of the free energy density so as to make it a function of $\mathbf{Q}$ and its gradient.

Before proceeding further I recall from Section 1.3 of Chapter 1 the main qualitative features of the definition of s and $\mathbf{n}$. The degree of orientation s is defined in terms of the statistical distribution of molecular orientations. Qualitatively, s measures on the macroscopic scale the degree of microscopic order; it may vary in space and ranges in the interval $[-\frac{1}{2}, 1]$. The lower bound on s represents the state of microscopic order in which all molecules in a material element, which can be identified with a point of $\mathscr{B}$, are orthogonal to $\mathbf{n}$, but otherwise disordered. The upper bound on s corresponds to a state of perfect microscopic order in which all molecules in an element are parallel to $\mathbf{n}$. Both these cases are ideal: they never occur in real substances.

When s vanishes the molecular orientation is completely disordered: the liquid crystal becomes an isotropic fluid on the macroscopic scale. In the isotropic phase $\mathbf{n}$ is deprived of any meaning since the orientation of one molecule is completely unrelated to the orientation of another. We interpret the occurrence of the value $s=0$ at a point of $\mathscr{B}$ as the sign that the liquid crystal is there in its isotropic phase, though no change occurs in the temperature.

In the new theory s and $\mathbf{n}$ are fields defined thus:

$$s: \mathscr{B} \to [-\tfrac{1}{2}, 1], \quad \mathbf{n}: \mathscr{B} \setminus \mathscr{S}(s) \to \mathbb{S}^2, \tag{6.46}$$

where

$$\mathscr{S}(s) := \{p \in \mathscr{B} \mid s(p) = 0\} \tag{6.47}$$

is the **singular set** of the pair $(s, \mathbf{n})$. If $\mathscr{S}(s) \neq \varnothing$, we say that the pair $(s, \mathbf{n})$ is **singular**, otherwise we say that it is **regular**. The singular set is the set where discontinuities of $\mathbf{n}$ may occur. Thus, within this theory, defects of liquid crystals are to be interpreted as transitions to the isotropic phase, which happen to be confined in space.

The theory that we employ in this chapter has been strongly motivated by the desire for a comprehensive treatment of defects with arbitrary space dimension. While point defects are described fairly well within the classical theory (*cf.* Section 6.1 above), this is not the case for both line and plane defects: the former need to be approximated by regular fields in order to keep Frank's energy finite, as in the solution of Cladis and Kléman illustrated in Section 6.5 below, and the latter is inconsistent with all orientation fields that have finite energy. More precisely, Corollary 6.7 in Section 6.1 says that every field $\mathbf{n}$ that minimizes Frank's energy functional subject to strong anchoring conditions may be discontinuous at most on sets of space dimension strictly less than 1.

Ericksen's new mathematical theory results from a long progress towards a deeper understanding of defects in liquid crystals. De Gennes (1969) proposed a theory to describe the change of liquid crystals into isotropic fluids that was modelled after the general theory of Landau for phase transitions. The degree of orientation came on the scene at that time. Fan (1971) employed a degree of orientation variable in space to tame the singularities of $\mathbf{n}$: he allowed the liquid crystal to become isotropic wherever the classical theory predicts a singularity of $\mathbf{n}$ with infinite energy.

Ericksen (1991) has pursued Fan's line of thought and has proposed a general theory for both the statics and the dynamics of liquid crystals with variable degree of orientation. Here we only explore some consequences of the equilibrium theory. The dynamical theory has been applied in several papers of M.C. Calderer to describe the evolution of defects which have apparently been observed in some polymeric liquid crystals (see *e.g.* Calderer (1990, 1991a and b, 1992)). We still lack explanation of how defects arise from regular orientation fields as time elapses. The reader is also referred to Macmillan's (1992a, b) recent contribution to the dynamics of liquid crystals.

6.2.1 Energy functional

Let $\mathscr{B}$ be the region in space occupied by the liquid crystal; denote by

$$\mathscr{F}_E[s,\mathbf{n}] := \int_{\mathscr{B}} \sigma_E(s, \nabla s, \mathbf{n}, \nabla\mathbf{n})\,dv \tag{6.48}$$

the energy functional, which now depends on both s and $\mathbf{n}$. The fields s and $\mathbf{n}$, which are defined as in (6.46), are taken within the class

$$\mathscr{C} := \{(s,\mathbf{n}) \mid s \in C^0(\bar{\mathscr{B}}) \cap C^1(\bar{\mathscr{B}} \setminus \mathscr{S}(s)), \quad \mathbf{n} \in C^1(\bar{\mathscr{B}} \setminus \mathscr{S}(s))\}, \tag{6.49}$$

where $\bar{\mathscr{B}}$ denotes the closure of $\mathscr{B}$ and $\mathscr{S}(s)$ is the singular set of the pair $(s, \mathbf{n})$.

The mapping $(s, \nabla s, \mathbf{n}, \nabla \mathbf{n}) \mapsto \sigma_E(s, \nabla s, \mathbf{n}, \nabla \mathbf{n})$, which gives the free energy per unit volume at a given temperature, is subject here to conditions that generalize those put forward within the theory that regards s as constant (*cf.* Section 3.1 of Chapter 3). Thus, frame-indifference and material symmetry require that

$$\sigma_E(s, \mathbf{Q}\nabla s, \mathbf{Q}\mathbf{n}, \mathbf{Q}\nabla\mathbf{n}\mathbf{Q}^T) = \sigma_E(s, \nabla s, \mathbf{n}, \nabla \mathbf{n}) \quad \text{for all } \mathbf{Q} \in \mathrm{G}(\mathscr{V}), \tag{6.50}$$

where $G(\mathscr{V})$ is the group defined by

$$G(\mathscr{V}) := \begin{cases} O(\mathscr{V}) & \text{for nematics,} \\ SO(\mathscr{V}) & \text{for cholesterics.} \end{cases} \tag{6.51}$$

In particular, the central inversion $-\mathbf{I}$, belongs to $G(\mathscr{V})$ for nematics, and so (6.50) implies that

$$\sigma_E(s, -\nabla s, -\mathbf{n}, \nabla \mathbf{n}) = \sigma_E(s, \nabla s, \mathbf{n}, \nabla \mathbf{n}). \tag{6.52}$$

Furthermore, σ_E must be even in $\mathbf{n}$, that is

$$\sigma_E(s, \nabla s, -\mathbf{n}, -\nabla \mathbf{n}) = \sigma_E(s, \nabla s, \mathbf{n}, \nabla \mathbf{n}); \tag{6.53}$$

σ_E must also be positive definite, that is

$$\sigma_E(s, \nabla s, \mathbf{n}, \nabla \mathbf{n}) > 0 \quad \text{for all } (s, \mathbf{n}) \in \mathscr{C} \setminus \mathscr{N} \tag{6.54}$$

and

$$\sigma_E(s, \nabla s, \mathbf{n}, \nabla \mathbf{n}) = 0 \quad \text{for all } (s, \mathbf{n}) \in \mathscr{N}, \tag{6.55}$$

where $\mathscr{N}$ is a class of natural configurations, different for nematics and cholesterics, whence one should start to produce the configuration $(s, \mathbf{n})$, expending the energy $\mathscr{F}_E[s, \mathbf{n}]$.

Henceforth in this chapter we confine attention to nematics. For them $\mathscr{N}$ consists of all pairs $(s, \mathbf{n})$ of $\mathscr{C}$ for which both s and $\mathbf{n}$ are constant fields: the former takes a single value s_1 depending on the temperature, while the latter can take any value on $\mathbb{S}^2$.

Following Ericksen (1991), we assume a special form for σ_E:

$$\sigma_E(s, \nabla s, \mathbf{n}, \nabla \mathbf{n}) = \sigma_0(s, \mathbf{n}) + \sigma_1(s, \nabla s, \mathbf{n}, \nabla \mathbf{n}) + \sigma_2(s, \nabla s, \mathbf{n}, \nabla \mathbf{n}), \tag{6.56}$$

where σ_1 is linear in ∇s and $\nabla \mathbf{n}$, and σ_2 is homogeneously quadratic in these fields.

Combining (6.52) and (6.53), we arrive at

$$\sigma_E(s, -\nabla s, \mathbf{n}, -\nabla\mathbf{n}) = \sigma_E(s, \nabla s, \mathbf{n}, \nabla\mathbf{n}), \tag{6.57}$$

whence for σ_E as in (6.56) we conclude that

$$\sigma_1 = 0. \tag{6.58}$$

Likewise, by the use of (6.50), (6.54) and (6.55) we show that σ_0 cannot depend on $\mathbf{n}$ and must be such that

$$\sigma_0(s_1) = 0 \quad \text{and} \quad \sigma_0(s) > 0 \quad \text{for all } s \neq s_1. \tag{6.59}$$

Furthermore, paralleling our reasoning in Section 3.2 of Chapter 3, one arrives at the following formula for σ_2:

$$\begin{aligned}\sigma_2(s, \nabla s, \mathbf{n}, \nabla\mathbf{n}) = {} & k_1(\operatorname{div}\mathbf{n})^2 + k_2(\mathbf{n}\cdot\operatorname{curl}\mathbf{n})^2 + k_3|\mathbf{n}\wedge\operatorname{curl}\mathbf{n}|^2 \\ & + (k_2 + k_4)(\operatorname{tr}(\nabla\mathbf{n})^2 - (\operatorname{div}\mathbf{n})^2) + h_1|\nabla s|^2 \\ & + h_2(\nabla s\cdot\mathbf{n})^2 + h_3(\operatorname{div}\mathbf{n})(\nabla s\cdot\mathbf{n}) + h_4\nabla s\cdot(\nabla\mathbf{n})\mathbf{n},\end{aligned} \tag{6.60}$$

where for a given temperature the k's and h's are functions of s. To explore the consequences of (6.54) and (6.55) on this formula, it is convenient to rewrite it at follows

$$\begin{aligned}\sigma_2(s, \nabla s, \mathbf{n}, \nabla\mathbf{n}) = {} & k_1'(\operatorname{div}\mathbf{n})^2 + k_2(\mathbf{n}\cdot\operatorname{curl}\mathbf{n})^2 + k_3'|\mathbf{n}\wedge\operatorname{curl}\mathbf{n}|^2 \\ & + (k_2 + k_4)(\operatorname{tr}(\nabla\mathbf{n})^2 - (\operatorname{div}\mathbf{n})^2) \\ & + k_5|\nabla s - (\nabla s\cdot\mathbf{n})\mathbf{n} - \lambda(\nabla\mathbf{n})\mathbf{n}|^2 \\ & + k_6(\nabla s\cdot\mathbf{n} - \mu\operatorname{div}\mathbf{n})^2,\end{aligned} \tag{6.61}$$

where

$$\begin{aligned} & k_1' := k_1 - \mu^2 k_6 = k_1 - \frac{h_3^2}{4(h_1 + h_2)}, \\ & k_3' := k_3 - \lambda^2 k_5 = k_3 - \frac{h_4^2}{4h_1}, \\ & k_5 := h_1, \quad k_6 := h_1 + h_2, \\ & \mu := -\frac{h_3}{2(h_1 + h_2)}, \quad \lambda := -\frac{h_4}{2h_1}. \end{aligned} \tag{6.62}$$

Making use of (6.61) and (6.62), Ericksen (1991) proved that σ_2 is strictly positive whenever the following inequalities hold:

$$k_1' > 0,\ k_2 > |k_4|,\ k_3' > 0,\ 2k_1' - k_2 - k_4 > 0,\ k_5 > 0,\ k_6 > 0. \tag{6.63}$$

6.2.2 The potential σ_0

Moving from different premisses, both De Gennes (1969) and Doi (1981) arrived at the same form of the potential σ_0, which is

$$\sigma_0(s) = \tfrac{1}{2}as^2 - \tfrac{1}{3}bs^3 + \tfrac{1}{4}cs^4 + d, \tag{6.64}$$

where a, b, c, and d are positive functions of the temperature. The qualitative features of this potential are illustrated by the graphs of Fig. 6.3 for five different temperatures.

For each liquid crystal there are three characteristic temperatures, namely T^*, T_c, and T^+. T^* and T^+ bound the range of temperatures within which the substance is a liquid crystal. When $T = T^*$, σ_0 possesses a minimum at $s = s_1$ and an inflection point at $s = 0$ (Fig. 6.3a).

When $T^* < T < T_c$, σ_0 possesses two minima: one at $s = 0$ and the other at $s = s_1$; the former minimum is relative while the latter is absolute (Fig. 6.3b). In this range of temperatures the equilibrium phase with higher degree of orientation corresponds to the minimum value of σ_0; thus, the oriented phase is **stable**, whereas the isotropic **phase** is **metastable**.

For $T = T_c$ the two minima at σ_0 have the same value (Fig. 6.3c). This temperature marks the transition to the isotropic phase of a nematic liquid crystal in any state homogeneous in space. Thus, T_c is to be identified with the temperature T_{NI} introduced in Section 1.1 of Chapter 1.

When $T_c < T < T^+$ the phase with $s = s_1$ becomes metastable, while the isotropic phase is stable (Fig. 6.3d).

Finally, for $T = T^+$, σ_0 possesses a minimum at $s = 0$ and an inflection point at $s = s_1$ (Figure 6.3e): the isotropic phase is stable, but the oriented equilibrium phase is neither stable nor metastable.

It is worth noting that the parameter s_1, which in the preceding discussion designates the oriented equilibrium phase, is indeed a function of the temperature: as T increases in the interval $[T^*, T^+]$ s_1 decreases, but it never vanishes.

If in (6.56) σ_0 prevails over σ_2, it tends to impose everywhere in $\mathscr{B}$ the value of s that makes it minimum. Thus, when $T^* < T < T_c$ one expects that the minimizers of the functional $\mathscr{F}_E$ above be the same as Frank's and that the new theory lead to conclusions not too different from those reached by the classical one. Such a heuristic argument suggests also that the classical theory should be

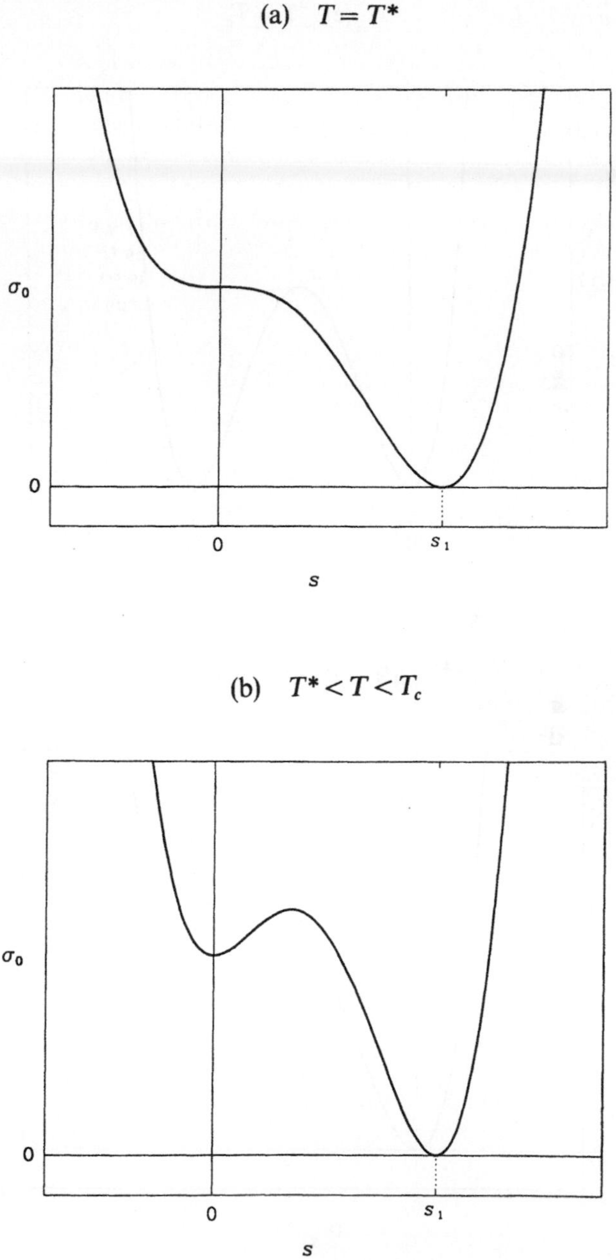

Figure 6.3

(c) $T = T_c$

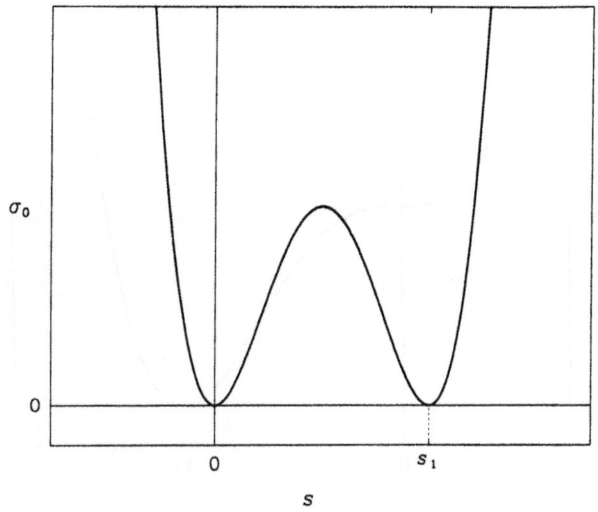

(d) $T_c < T < T^+$

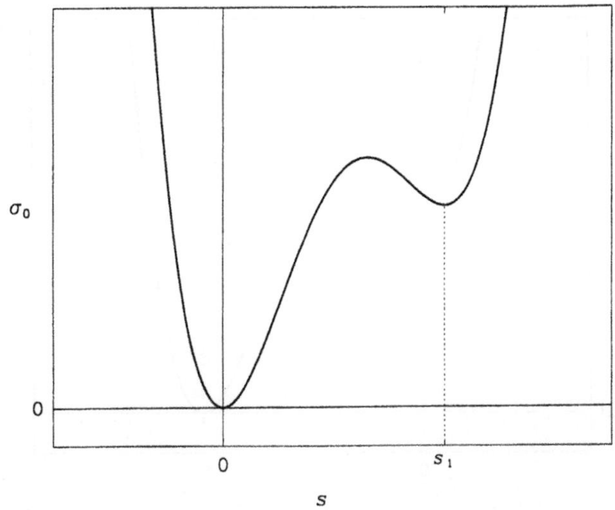

Figure 6.3 *(cont'd)*

(e) $T = T^+$

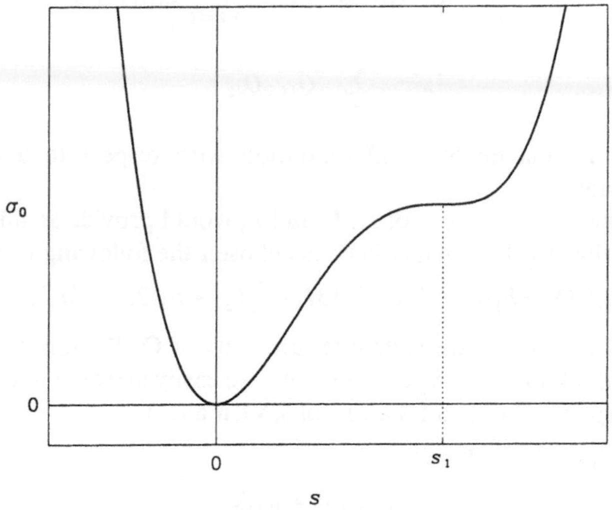

Figure 6.3 *(cont'd)*

more successful when T is closer to T^*. On the other hand, when $T = T_c$, σ_0 though prevailing on σ_2 *cannot* dictate *one* value of s because it attains its minimum at *two* distinct values of s, namely $s = 0$ and $s = s_1$. Thus, the consequences of the new theory are likely to differ more from those of the classical one when T is closer to T_c rather than to T^*.

6.2.3 Consistency conditions

The formula we have recorded in (6.60) above for σ_2 has been derived under the assumption that s and $\mathbf{n}$ are independent from one another. If, on the other hand, we regard them as meaningful just in that they are related to $\mathbf{Q}$ through (6.45), we should replace σ_2 with a function σ_2' of $\mathbf{Q}$ and its gradient $\nabla\mathbf{Q}$, which is quadratic in this latter and isotropic. Following Ericksen (1991) we write in Cartesian coordinates the three independent quadratic invariants of $\nabla\mathbf{Q}$, namely

$$I_1 := \sum_{i,j,k=1}^{3} Q_{ij,i}Q_{kj,k}, \quad I_2 := \sum_{i,j,k=1}^{3} Q_{ij,k}Q_{ij,k},$$
$$I_3 := \sum_{i,j,k=1}^{3} Q_{ij,k}Q_{ik,j}, \tag{6.65}$$

where a comma denotes differentiation with respect to a space co-ordinate.

Any linear combination of I_1, I_2 and I_3 would provide an admissible function σ_2'; Ericksen (1991) has chosen the following one:

$$\sigma_2'(\mathbf{Q}, \nabla\mathbf{Q}) = b_1(I_3 - I_1) + b_2(3I_1 - \tfrac{1}{2}I_2) + b_3(2I_2 - 3I_1), \tag{6.66}$$

where b_1, b_2 and b_3 are isotropic functions of $\mathbf{Q}$. Writing $\mathbf{Q}$ as in (6.45) and taking b_1, b_2, b_3 as constants, we easily arrive at a formula which expresses σ_2' as a function of $s, \nabla s, \mathbf{n}$ and $\nabla\mathbf{n}$:

$$\begin{aligned}\sigma_2'(\mathbf{Q}, \nabla\mathbf{Q}) &= \sigma_2^0(s, \nabla s, \mathbf{n}, \nabla\mathbf{n})\\ &:= (b_2 - b_3)(\nabla s\cdot\mathbf{n})^2 + b_3|\nabla s|^2\\ &\quad + 2(2b_2 - b_1 - 2b_3)s(\operatorname{div}\mathbf{n})(\nabla s\cdot\mathbf{n})\\ &\quad + 2(b_1 - b_2 + b_3)s\nabla s\cdot(\nabla\mathbf{n})\mathbf{n}\\ &\quad + (2b_2 + b_3)s^2[(\operatorname{div}\mathbf{n})^2 + |\mathbf{n}\wedge\operatorname{curl}\mathbf{n}|^2]\\ &\quad + (4b_3 - b_2)s^2(\mathbf{n}\cdot\operatorname{curl}\mathbf{n})^2\\ &\quad + (b_1 - b_2 + 4b_3)s^2[\operatorname{tr}(\nabla\mathbf{n})^2 - (\operatorname{div}\mathbf{n})^2].\end{aligned} \tag{6.67}$$

Comparing (6.67) and (6.60) we see that the former formula reduces to the latter if we set

$$\begin{gathered}k_1 = k_3 = (2b_2 + b_3)s^2, \quad k_2 = k_3 = (4b_3 - b_2)s^2, \quad k_4 = b_1s^2,\\ h_1 = b_3, \quad h_2 = b_2 - b_3, \quad h_3 = 2(2b_2 - b_1 - 2b_3)s,\\ h_2 = 2(b_1 - b_2 + b_3)s.\end{gathered} \tag{6.68}$$

Furthermore, when s does not vanish, inequalities (6.63) imply the following inequalities for the constants b_1, b_2 and b_3:

$$\begin{gathered}b_2 > 0, \quad b_3 > 0, \quad 4b_3 - b_2 > |b_1|,\\ b_2(2b_2 + b_3) > (b_1 - 2b_2 + 2b_3)^2\\ b_3(2b_2 + b_3) > (b_1 - b_2 + b_3)^2.\end{gathered} \tag{6.69}$$

Equations (6.68) require, in particular, that k_1 and k_3 be equal, which is contrary to most experimental evidence. To avoid this disagreement one should allow b_1, b_2 and b_3 in (6.66) to depend on s. Here we do not pursue this line of thought any farther; rather, we consider an approximation which is compatible with (6.68) and parallels the one-constant approximation to Frank's formula introduced in Section 3.3 of Chapter 3.

6.2.4 One-constant approximation

Setting

$$k_1 = k_2 = k_3 = k_E s^2, \quad k_4 = 0, \quad h_1 = k_E k, \quad h_2 = h_3 = h_4 = 0, \quad (6.70)$$

where both k_E and k are positive constants, by formula (3.25) of Chapter 3, we give the function σ_2 in (6.60) above the following form:

$$\sigma_2(s, \nabla s, \mathbf{n}, \nabla \mathbf{n}) = k_E(k|\nabla s|^2 + s^2|\nabla \mathbf{n}|^2). \quad (6.71)$$

A glance at (6.68) suffices to show that (6.67) and (6.71) agree, provided that

$$b_1 = 0, \quad b_2 = b_3 = \tfrac{1}{3}k_E, \quad \text{and} \quad k = \tfrac{1}{3}. \quad (6.72)$$

In this chapter we often write σ_E as

$$\sigma_E(s, \nabla s, \mathbf{n}, \nabla \mathbf{n}) = k_E(k|\nabla s|^2 + s^2|\nabla \mathbf{n}|^2) + \sigma_0(s), \quad (6.73)$$

bearing in mind that it may be regarded as a quadratic form of $\nabla \mathbf{Q}$ only when $(6.72)_4$ is satisfied.

Equation (6.73) provides the simplest formula quadratic in both ∇s and $\nabla \mathbf{n}$, in which both s and $\mathbf{n}$ are taken as independent state variables, regardless of their relation to the order tensor $\mathbf{Q}$. This formula exhibits the essential feature of the new model: the elastic energy associated with the change in space of $\mathbf{n}$ loses strength wherever s approaches zero, at the expense of the energy required to lower the degree of order in the material. We take k as an adjustable parameter which balances one competing energy against the other.

6.3 New defects

We now describe how the new theory treats defects with space dimension 1 or higher, the defects that the classical theory fails to explain. In this section our development parallels that of Section 6.1:

we attempt to establish the existence of minimizers for the functional $\mathscr{F}_E$ introduced in (6.48) and their degree of regularity.

6.3.1 Existence of minimizers

Only Lin and Poon (1993) have dealt so far with the functional $\mathscr{F}_E$ when σ_E is the most general quadratic form in ∇s and $\nabla \mathbf{n}$ (*cf.* (6.56) and (6.60)). We first state their existence theorem; apart from it, our study below will always concern the special form of σ_E given in (6.73).

Let $\mathscr{B}$ be a bounded regular region in the three-dimensional Euclidean space $\mathscr{E}$; as above we take it to be with no edge and no vertex. Let s_0 and $\mathbf{n}_0$ be fields prescribed thus on the whole of $\partial\mathscr{B}$:

$$s_0:\partial\mathscr{B}\rightarrow[-\tfrac{1}{2},1], \quad \mathbf{n}_0:\partial\mathscr{B}\backslash\mathscr{S}(s_0)\rightarrow\mathbb{S}^2, \tag{6.74}$$

where

$$\mathscr{S}(s_0):=\{p\in\partial\mathscr{B}\,|\,s_0(p)=0\} \tag{6.75}$$

is the singular set of the pair $(s_0, \mathbf{n}_0)$. The class of admissible minimizers for $\mathscr{F}_E$ is defined by

$$\begin{aligned}\mathscr{A}(s_0,\mathbf{n}_0):=\{(s,\mathbf{n})\,|\,s\in W^{1,2}(\mathscr{B},[-\tfrac{1}{2},1]),\mathbf{u}\in W^{1,2}(\mathscr{B},\mathscr{V}),\mathbf{u}:=s\mathbf{n},\\ s_0 \text{ and } \mathbf{u}_0:=s_0\mathbf{n}_0 \text{ are the traces of } s \text{ and } \mathbf{u} \text{ on } \partial\mathscr{B}\},\end{aligned} \tag{6.76}$$

where $W^{1,2}$ denotes as above the space of all L^2-mappings with first weak derivatives in L^2.

The theorems reviewed in this section do not require the potential σ_0 introduced above to be as in (6.64), though this seems to be the form most appropriate for it. Some authors assume σ_0 to possess qualitative features that (6.64) does not display, while others assume for σ_0 a form far more general than it.

Lin and Poon (1993) proved the following theorem:

Theorem 6.16 Let σ_E be the function defined by

$$\sigma_E(s,\nabla s,\mathbf{n},\nabla\mathbf{n}):=\sigma_0(s)+\sigma_2(s,\nabla s,\mathbf{n},\nabla\mathbf{n}). \tag{6.77}$$

Assume that σ_0, which maps $]-\tfrac{1}{2},1[$ into $\mathbb{R}$, is of class C^1 and such that

$$\lim_{s\rightarrow-(1/2)^+}\sigma_0(s)=\lim_{s\rightarrow1^-}\sigma_0(s)=+\infty \tag{6.78}$$

and

$$\lim_{s\rightarrow0}\frac{\sigma_0(s)}{s^2}\neq0. \tag{6.79}$$

Assume further that σ_2 is as in (6.61), where the functions k_1', k_2, k_3' and k_4 are now defined by

$$k_i' := \alpha_i s^2 \quad \text{for } i = 1,3 \quad \text{and} \quad k_i := \alpha_i s^2 \quad \text{for } i = 2,4, \tag{6.80}$$

with $\alpha_1, \alpha_2, \alpha_3$ and α_4 constants that satisfy

$$\alpha_1 > \alpha_2 + \alpha_4, \quad \alpha_2 > 0, \quad \alpha_3 > \alpha_2 + \alpha_4, \quad \alpha_2 \geqslant |\alpha_4|, \tag{6.81}$$

while both k_5 and k_6 are positive constants. Then for any given pair $(s_0, \mathbf{n}_0)$ as in (6.74) there are pairs $(s^*, \mathbf{n}^*)$ in $\mathscr{A}(s_0, \mathbf{n}_0)$ where $\mathscr{F}_E$ attains its minimum:

$$\mathscr{F}_E[s^*, \mathbf{n}^*] = \inf_{(s,\mathbf{n})\in\mathscr{A}(s_0,\mathbf{n}_0)} \mathscr{F}_E[s, \mathbf{n}]. \tag{6.82}$$

The proof of Theorem 6.16 rests upon the discovery of a null Lagrangian for the functional $\mathscr{F}_E$. Let the function σ_L^* be defined by

$$\sigma_L^*(s, \nabla s, \mathbf{n}, \nabla\mathbf{n}) := s^2\{\operatorname{tr}(\nabla\mathbf{n})^2 - (\operatorname{div}\mathbf{n})^2\} + 2s\nabla s\cdot\{(\nabla\mathbf{n})\mathbf{n} - (\operatorname{div}\mathbf{n})\mathbf{n}\}. \tag{6.83}$$

A direct computation shows that

$$\sigma_L^*(s, \nabla s, \mathbf{n}, \nabla\mathbf{n}) = \operatorname{div}\{s^2[(\nabla\mathbf{n})\mathbf{n} - (\operatorname{div}\mathbf{n})\mathbf{n}]\}, \tag{6.84}$$

whence it follows that the integral over $\mathscr{B}$ of σ_L^* can be reduced to a surface integral on $\partial\mathscr{B}$ which depends only on the traces of both s and $\mathbf{n}$. In a preparatory lemma Lin and Poon (1993) showed that there are two positive numbers, λ_1 and $\lambda_2 > \lambda_1$, depending on the constants $\alpha_1, \alpha_2, \alpha_3, \alpha_4$ and k_5, k_6, such that

$$\lambda_1(|\nabla s|^2 + s^2|\nabla\mathbf{n}|^2) \leqslant \sigma_2^*(s, \nabla s, \mathbf{n}, \nabla\mathbf{n}) \leqslant \lambda_2(|\nabla s|^2 + s^2|\nabla\mathbf{n}|^2), \tag{6.85}$$

where

$$\sigma_2^* := \sigma_2 + \tfrac{1}{2}(\alpha - \alpha_2 - \alpha_4)\sigma_L^* \tag{6.86}$$

and α is any appropriately small positive number.

Henceforth we assume that σ_E is given the special form in (6.73), and so $\mathscr{F}_E$ will be the functional

$$\mathscr{F}_F[s, \mathbf{n}] = \int_{\mathscr{B}} \{k_E(k|\nabla s|^2 + s^2|\nabla\mathbf{n}|^2) + \sigma_0(s)\}\, dv, \tag{6.87}$$

where both k_E and k are positive moduli.

The existence of minimizers for this functional has been established under slightly different hypotheses in the papers of Ambrosio (1990a) and Lin (1991a and b). They employ different techniques, but their conclusions are rather similar. Here we mainly follow Ambrosio (1990a).

Theorem 6.17. Let the mappings s_0 and $\mathbf{n}_0$ be given as in (6.74). For any continuous function σ_0 of $[-\frac{1}{2}, 1]$ into $\mathbb{R}$ there are pairs $(s^*, \mathbf{n}^*)$ in $\mathscr{A}(s_0, \mathbf{n}_0)$ where the functional $\mathscr{F}_E$ in (6.87) attains its minimum.

Ambrosio (1990a) also proved that every minimizer $(s^*, \mathbf{n}^*)$ of $\mathscr{F}_E$ in $\mathscr{A}(s_0, \mathbf{n}_0)$ is such that

$$\lim_{r \to 0^+} \frac{1}{r} \int_{B_r(x)} \{k_E(k|\nabla s|^2 + s^2|\nabla \mathbf{n}|^2) + \sigma_0\}\, dv = 0 \tag{6.88}$$

for all $x \in \mathscr{B}$. As above, $B_r(x)$ denotes the ball of radius r and centre at x. Moreover, for any subset $\mathscr{K}$ with closure included in $\mathscr{B}$, there is a constant C depending only on $\mathscr{K}$ such that

$$\frac{1}{r} \int_{B_r(x)} \{k_E(k|\nabla s|^2 + s^2|\nabla \mathbf{n}|^2) + \sigma_0(s)\}\, dv < C \tag{6.89}$$

for all $x \in \mathscr{K}$.

These conclusions are most remarkable in that they have no analogue in the classical theory, even in the one-constant approximation to Frank's energy functional, as a glance at (6.29) will suffice to show.

6.3.2 Regularity of minimizers

To appreciate the full extent of the new theory, and to find out whether it provides a complete description of the variety of defects occurring in liquid crystals, we address the question of the regularity of energy minimizers. Both Ambrosio (1990b) and Lin (1991a and b) have studied this problem. The former author confined attention to the behaviour of minimizers away from their defects, while the latter moved a few steps forward and gave the first estimate of the space dimension of the singular set in terms of the parameter k.

For clarity, before bringing together the outcomes of these studies we recall an elementary definition.

Definition 6.2. Let u be a function of a closed subset $\mathscr{K}$ of $\mathscr{E}$ into $\mathbb{R}$ and let α be given in $]0, 1[$. We say that u is **Hölder continuous**

with exponent α in $\mathscr{K}$ if there is a positive constant c such that

$$|u(x) - u(y)| < c|x - y|^{\alpha} \quad \text{for all } x, y \in \mathscr{K}. \tag{6.90}$$

We denote by $C^{0,\alpha}$ the class of all functions that satisfy (6.90). If, in addition, u is differentiable in $\mathscr{K}$, we say that it belongs to the class $C^{1,\alpha}$ whenever it belongs to $C^{0,\alpha}$ together with its first partial derivatives. If (6.90) is satisfied with $\alpha = 1$, the function u is said to be **Lipschitz continuous**. The above definition can easily be extended to both continuous and differentiable mappings into the translations space $\mathscr{V}$.

Theorem 6.18. Let the potential σ_0 in (6.87) be a function of class C^1 such that

$$\sigma_0'(0) = 0. \tag{6.91}$$

If $(s^*, \mathbf{n}^*)$ is a minimizer of $\mathscr{F}_E$ in $\mathscr{A}(s_0, \mathbf{n}_0)$, for any compact subset $\mathscr{K}$ of $\mathscr{B}$ there exists $\alpha \in]0, 1[$, such that both s^* and $\mathbf{u}^* := s^*\mathbf{n}^*$ are Hölder continuous with exponent α in $\mathscr{K}$.

When the parameter k falls in the interval $]0, 1[$, with a slightly heavier assumption on σ_0 we can reach a much neater conclusion.

Theorem 6.19. Let σ_0 be of class C^2 and let k be given in $]0, 1[$. If $(s^*, \mathbf{n}^*)$ is a pair that minimizes $\mathscr{F}_E$ in $\mathscr{A}(s_0, \mathbf{n}_0)$ then both s^* and $\mathbf{u}^* := s^*\mathbf{n}^*$ are Lipschitz continuous in any compact subset $\mathscr{K}$ of $\mathscr{B}$.

In words, we could say that these regularity theorems, which are all Lin's, establish Hölder and Lipschitz continuity of both s^* and $\mathbf{u}^*$ locally in $\mathscr{B}$. In particular, this amounts to saying that constants like c in (6.90) would depend on the closed subset of $\mathscr{B}$ to which we restrict attention.

6.3.3 Singular set

We learned in Section 6.1 that within the classical theory a defect in the orientation field $\mathbf{n}$ that minimizes the free energy can arise only wherever the function $S_{\mathbf{n}}$ in (6.15) does not vanish. From Section 6.2 we know that within the new theory $\mathbf{n}$ is not even defined in the set $\mathscr{S}(s)$ where s vanishes. Furthermore, the theorems above on the regularity of a minimizing pair $(s^*, \mathbf{n}^*)$ tell us that $\mathbf{n}^*$ cannot be singular away from $\mathscr{S}(s^*)$. Thus the **nodal set** of s^* accommodates all singularities of $\mathbf{n}^*$.

It is crucial here to estimate the Hausdorff dimension of $\mathscr{S}(s^*)$, as it was for the singular set $\mathscr{S}(\mathbf{n}^*)$ defined in Section 6.1. Below we restrict attention to the minimizing pairs $(s^*, \mathbf{n}^*)$ for which s^* is negative nowhere in $\mathscr{B}$. Minimizers like these actually exist whenever the potential σ_0 possesses a peculiar feature, which the function in (6.64) also has.

Theorem 6.20. Let σ_0 be a function of class C^1 such that $\sigma_0'(s) \leqslant 0$ whenever $s \leqslant 0$. If the function $s_0: \partial\mathscr{B} \to [-\frac{1}{2}, 1]$ is such that $s_0(x) \geqslant 0$ for all $x \in \partial\mathscr{B}$, then for each pair $(s^*, \mathbf{n}^*)$ that minimizes $\mathscr{F}_E$ in $\mathscr{A}(s_0, \mathbf{n}_0)$ the function s^* is not negative in $\mathscr{B}$.

Theorem 6.21. Suppose that the field s_0 in $(6.74)_1$ does not vanish identically on $\partial\mathscr{B}$. If $(s^*, \mathbf{n}^*)$ is any minimizing pair of $\mathscr{F}_E$ in $\mathscr{A}(s_0, \mathbf{n}_0)$, then the Hausdorff dimension $d_{\mathscr{S}(s^*)}$ of its singular set is such that

$$d_{\mathscr{S}(s^*)} \leqslant 2 \quad \text{for } 0 < k \leqslant 1 \quad \text{and} \quad d_{\mathscr{S}(s^*)} \leqslant 1 \quad \text{for } k > 1.$$

Thus, according to the new theory both plane and line defects *may* occur in the orientation fields that minimize the free energy: they are the *new defects*, those incompatible with the classical theory, which accounts only for point defects. We shall see in the two following sections how plane and line defects *do* actually arise in the solutions of specific variational problems posed within the new theory.

Remark 1. The potential σ_0 (6.87) does not affect the space dimension of the singular set for an energy minimizer.

Theorem 6.21, which is proved in Lin (1991b), does not have the last word on the dimension of the singular set of a minimizer. It has recently been improved by Hardt and Lin (1993).

Theorem 6.22. If the hypotheses of Theorem 6.21 hold, then the singular set of every minimizer consists only of isolated points for all $k > 1$.

6.3.4 Projective plane

On several occasions in this book we have remarked that $\mathbb{S}^2$ is not the most appropriate manifold to describe the orientation of a liquid crystal, because $\mathbf{n}$ and its opposite have precisely the same physical meaning. It might be sensible to take $\mathbf{n}$ in the **projective plane** $\mathbb{P}^2$, the manifold that results from $\mathbb{S}^2$ when all antipodal points are identified.

Hardt and Lin (1993) asked whether the space dimension of the singular set of a minimizer would be affected by this basic change, in the classical theory as well as in the new one.

They showed that $\mathscr{S}(\mathbf{n}^*)$ is again a closed subset of $\mathscr{B}$ of one-dimensional Hausdorff measure zero, as in Section 6.1. This actually means that within the classical theory one may locally orient a minimizing field into $\mathbb{P}^2$ so as to obtain a minimizing field into $\mathbb{S}^2$. In other words, the classical theory is not sensitive to the change of $\mathbb{S}^2$ into $\mathbb{P}^2$.

This is not the case with the new theory. Hardt and Lin (1993) also showed that when the projective plane takes the place of the unit sphere, for $k > 1$, the estimate in Theorem 6.21 applies, while that in Theorem 6.22 does not.

Remark 2. There is no indication as to whether the conclusions about the regularity of energy minimizers reached above apply also when $\mathscr{F}_E$ is given a form more general than (6.87).

6.3.5 Neglecting σ_0

Lin (1989) studied variational problems for $\mathscr{F}_E$ subject on $\partial\mathscr{B}$ to the condition

$$\left.\frac{\partial s}{\partial \mathbf{v}}\right|_{\partial\mathscr{B}} = \mathbf{0}, \qquad (6.92)$$

where $\mathbf{v}$ is the outward unit normal of $\mathscr{B}$. Here we prefer to prescribe s on $\partial\mathscr{B}$ for two reasons. First, because we take the course of prescribing s wherever on $\partial\mathscr{B}$ we are to prescribe $\mathbf{n}$ also, since this latter would be undefined if the former were to vanish. Second, because (6.92) is not compatible with neglecting σ_0 in (6.87), which will be done in the last two sections of this chapter to simplify the analysis. In fact, if (6.92) is required for s on $\partial\mathscr{B}$ and σ_0 is neglected in (6.87), $\mathscr{F}_E$ will attain its minimum for $s \equiv 0$ and arbitrary $\mathbf{n}$, that is when $\mathscr{B}$ consists only of the isotropic phase.

Neglecting σ_0 is for us just a matter of convenience. There is indeed evidence that this is not always legitimate for thermotropic liquid crystals away from their clearing point. Our choice has rather elegant mathematical consequences which are illustrated below. The first is a uniqueness theorem (*cf.* Lin (1991b)).

Theorem 6.23. When $\sigma_0 = 0$ in (6.87) there is precisely one minimizer of $\mathscr{F}_E$ in $\mathscr{A}(s_0, \mathbf{n}_0)$ if $0 < k < 1$.

It is worth noting that the energy minimizer may well not be unique for $k \geqslant 1$.

Furthermore, when $\sigma_0 = 0$, a recursive argument building upon Theorems 6.18 and 6.19 shows that the minimizers of $\mathscr{F}_E$ are at least of class C^∞ away from defects; actually, they are even analytic as the minimizers of $\mathscr{F}_F$ (*cf.* Section 6.1).

6.4 Plane defects

Now we study within the new theory a variational problem, which was first considered by Ambrosio and Virga (1991), who neglected the potential σ_0 in the energy functional, and it was then studied by Roccato and Virga (1992) under a less restrictive hypothesis. Here I bring together the conclusions of both these papers.

6.4.1 Variational problem

Let $\mathscr{B}$ be the region in the Euclidean space $\mathscr{E}$ bounded by two parallel plates, 2ℓ apart:

$$\mathscr{B} := \{p \in \mathscr{E} \mid -\ell < (p-o)\cdot\mathbf{e} < \ell, 0 < (p-o)\cdot\mathbf{e}_i < \ell_i, i = 1, 2\}, \quad (6.93)$$

where o is a point of $\mathscr{E}$ and $\mathbf{e}, \mathbf{e}_1$ and $\mathbf{e}_2$ are mutually orthogonal unit vectors. The degree of orientation s and the optic axis are given in this region by the fields defined in (6.46). We denote by x the co-ordinate of a point p along $\mathbf{e}$: that is, $x(p) := (p-o)\cdot\mathbf{e}$. The plates that bound $\mathscr{B}$ lie at $x = -\ell$ and $x = \ell$; we call them $\mathscr{S}^-$ and $\mathscr{S}^+$, respectively.

The total free energy is the functional $\mathscr{F}_E$ in (6.87) above. $\mathscr{F}_E[s, \mathbf{n}]$ is well defined if $s^2|\nabla\mathbf{n}|^2$ is given a definite value in the singular set $\mathscr{S}(s)$ of the pair $(s, \mathbf{n})$: we agree to set

$$s^2|\nabla\mathbf{n}|^2 = 0 \quad \text{in } \mathscr{S}(s). \quad (6.94)$$

We prescribe $\mathbf{n}$ on $\mathscr{S}^-$ and $\mathscr{S}^+$ thus:

$$\mathbf{n}|_{\mathscr{S}^-} = \cos\alpha_0\mathbf{e}_1 - \sin\alpha_0\mathbf{e}_2, \quad \mathbf{n}|_{\mathscr{S}^+} = \cos\alpha_0\mathbf{e}_1 + \sin\alpha_0\mathbf{e}_2, \quad (6.95)$$

where $\alpha_0 \in]-\pi/2, \pi/2[$ (see Fig. 6.4).

We prescribe s likewise:

$$s|_{\mathscr{S}^-} = s|_{\mathscr{S}^+} = s_1, \quad (6.96)$$

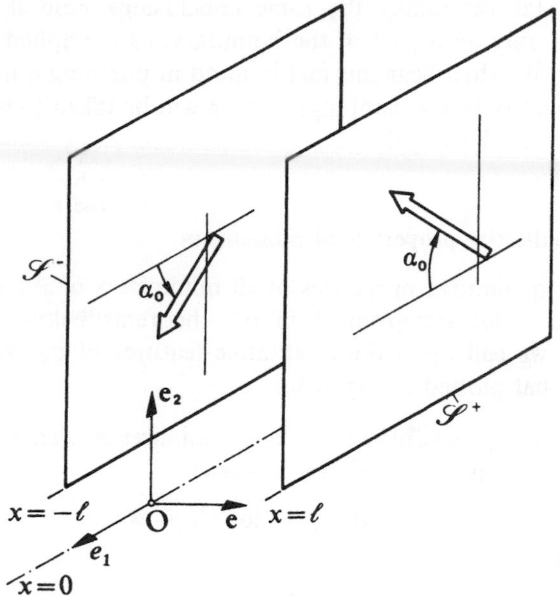

Figure 6.4

where s_1 is the minimizer of σ_0 in (6.64) when $T^* < T < T_c$ (*cf.* also Fig. 6.3b).

The reader must be aware of the fact that prescribing s on $\partial\mathscr{B}$ is not always justified because there may be no way to force s to take the value that we please at a solid surface. In particular, this seems unnatural when σ_0 is not neglected in (6.73): in such a case one might prefer not to prescribe s on $\partial\mathscr{B}$ and leave σ_0 to drive s in $\mathscr{B}$ so as to minimize the total free energy. Alternatively, one might introduce an anchoring energy, as we did in Section 5.6 of Chapter 5, and let it depend also on s. There are, however, also good reasons in favour of conditions (6.96). First, they make our development much easier. Second, when there is experimental evidence that σ_0 is the leading term of the energy density (6.73) (*cf.* Goossens (1985)), one should presume that s takes the value that minimizes σ_0 at least where $\mathbf{n}$ is prescribed.

From Section 6.3 we already know a great deal about the minimizers of $\mathscr{F}_E$ subject to (6.95) and (6.96). Though the theorems in Section 6.3 are stated under the assumption that both s and $\mathbf{n}$ are prescribed on the *whole* boundary of the region occupied by the liquid crystal, essentially the same conclusions hold if they are prescribed only on a part of the boundary, as is implied by (6.95) and (6.96). We shall bear this fact in mind in pursuing our analysis. In particular, both the mappings s and $\mathbf{n}$ will be taken to be of class C^2 away from the singular set.

6.4.2 Qualitative properties of minimizers

The main qualitative properties of all minimizers of $\mathscr{F}_E$ subject to (6.95) and (6.96) are grouped in the theorems below. In these, whenever we call upon the qualitative features of σ_0 we take its graph as that plotted in Fig. 6.3b.

Theorem 6.24. Let $(s, \mathbf{n})$ be a pair that minimizes $\mathscr{F}_E$. Then s depends only on x, the space co-ordinate along $\mathbf{e}$:

$$s(p) = \sigma(x(p)) \quad \text{for all } p \in \mathscr{B}. \tag{6.97}$$

Moreover,

$$\mathbf{n} = \cos\alpha(x)\mathbf{e}_1 + \sin\alpha(x)\mathbf{e}_2 \tag{6.98}$$

away from the singular set $\mathscr{S}(s)$.

The proof of Theorem 6.24 is omitted here. The reader can retrace it in Section 9 of Ambrosio and Virga (1991): there the potential σ_0 was set equal to zero, but this conclusion is not affected by σ_0.

In particular, Theorem 6.24 tells us that $\mathbf{n}$ also depends only on x and that in $\mathscr{B}\setminus\mathscr{S}(s)$ it is parallel to the plates $\mathscr{S}^-$ and $\mathscr{S}^+$. Furthermore, the singular set $\mathscr{S}(s)$ is the union of rectangles:

$$\mathscr{S}(s) = \bigcup \{o + x\mathbf{e} + x_1\mathbf{e}_1 + x_2\mathbf{e}_2 \,|\, x \in S(\sigma),\ x_1 \in [0, \ell_1], x_2 \in [0, \ell_2]\}, \tag{6.99}$$

where $S(\sigma)$ is the nodal set of σ

$$S(\sigma) := \{x \in [-\ell, \ell] \,|\, \sigma(x) = 0\}. \tag{6.100}$$

We say that a pair (σ, α) is **regular** if $S(\sigma) = \varnothing$, and that it is **singular** if $S(\sigma) \neq \varnothing$.

The functions α and σ inherit the following boundary conditions from $\mathbf{n}$ and s:

$$\alpha(-\ell) = -\alpha_0, \quad \alpha(\ell) = \alpha_0, \tag{6.101}$$

$$\sigma(-\ell) = \sigma(\ell) = s_1. \tag{6.102}$$

Let F be the functional defined by

$$F[\sigma,\alpha]:=k_E\ell_1\ell_2\int_{-\ell}^{\ell}\{k\sigma'^2+s^2\alpha'^2+\psi(\sigma)\}\,dx, \qquad (6.103)$$

where

$$\psi:=\frac{\sigma_0}{k_E}. \qquad (6.104)$$

The class of admissible functions for F is

$$\mathscr{A}:=\{(\sigma,\alpha)\,|\,\sigma\in AC(-\ell,\ell),\ \alpha\in AC_{\text{loc}}(]-\ell,\ell[\setminus S(\sigma))\}. \qquad (6.105)$$

Here $AC(-\ell,\ell)$ is the class of functions absolutely continuous on $]-\ell,\ell[$, while $AC_{\text{loc}}(]-\ell,\ell[\setminus S(\sigma))$ is the class of functions absolutely continuous in any compact subset of $]-\ell,\ell[\setminus S(\sigma)$. Henceforth a prime denotes differentiation with respect to x. In accordance with (6.94) we set

$$\sigma^2\alpha'^2=0 \quad \text{in } S(\sigma), \quad \text{for all } (\sigma,\alpha)\in\mathscr{A}. \qquad (6.106)$$

By Theorem 6.24, the minimum of $\mathscr{F}_E$ subject to (6.95) and (6.96) is the same as the minimum of F subject to (6.101) and (6.102).

Theorem 6.25. If the pair (σ,α) minimizes F in $\mathscr{A}$ then

$$\min_{[-\ell,\ell]}\sigma\geqslant 0, \quad \max_{[-\ell,\ell]}\sigma=s_1. \qquad (6.107)$$

Proof. The Euler equations for F in $]-\ell,\ell[\setminus S(\sigma)$ are

$$k\sigma''=\sigma\alpha'^2+\tfrac{1}{2}\dot{\psi}(\sigma), \qquad (6.108)$$

where a superimposed dot denotes differentiation with respect to σ, and

$$(\sigma^2\alpha')'=0. \qquad (6.109)$$

Since the right-hand side of (6.108) is negative for all $\sigma\in]-\frac{1}{2},1[$, no solution of (6.108) possesses a negative isolated minimum in $]-\ell,\ell[$. Thus, by (6.102), $(6.107)_1$ is proved. Similarly, since the right-hand side of (6.108) is positive for all $\sigma\in]s_1,1[$, no solution of (6.108) possesses in $]-\ell,\ell[$ an isolated maximum greater than s_1. The proof of (6.107) is thus complete.

Remark. Theorem 6.25 rests upon an elementary *maximum principle*. It can be shown that the same argument applies also if the minimizers of $\mathscr{F}_E$ do not enjoy the symmetry established in Theorem 6.24.

Whenever σ_0 is decreasing in $]-\frac{1}{2},0[$ and increasing in $]s_1,1[$, all minimizing s range in the interval $[0,s_1]$, provided that the boundary data also range in this interval. Thus, in solving a variational problem for $\mathscr{F}_E$ we need not worry about the precise form of σ_0 outside $[0,s_1]$, because it does not affect the energy minimizers.

Theorem 6.26. Let the restriction to $[0,s_1]$ of the function ψ in (6.104) be of class C^2. If the pair (σ,α) minimizes F in $\mathscr{A}$ and $S(\sigma)=\varnothing$, then σ is an even function, while α is odd. Moreover, σ' vanishes only at $x=0$.

Proof. If σ does not vanish in $]-\ell,\ell[$, equation (6.109) implies that there is a constant c such that

$$\alpha' = \frac{c}{\sigma^2}. \tag{6.110}$$

Integrating (6.110) and making use of $(6.101)_1$, we arrive at

$$\alpha(x) = -\alpha_0 + c\int_{-\ell}^{x} \frac{1}{\sigma^2(\xi)}\,d\xi; \tag{6.111}$$

thus, by $(6.101)_2$, c is subject to

$$2\alpha_0 = c\int_{-\ell}^{\ell} \frac{1}{\sigma^2(x)}\,dx. \tag{6.112}$$

Inserting (6.110) into (6.108), we get

$$k\sigma'' = \frac{c^2}{\sigma^3} + \tfrac{1}{2}\dot{\psi}(\sigma). \tag{6.113}$$

By solving equation (6.113) subject to (6.102) and (6.112), we determine both c and σ; equation (6.111) then gives α. Thus we obtain all regular minimizers of F, if there are any.

We first note that if $\alpha_0 \neq 0$ then $s \equiv s_1$ cannot be a solution of both (6.112) and (6.113) because the latter equation would then imply that $c=0$, which contradicts the former equation.

Let $\hat{\sigma}$ be a solution of (6.112), (6.113) and (6.102) that minimizes F. Then $\hat{\sigma}'(-\ell)$ is negative because (6.113) and $(6.102)_1$ imply that $\hat{\sigma}''(-\ell)>0$, and if it were the case that $\hat{\sigma}'(\ell)\geqslant 0$ then $(6.107)_2$ would be violated. Furthermore, by $(6.107)_1$ there exists $x_0\in]-\ell,\ell[$ such that $\hat{\sigma}(x_0)>0$, $\hat{\sigma}'(x_0)=0$, and $\hat{\sigma}'(x)<0$ for all $x\in]-\ell,x_0[$. Now let $\ell_0 := \min\{\ell, 2x_0+\ell\}$ and consider in $[x_0,\ell_0]$ the initial-value problem for (6.113) whose data are $\sigma(x_0) = \hat{\sigma}(x_0)$ and $\sigma'(x_0) = 0$. Since equation (6.113) does not depend explicitly on x, the only solution of this problem is the function defined by $\sigma(x) = \hat{\sigma}(2x_0 - x)$ for all $x \in [x_0, \ell_0]$; this is also the continuation of $\hat{\sigma}$ in $[x_0, \ell_0]$ because $\hat{\sigma}$ solves (6.113) also in this interval. Thus we conclude that

$$\hat{\sigma}(x) = \hat{\sigma}(2x_0 - x) \quad \text{for all } x \in [x_0, \ell_0], \tag{6.114}$$

$$\hat{\sigma}' < 0 \quad \text{in }]-\ell, x_0[\quad \text{and} \quad \hat{\sigma}' > 0 \quad \text{in } [x_0, \ell_0]. \tag{6.115}$$

It follows from (6.114) and (6.115) that $x_0 = 0$; otherwise, either $(6.102)_2$ or $(6.107)_2$ would be violated.

To complete the proof of the theorem it remains to be noted that (6.111) gives an odd function when σ is even.

6.4.3 Minimizers

To find an explicit solution of the variational problem stated above, we abandon here the level of generally kept thus far. To be precise, we assume that

$$\psi = 0. \tag{6.116}$$

Under this assumption we show that when the pair (σ, α) minimizes F in $\mathscr{A}$ the nodal set of σ must be either a singleton or the empty set.

Lemma 6.27. If the pair (σ, α) minimizes F in $\mathscr{A}$, then $S(\sigma)$ is connected.

Proof. Since σ is continuous, $S(\sigma)$ is a closed subset of $]-\ell, \ell[$. Let $m_\sigma := \min S(\sigma)$ and $M_\sigma := \max S(\sigma)$. We suppose for a contradiction that the set $\{x \in [m_\sigma, M_\sigma] \mid \sigma(x) \neq 0\}$ is not empty. We define the function $\tilde{\sigma}$ by

$$\tilde{\sigma}(x) := \begin{cases} \sigma(x) & \text{if } x \in]-\ell, \ell[\setminus [m_\sigma, M_\sigma], \\ 0 & \text{otherwise.} \end{cases} \tag{6.117}$$

An easy computation shows that

$$F[\sigma, \alpha] > F[\tilde{\sigma}, \alpha]. \tag{6.118}$$

Thus, (σ, α) does not minimize F, which is a contradiction.

Lemma 6.28. If the pair (σ, α) minimizes F in $\mathscr{A}$, then $S(\sigma)$ cannot be an interval.

Proof. Let $S(\sigma)=[m_\sigma, M_\sigma]$ with $m_\sigma < M_\sigma$. We define a piecewise linear function $\lambda:[-\ell,\ell]\to[-\ell,\ell]$ such that $\lambda^{-1}([m_\sigma, M_\sigma])=[m'_\sigma, M'_\sigma]$ with $m'_\sigma > m_\sigma$ and $M'_\sigma < M_\sigma$. The derivative of λ is the piecewise constant function defined by

$$\lambda'(x):=\begin{cases}\lambda_1 & \text{if } x\in[-\ell, m'_\sigma[,\\ \lambda_2 & \text{if } x\in]m'_\sigma, M'_\sigma[,\\ \lambda_3 & \text{if } x\in]M'_\sigma,\ell],\end{cases} \tag{6.119}$$

where $\lambda_1, \lambda_3 < 1$ and $\lambda_2 > 1$. It is plain that the pair $(\sigma\circ\lambda, \alpha\circ\lambda)$ belongs to $\mathscr{A}$ and that $S(\sigma\circ\lambda)=[m'_\sigma, M'_\sigma]$. It is easily seen that

$$F[\sigma\circ\lambda,\alpha\circ\lambda]=k_E\ell_1\ell_2\left\{\lambda_1\int_{-\ell}^{m_\sigma}\{k\sigma'^2+\sigma^2\alpha'^2\}\,dx\right.$$

$$\left.+\lambda_2\int_{M_\sigma}^{\ell}\{k\sigma'^2+\sigma^2\alpha'^2\}\,dx\right\}<F[\sigma,\alpha]. \tag{6.120}$$

Thus (σ,α) does not minimize F in $\mathscr{A}$, unless $m_\sigma = M_\sigma$.

Bringing together Lemmas 6.27 and 6.28 we prove the following theorem.

Theorem 6.29. If $(\sigma,\alpha)\in\mathscr{A}$ minimizes F when (6.116) is satisfied then either $S(\sigma)=\varnothing$ or $S(\sigma)=\{x_0\}$ with $x_0\in]-\ell,\ell[$.

We are now in a position to determine the minimizers of F in $\mathscr{A}$.

Theorem 6.30. When $\psi=0$ there is only one minimizer of F in $\mathscr{A}$ for all values of k. If $0<k\leqslant(2\alpha_0/\pi)^2$, the minimizer is singular, while it is regular if $k>(2\alpha_0/\pi)^2$.

Proof. We first seek regular minimizers of F. When $\psi=0$ equation (6.113) becomes

$$k\sigma''=\frac{c^2}{\sigma^3}. \tag{6.121}$$

Hence σ is convex because, by $(6.107)_1$, it is positive. It is easily seen that the only convex solution of (6.121) in $]-\ell,\ell[$ that satisfies (6.102) is

$$\sigma(x)=s_1\sqrt{y\left(\frac{x}{\ell}\right)^2+1-y} \tag{6.122}$$

with

$$c^2 = k\left(\frac{s_1^2}{\ell}\right)^2 y(1-y), \tag{6.123}$$

where y is a parameter in $]0,1[$. As y varies, (6.122) and (6.123) represent a family of solutions of (6.121). The role of this parameter becomes clear as we evaluate at $x=0$ the function in (6.122):

$$y = 1 - \frac{\sigma^2(0)}{s_1^2}. \tag{6.124}$$

By the use of (6.122), we calculate the following integral:

$$\int_{-\ell}^{\ell} \frac{1}{\sigma^2(x)}\,dx = \frac{2\ell}{s_1^2}\frac{1}{\sqrt{y(1-y)}}\arctan\frac{\sqrt{y}}{\sqrt{1-y}}. \tag{6.125}$$

Inserting both (6.125) and (6.123) into (6.112), we get an equation for y whose roots determine the regular minimizers of F subject to (6.101) and (6.102):

$$\frac{\alpha_0}{\sqrt{k}} = \arctan\frac{\sqrt{y}}{\sqrt{1-y}}, \tag{6.126}$$

There is precisely one solution of this equation, provided that

$$\sqrt{k} > \frac{2\alpha_0}{\pi}. \tag{6.127}$$

Equation (6.121) implies that

$$k\sigma'^2 = a - \frac{c^2}{s^2}, \tag{6.128}$$

where a is an arbitrary constant. Making use of (6.122) and (6.123) in (6.128) we relate a and y:

$$a = \frac{ks_1^2}{\ell^2}y. \tag{6.129}$$

Inserting (6.110) and (6.128) into (6.103), with the aid of (6.129), we conclude that when $\psi = 0$, if there is a minimizer (σ, α) of F such that $S(\sigma) = \varnothing$, then

$$\min\{F[\sigma,\alpha] \mid (\sigma,\alpha)\in\mathscr{A}, S(\sigma)=\varnothing\} = k_E\ell_1\ell_2\frac{2ks_1^2}{\ell}y, \tag{6.130}$$

where y is the solution of (6.126).

We now seek minimizers of F in $\mathscr{A}$ such that $S(\sigma)=\{x_0\}$ with $x_0\in]-\ell,\ell[$. Let σ be given with $S(\sigma)=\{x_0\}$. The function α that minimizes $\int_{-\ell}^{\ell}\sigma^2\alpha'^2dx$ subject to (6.101) is constant on both the intervals $]-\ell,x_0[$ and $]x_0,\ell[$, and it is discontinuous at $x=x_0$. When $\psi=0$ equation (6.108) thus becomes

$$k\sigma''=0 \quad \text{in }]-\ell,\ell[\setminus\{x_0\}, \tag{6.131}$$

and so σ minimizes F only if it is piecewise linear. Hence the boundary conditions (6.101) and (6.102) determine the solution:

$$\sigma(x)=\begin{cases} -s_1\dfrac{x-x_0}{x_0+\ell} & \text{if } x\in[-\ell,x_0], \\[2ex] s_1\dfrac{x-x_0}{\ell-x_0} & \text{if } x\in[x_0,\ell], \end{cases} \tag{6.132}$$

$$\alpha(x)=\begin{cases} -\alpha_0 & \text{if } x\in[-\ell,x_0[, \\ \alpha_0 & \text{if } x\in]x_0,\ell]. \end{cases} \tag{6.133}$$

If for $\psi=0$ we evaluate F on this pair, we get

$$F[\sigma,\alpha]=k_E\ell_1\ell_2\frac{2ks_1^2\ell}{\ell^2-x_0^2}, \tag{6.134}$$

which attains its minimum when $x_0=0$. Hence, if there is a minimizer (σ,α) of F such that $S(\sigma)\neq\varnothing$, then

$$\min\{F[\sigma,\alpha]\,|\,(\sigma,\alpha)\in\mathscr{A}, S(\sigma)\neq\varnothing\}=k_E\ell_1\ell_2\frac{2ks_1^2}{\ell}. \tag{6.135}$$

Comparing (6.135) and (6.130), and recalling that $y\in]0,1[$, we conclude that if (6.127) is satisfied then

$$\min\{F[\sigma,\alpha]\,|\,(\sigma,\alpha)\in\mathscr{A}\}=\min\{F[\sigma,\alpha]\,|\,(\sigma,\alpha)\in\mathscr{A}, S(\sigma)=\varnothing\}.$$

On the other hand, if (6.127) is not satisfied, then

$$\min\{F[\sigma,\alpha]\,|\,(\sigma,\alpha)\in\mathscr{A}\}=\min\{F[\sigma,\alpha]\,|\,(\sigma,\alpha)\in\mathscr{A}, S(\sigma)\neq\varnothing\}.$$

Actually, in this case there is no regular solution of the equilibrium equations.

Theorem 6.30 shows that a bifurcation occurs at $k=(2\alpha_0/\pi)^2$. The orientation field that minimizes the free energy in $\mathscr{B}$ exhibits a defect at the mid plane of the cell when $k\leqslant(2\alpha_0/\pi)^2$, while it is everywhere smooth when $k>(2\alpha_0/\pi)^2$, and it displays a non-uniform twist between the plates. Since $|\alpha_0| < \pi/2$ this conclusion agrees with Theorem 6.21 of Section 6.3 and shows that plane defects can actually occur in minimizers of $\mathscr{F}_E$.

6.4.4 An approximation for ψ

The reader might think that the plane defect described above is indeed an artifact due to the absence of the potential σ_0 in $\mathscr{F}_E$. A step towards understanding the effect of σ_0 on the occurrence of defects in the energy minimizers was taken in Section 3 of Virga (1991) by using an approximation of the function in (6.64) which is appropriate only when the two wells of σ_0 have the same depth. This line of thought was pursued in Roccato and Virga (1992) by the use of an approximation to σ_0 that also fits the case when the two wells have different depths.

We approximate the function ψ in (6.104) by a function that takes only three values in the interval $[0, s_1]$:

$$\psi(s) = \begin{cases} 0 & \text{for } s = 0, \\ h & \text{for } s \in]0, s_1[, \\ -a & \text{for } s = s_1, \end{cases} \tag{6.136}$$

where h and a are both positive constants. We aim at understanding in a simple case how σ_0 interferes with the occurrence of singular minimizers of $\mathscr{F}_E$, and so we choose a special class of admissible functions. Let $\mathscr{C}$ be the class of all pairs (σ, α) in $\mathscr{A}$ for which there exists $\eta \in [0, 1]$ such that

$$\sigma(x) = s_1 \quad \text{for all } x \in [-\ell, -\eta\ell] \cup [\eta\ell, \ell] \tag{6.137}$$

(*cf.* Section 4 of Roccato and Virga (1992) for more details about this class).

We set

$$z_c := \left(\frac{\pi\alpha_0 s_1}{4\ell}\right)^2 \tag{6.138}$$

and let k_c be the function defined by

$$k_c(z) := \begin{cases} \left(\dfrac{2\alpha_0}{\pi}\right)^2 & \text{for } z \in [0, z_c], \\ \left(\dfrac{\alpha_0^2 s_1}{2\ell}\right)^2 \dfrac{1}{z} & \text{for } z \in [z_c, \infty[. \end{cases} \tag{6.139}$$

Roccato and Virga (1992) proved the following theorem.

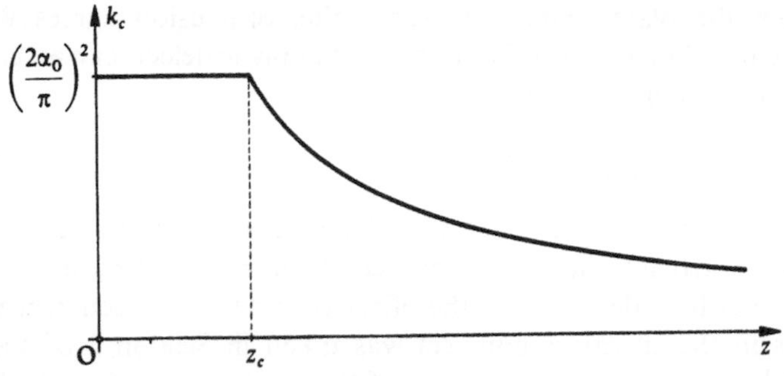

Figure 6.5

Theorem 6.31. For every value of k there is precisely one minimizer of F in $\mathscr{C}$ subject to (6.101) and (6.102). It is singular if, and only if,

$$k \leqslant k_c(h+a), \tag{6.140}$$

where k_c is the function defined in (6.139).

Figure 6.5 shows the graph of k_c. We see that the parameters that really matter in deciding whether the minimizer of F in $\mathscr{C}$ is singular or not are k and $z := h + a$, this latter being proportional, for the special ψ employed here, to the depth of the potential well at $s = s_1$.

It is worth noting that, no matter low large z is, the minimizer of F in $\mathscr{C}$ is singular for k sufficiently small. Furthermore, when $z \leqslant z_c$ the critical value of k that marks the onset of the bifurcation is the same as that in Theorem 6.30.

Suppose that, for fixed z, k decreases: when it crosses $k_c(z)$ a defect develops at the mid plane between the plates $\mathscr{S}^-$ and $\mathscr{S}^+$; the orientation of $\mathbf{n}$ becomes constant on both sides of the defect, giving rise to a texture with two adjacent domains.

We are now assured that plane defects, far from being merely consistent with the new theory, do actually occur in special problems also when the potential σ_0 is not neglected.

6.4.5 Instability of the uniform twist

The problem considered in this section has received much attention in applications; the domain $\mathscr{B}$ in (6.93), on the boundary of which $\mathbf{n}$ is required to satisfy (6.95), is indeed the same twisted nematic cell studied in Section 4.5 of Chapter 4. There we applied the classical theory, and we found that in the absence of any external field the free energy attains its minimum when the orientation is uniformly twisted between the plates $\mathscr{S}^-$ and $\mathscr{S}^+$, that is when $\mathbf{n}$ is as in (6.98) with the angle α given by the function

$$\alpha^*(x) := \frac{\alpha_0}{\ell} x \quad \text{for } x \in [-\ell, \ell]. \tag{6.141}$$

In the classical theory the degree of orientation s is taken as constant throughout the material. We now wonder whether there is $s^* \in]-\frac{1}{2}, 1[$ such that taking $\sigma^* \equiv s^*$ and $\alpha = \alpha^*$ we obtain an equilibrium configuration for the functional F in (6.103).

Since α^* is continuous the set $S(\sigma^*)$ must be empty, and so $s^* \neq 0$. Inserting both σ^* and α^* into (6.108) and (6.109), which are the Euler equations for F, we see that the latter equation is satisfied by any value of s^*, while the former becomes an equation for it:

$$s^* \left(\frac{\alpha_0}{\ell}\right)^2 + \frac{1}{2}\dot{\psi}(s^*) = 0. \tag{6.142}$$

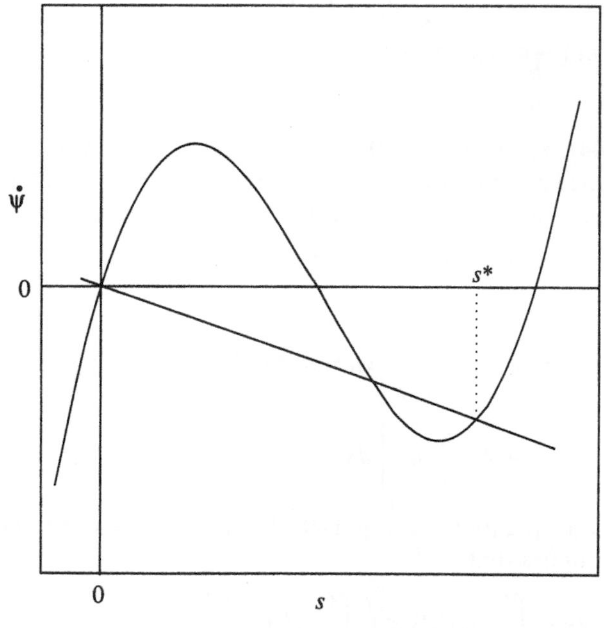

Figure 6.6

The roots of this equation correspond to the intersections between the graphs in Fig. 6.6. If $(\alpha_0/\ell)^2$ is sufficiently small there are two positive roots of (6.142). It is not difficult to show that F takes the smaller value on the larger root; thus we choose this latter as the equilibrium value of σ^* when $\alpha=\alpha^*$.

If follows from (6.142) that in $\mathscr{B}$ there is no uniform twist compatible with the new theory when ψ is set equal to zero. Our analysis above would then be demeaned if the uniform twist just found were always stable when $\psi\neq 0$. The following theorem shows that this is not the case.

Theorem 6.32. Let the pair (σ^*,α^*) represent the uniform twist in the cell $\mathscr{B}$. If

$$\ddot{\psi}(s^*)<6\left(\frac{\alpha_0}{\ell}\right)^2 \tag{6.143}$$

then (σ^*,α^*) is not a minimizer of F in $\mathscr{A}$.

Proof. Let ε_0 be a given positive number and let u and v be functions of class C^1 in $[-\ell,\ell]$ such that v satisfies

$$v(-\ell)=v(\ell)=0. \tag{6.144}$$

For all $\varepsilon\in[-\varepsilon_0,\varepsilon_0]$ define

$$\sigma_\varepsilon:=\sigma^*+\varepsilon u,\quad \alpha_\varepsilon:=\alpha^*+\varepsilon v. \tag{6.145}$$

By (6.144), α_ε satisfies (6.101) for all $\varepsilon\in[-\varepsilon_0,\varepsilon_0]$, while σ_ε is unconstrained at the end-points of the interval $[-\ell,\ell]$, as σ^*. Inserting (6.145) into (6.103), and making use of (6.144), we arrive at

$$F[\sigma_\varepsilon,\alpha_\varepsilon]=F[\sigma^*,\alpha^*]+\varepsilon^2F_2[u,v]+o(\varepsilon^2),$$

where

$$F_2[u,v]:=k_E\ell_1\ell_2\int_{-\ell}^{\ell}\left\{ku'^2+s^{*2}v'^2+\left(\left(\frac{\alpha_0}{\ell}\right)^2+\tfrac{1}{2}\ddot{\psi}(s^*)\right)u^2 + 4s^*\frac{\alpha_0}{\ell}uv'\right\}dx. \tag{6.146}$$

Hölder's inequality (*cf. e.g.* p. 140 of Hardy, Littlewood and Pólya (1952)) implies that

$$\int_{-\ell}^{\ell}uv'\,dx\leqslant\int_{-\ell}^{\ell}|uv'|\,dx\leqslant\left(\int_{-\ell}^{\ell}u^2\,dx\right)^{1/2}\left(\int_{-\ell}^{\ell}v'^2\,dx\right)^{1/2}. \tag{6.147}$$

Thus, setting

$$y_1 := \left(\int_{-\ell}^{\ell} u^2 dx \right)^{1/2}, \quad y_2 := \left(\int_{-\ell}^{\ell} u'^2 dx \right)^{1/2}, \quad y_3 := \left(\int_{-\ell}^{\ell} v'^2 dx \right)^{1/2},$$

we have that

$$\frac{1}{k_E \ell_1 \ell_2} F_2[u, v] \leqslant f(y_1, y_2, y_3),$$

where

$$f(y_1, y_2, y_3) := \left(\left(\frac{\alpha_0}{\ell} \right)^2 + \tfrac{1}{2} \ddot{\psi}(s^*) \right) y_1^2 + 4s^* \frac{|\alpha_0|}{\ell} y_1 y_3 + s^{*2} y_3^2 + k y_2^2. \tag{6.148}$$

We now show that when (6.143) is satisfied the function f is *not* positive definite, and so neither is F_2. Since u is not prescribed both at $x = -\ell$ and at $x = \ell$, y_1 and y_2 are independent from one another, and so f is a quadratic form in three variables. The eigenvalues of this form are $\lambda_2 = k$ and the roots λ_1 and λ_3 of the following equation:

$$\lambda^2 - \left(s^{*2} + \left(\frac{\alpha_0}{\ell} \right)^2 + \tfrac{1}{2} \ddot{\psi}(s^*) \right) \lambda + s^{*2} \left(\tfrac{1}{2} \ddot{\psi}(s) - 3 \left(\frac{\alpha_0}{\ell} \right)^2 \right) = 0. \tag{6.149}$$

It is clear from (6.149) that λ_1 and λ_3 have opposite signs if (6.143) is satisfied.

6.5 Disclinations

Line defects are often called **disclinations**. We have already encountered some in Section 3.7 of Chapter 3 among the special solutions to the equilibrium equations of the classical theory. They are *Frank's disclinations*, classified according to their Frank's index (*cf.* equation (3.181) of Chapter 3). All Frank's disclinations, though resembling defects actually observed, possess infinite energy. How to reconcile these disclinations to a variational theory has long been a challenging problem. Here we first recall how this problem was solved within the classical theory for disclinations of index 2, and then we see how the new theory casts more light on it.

6.5.1 Fluting

Let $\mathscr{B}$ be a circular cylinder of radius R and height H: in cylindrical co-ordinates

$$\mathscr{B} = \{p = o + r\mathbf{e}_r + z\mathbf{e}_z | r \in [0, R[, z \in]0, H[\}, \tag{6.150}$$

where o is the centre of one base of the cylinder. Let $\mathscr{S}_0$ denote the lateral boundary of $\mathscr{B}$. We assume that $\mathbf{n}$ is prescribed on $\mathscr{S}_0$ as the outward unit normal:

$$\mathbf{n}|_{\mathscr{S}_0} = \mathbf{e}_r. \tag{6.151}$$

We consider Frank's functional $\mathscr{F}_F$ within the one-constant approximation:

$$\mathscr{F}_F = k_F \int_{\mathscr{B}} |\nabla \mathbf{n}|^2 \, dv. \tag{6.152}$$

Cladis and Kléman (1972) studied the problem of minimizing $\mathscr{F}_F$ subject to (6.151). They looked for minimizers in the class of axisymmetric fields

$$\mathbf{n} = \cos\varphi \mathbf{e}_r + \sin\varphi \mathbf{e}_z, \tag{6.153}$$

where φ is a real-valued function of class C^1 that depends only on the radial co-ordinate r and obeys the boundary condition

$$\varphi(R) = 0. \tag{6.154}$$

In this class of orientation fields the energy functional in (6.152) reads

$$\mathscr{F}_F[\mathbf{n}] = 2\pi H k_F F[\varphi], \tag{6.155}$$

where

$$F[\varphi] := \int_0^R \left(\varphi'^2 + \frac{\cos^2\varphi}{r^2}\right) r \, dr, \tag{6.156}$$

a prime denoting differentiation with respect to r. Cladis and Kléman proved that the minimizer of F subject to (6.154) is

$$\varphi_f(r) := \frac{\pi}{2} - 2 \arctan\left(\frac{r}{R}\right); \tag{6.157}$$

an easy computation shows that

$$F[\varphi_f] = 2. \tag{6.158}$$

The function φ_f decreases from $\pi/2$ to 0 as r ranges from 0 to R.

The orientation $\mathbf{n}_f$ that corresponds to φ_f through equation (6.153) is a field of class C^1 in the whole of $\mathscr{B}$ and coincides with the planar field $\mathbf{e}_r$ only on $\mathscr{S}_0$. This is the reason why Cladis and Kléman's solution is sometimes referred to as the solution *escaping into the third dimension.* The integral lines of $\mathbf{n}_f$ exhibit a vertical *fluting*, but possess no actual defect. Cladis and Kléman also studied this variational problem in the case when the one-constant approximation to Frank's energy functional does not apply. Their analysis becomes more involved in this case, but their conclusions are qualitatively the same as that above.

Meyer (1973), apparently unaware of Cladis and Kléman's work, treated likewise all Frank's disclinations with even Frank's index.

6.5.2 Variational problem

We now formulate within the new theory a variational problem which parallels the one solved by Cladis and Kléman (1972) and Meyer (1973). Here the energy functional depends also on the degree of orientation s and we take it in the form

$$\mathscr{F}_E[s, \mathbf{n}] = k_E \int_{\mathscr{B}} \{k|\nabla s|^2 + s^2|\nabla \mathbf{n}|^2\}\, dv, \tag{6.159}$$

where we have once again neglected the potential σ_0. In Section 6.4 we made an attempt to relax this assumption. Here our analysis would simply fall apart if σ_0 were not neglected.

I am no longer trying to justify the special functional in (6.159): I acknowledge that the only reason to give it this form is its simplicity.

When s is taken as constant in (6.159), $\mathscr{F}_E$ reduces to $\mathscr{F}_F$ in (6.152), modulo an inessential constant.

Besides (6.151), we impose on $\mathscr{F}_E$ the boundary condition

$$s|_{\mathscr{S}_0} = s_0, \tag{6.160}$$

where s_0 is a positive constant.

Now the planar field $\mathbf{e}_r$ becomes a legitimate candidate to minimize the free energy, provided it is associated with a suitable field s that vanishes along the axis of $\mathscr{B}$. Thus, it is conceivable that Frank's disclinations may be tamed so as to become even energy minimizers.

Following Mizel, Roccato and Virga (1991), we seek minimizers of $\mathscr{F}_E$ among axisymmetric pairs $(s, \mathbf{n})$. Specifically, we assume that s depends only on the radial co-ordinate r and that $\mathbf{n}$ is given the form (6.153). In this class of functions $\mathscr{F}_E$ can be reduced to a simpler functional:

$$\mathscr{F}_E[s, \mathbf{n}] = 2\pi H k_E[s, \varphi], \tag{6.161}$$

where

$$F[s, \varphi] := \int_0^R \left\{ k s'^2 + s^2 \left(\varphi'^2 + \frac{\cos^2 \varphi}{r^2} \right) \right\} r\, dr. \tag{6.162}$$

A prime still denotes differentiation with respect to r, and φ is a real-valued function on $[0, R]$ subject to (6.154), while s is a real-valued function on $[0, R]$ subject to

$$s(R) = s_0. \tag{6.163}$$

When s is a constant different from zero the functional in (6.162) is proportional to the functional in (6.156). To compare the energies given by $\mathscr{F}_F$ and $\mathscr{F}_E$ we assume that they are equal when $s \equiv s_0$ in $\mathscr{B}$, and so s_0 must be such that

$$k_E s_0^2 = k_F. \tag{6.164}$$

The singular set $\mathscr{S}(s)$ is a subset of $\mathscr{B}$ generated by a subset of $[0, R]$:

$$\mathscr{S}(s) = \{ p = o + r\mathbf{e}_r + z\mathbf{e}_z \mid r \in S(s), z \in [0, H] \}, \tag{6.165}$$

where

$$S(s) := \{ r \in [0, R] \mid s(r) = 0 \}. \tag{6.166}$$

With a slight abuse of language, we also call $S(s)$ the **singular set**.

The regularity for the minimizers of $\mathscr{F}_E$ established in Section 6.3 can easily be formulated also for the minimizers of F. Thus the minimizing s is Lipschitz continuous on $[0, R]$ for $0 < k < 1$ and just Hölder continuous for $k \geqslant 1$, while both the minimizing s and the minimizing φ are analytic away from the singular set $S(s)$.

Keeping these facts in mind, we state the following problem.

Variational Problem (VP). Find the minimizers of F in the class

$$\mathscr{C}(s_0) := \{ (s, \varphi) \mid s \in AC(]0, R[), \varphi \in AC_{\mathrm{loc}}(]0, R[\setminus S(s)), \\ s(R) = s_0 > 0, \varphi(R) = 0 \}. \tag{6.167}$$

6.5.3 Qualitative properties of minimizers

We now derive by direct inspection of the functional F a few qualitative properties of the pairs that solve (VP).

First, note that if (s, φ) is one solution of (VP) then $(s, -\varphi)$ is another one. There is no loss in restricting the range of φ to the positive real line, as we do henceforth. Thus no ambiguity arises.

Theorem 6.33. If the pair (s, φ) solves (VP), then

(a) $s(r) \geqslant 0$, (b) $\varphi(r) \leqslant \pi/2$ for all $r \in [0, R]$;
(c) $s'(r) \geqslant 0$, (d) $\varphi'(r) \leqslant 0$ for almost all $r \in [0, R]$.

Proof. To prove (a) we simply show that a function s that is negative somewhere in $[0, R]$ cannot minimize F. In fact, if we replace it by the function s^+ defined by $s^+(r) := \max\{s(r), 0\}$ for all $r \in [0, R]$, we lower the value of F. To prove (c), we employ a similar argument. If s is replaced by $\hat{s}$, where $\hat{s}(r) := \inf\{s(\bar{r}) \mid r \leqslant \bar{r} \leqslant R\}$, we again lower the value of F. The proofs of parts (b) and (d) closely parallel those of (a) and (c).

The following theorem states a sharp dichotomy for the singular set of a minimizing pair.

Theorem 6.34. If (s, φ) solves (VP), then either $S(s) = \varnothing$ or $S(s) = \{0\}$.

Proof. We first prove that if (s, φ) solves (VP), then either $S(s) = \varnothing$ or $S(s) = [0, r_0]$ with $0 \leqslant r_0 < R$. To show this it suffices to see that if $S(s) \neq \varnothing$ and $r_0 := \max\{r \in [0, R] \mid s(r) = 0\}$, then on replacing s by a function that vanishes in $[0, r_0]$, we lower the value of F.

The regularity of minimizers guarantees that all the pairs that solve (VP) obey the Euler equations of F away from the singular set. These equations are:

$$k(rs')' = sr\left(\varphi'^2 + \frac{\cos^2\varphi}{r^2}\right), \tag{6.168}$$

$$(rs^2\varphi')' = -\frac{s^2}{r}\cos\varphi\sin\varphi, \quad \text{in }]0, R[\setminus S(s); \tag{6.169}$$

they are subject to (6.154) and (6.163).

Recall that when φ is given, for each $r_1 \in]0, R[$ the initial-value problem $s(r_1) = s_1$, $s'(r_1) = s'_1$ for equation (6.168) possesses exactly one solution in $[r_1, R]$.

Now let $S(s)$ be the singular set of a minimizer and assume that it is not empty. It must be an interval $[0, r_0]$ with $0 \leqslant r_0 < R$. We show that assuming $0 < r_0 < R$ leads to a contradiction.

Let $\tau: [0, R] \rightarrow \mathbb{R}^+$ be a function of class C^1 such that

$$\tau(0) = \tau(R) = 0. \tag{6.170}$$

For a given $\varepsilon \in \mathbb{R}^+$, let ρ_ε be the function defined by

$$\rho_\varepsilon(r) := r + \varepsilon\tau(r) \quad \text{for all } r \in [0, R]. \tag{6.171}$$

For every function τ there exists ε_τ such that ρ_ε is a C^1-diffeomorphism of $[0, R]$ onto $[0, R]$ for all $\varepsilon \in [-\varepsilon_\tau, \varepsilon_\tau]$. Once a function τ has been selected, if the pair (s, φ) belongs to the class $\mathscr{C}(s_0)$ in (6.167) then for all $\varepsilon \in [-\varepsilon_\tau, \varepsilon_\tau]$ so also does the pair $(s_\varepsilon, \varphi_\varepsilon)$, where

$$s_\varepsilon(r) := s(\rho_\varepsilon(r)), \quad \varphi_\varepsilon(r) := \varphi(\rho_\varepsilon(r)) \quad \text{for all } r \in [0, R]. \tag{6.172}$$

If (s, φ) is a minimizer of F, then the variation

$$\delta F(s, \varphi)[\tau] := \frac{d}{d\varepsilon} F[s_\varepsilon, \varphi_\varepsilon]|_{\varepsilon=0} \tag{6.173}$$

must vanish for all τ of class C^1 that obey (6.170). When s vanishes on $[0, r_0]$ with $r_0 \geqslant 0$, a tedious but easy computation shows that

$$\delta F(s, \varphi)[\tau] = \int_{r_0}^{R} (r\tau' - \tau)\left(ks'^2 + s^2\left(\varphi'^2 - \frac{\cos^2\varphi}{r^2}\right)\right) dr. \tag{6.174}$$

Integrating by parts in (6.174), we arrive at

$$\delta F(s, \varphi)[\tau] = -\lim_{r \to r_0^+} (\tau(r) r w(r)) - \int_{r_0}^{R} \tau(rw' + 2w) dr, \tag{6.175}$$

where we have set

$$w := ks'^2 + s^2\left(\varphi'^2 - \frac{\cos^2\varphi}{r^2}\right). \tag{6.176}$$

Since $r_0 > 0$, the right-hand side of (6.175) vanishes for all functions τ if, and only if,

$$\lim_{r \to r_0^+} w(r) = 0 \tag{6.177}$$

and

$$rw' + 2w = 0 \quad \text{in }]r_0, R[. \tag{6.178}$$

The solution of (6.178) is

$$w(r) = \frac{c}{r^2}, \tag{6.179}$$

where c is an arbitrary constant. Equation (6.177) then requires $c = 0$, and so $w \equiv 0$ in $]r_0, R[$. Since s is continuous and $s(r_0) = 0$, taking the limit as $r \to r_0^+$, from (6.176) we arrive at

$$\lim_{r \to r_0^+} s'(r) = 0. \tag{6.180}$$

Since $s' \equiv 0$ on $[0, r_0]$, we conclude that $s'(r_0) = 0$. Equation (6.168), subject to the conditions $s(r_0) = 0$ and $s'(r_0) = 0$, has only the solution $s = 0$ in $[r_0, R]$ for every function φ of class $W^{1,2}$. Thus the singular set would be the whole interval $[0, R]$, which contradicts the assumption that $r_0 < R$.

Remark. We have just proved that r_0 must be zero. The variation of F in (6.175) still vanishes for all τ only if

$$w = 0 \quad \text{on the whole of }]0, R[. \tag{6.181}$$

6.5.4 Hamilton–Jacobi method

In Mizel, Roccato and Virga (1991) Euler equations play no role in distinguishing regular minimizers of F from singular ones. Another method is instead employed: the **Hamilton–Jacobi method**, which in control theory is usually referred to as the dynamic programming method. The differential equations on which it is based hold only for the minimizers of F, unlike Euler equations, which hold for all equilibrium configurations. It is explained in Section 5 of Mizel, Roccato and Virga (1991) how to apply this method to the functional in (6.162). I collect in the following theorems the main conclusions reached there.

Theorem 6.35. Let $k > 0$ be given. There is only one solution (s^*, φ^*) to (VP). The functions s^* and φ^* solve the equations

$$\varphi'(r) = \frac{\dot{B}_k(\varphi(r))}{2r}, \tag{6.182}$$

$$s'(r) = \frac{s(r) B_k(\varphi(r))}{kr}, \tag{6.183}$$

subject to (6.154) and (6.163), where $\dot{B}_k$ denotes the derivative of the function B_k that solves on $[0, \pi/2]$ the problem

$$\dot{B}(\varphi) = -2\sqrt{\cos^2\varphi - \frac{B^2(\varphi)}{k}}, \quad B\left(\frac{\pi}{2}\right) = 0. \tag{6.184}$$

For every positive value of k the differential problem in (6.184) has precisely one solution B_k of class C^2 on $]0, \pi/2[$ that is continuous up to the end-points of the interval $[0, \pi/2]$.

The main qualitative features of the solutions to equation $(6.184)_1$ can be proved through the phase-plane analysis illustrated below. Here we remark how one value taken by the solution B_k to the problem (6.184) is especially relevant to our development.

Theorem 6.36. For every $k > 0$ the minimum of F in $\mathscr{C}(s_0)$ is

$$\min_{(s,\varphi)\in\mathscr{C}(s_0)} F[s,\varphi] = s_0^2 B_k(0). \tag{6.185}$$

Thus, it is desirable to learn more about the function $k \mapsto B_k(0)$.

Theorem 6.37. The solution B_k to the problem (6.184) satisfies

$$\begin{aligned} B_k(0) &= \sqrt{k} \quad \text{for all} \quad k \in]0, 1], \\ B_k(0) &< \sqrt{k} \quad \text{for all} \quad k \in]1, \infty[. \end{aligned} \tag{6.186}$$

Furthermore,

$$\lim_{k\to\infty} B_k(0) = 2. \tag{6.187}$$

Figure 6.7 shows the graph of $B_k(0)$ versus k; the dashed line is the graph of the function $k \mapsto \sqrt{k}$.

It is worth noting that by (6.158), (6.164) and (6.185), we conclude

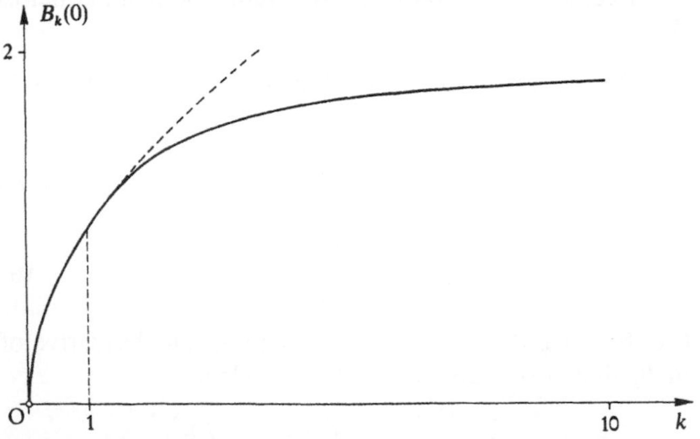

Figure 6.7

that

$$\min_{(s,\varphi)\in\mathscr{C}(s_0)} F[s,\varphi] < F[\varphi_f], \tag{6.188}$$

and so the fluted solution of Cladis and Kléman and Meyer possesses more energy than the pair that minimizes the functional in (6.162).

6.5.5 Phase-plane analysis

The following theorem tells us a great deal about the pair (s^*, φ^*) that minimizes F in $\mathscr{C}(s_0)$.

Theorem 6.38. For all $k > 0$ both s^* and φ^* depend on r through r/R. If $0 < k \leqslant 1$ then

$$s^*(r) = s_0\left(\frac{r}{R}\right)^{1/\sqrt{k}} \quad \text{and} \quad \varphi^*(r) = 0 \quad \text{for all } r\in[0, R]. \tag{6.189}$$

If $k > 1$ then $S(s^*) = \varnothing$ and $\varphi^{*\prime}(r) < 0$ for all $r\in[0, R]$; furthermore,

$$\lim_{r\to 0^+} \varphi^*(r) = \frac{\pi}{2}. \tag{6.190}$$

Thus, for $0 < k \leqslant 1$ the field $\mathbf{n}^*$ that minimizes $\mathscr{F}_E$ in (6.161) displays precisely Frank's disclination with Frank's index 2, whereas for $k > 1$ it is again a field with no defect, fluted along the axis of the cylinder.

Both s^* and φ^* have been computed for different values of k. Their graphs are plotted in Figs. 6.8 and 6.9. It is worth noting that while φ^* looks much like φ_f for all $k > 1$, s^* approaches 0 more rapidly near the axis of the cylinder as k gets closer to 1. Qualitatively, the size of the **core** where the liquid crystal is nearly isotropic becomes larger as k approaches 1 from above, while the orientation does not change dramatically. These configurations are precursors of the disclination that takes over when $k \leqslant 1$.

Our analysis refines that of Dafermos (1970), whose idea was to insert a variable core of isotropic phase around Frank's disclinations to make their energy finite.

The proof of both Theorems 6.37 and 6.38 rests mainly upon a phase-plane analysis of equation $(6.184)_1$. We do not go here into the details of this analysis; I shall be content to present briefly a few illustrations, just to give the reader the flavour of it.

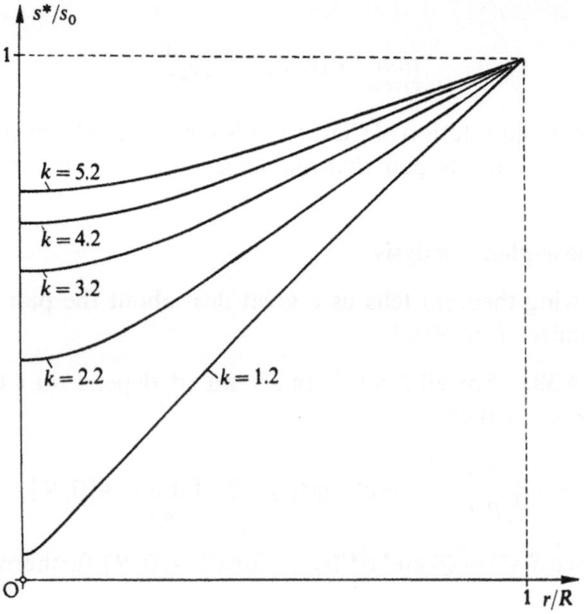

Figure 6.8

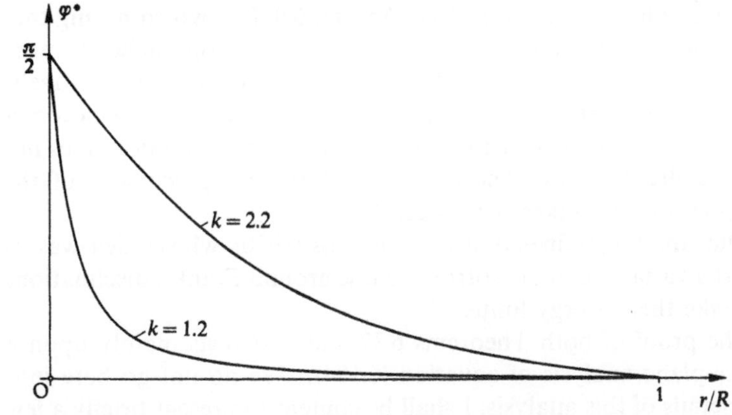

Figure 6.9

The terminology that we employ is the following. An **integral trajectory** is a curve that represents a solution of $(6.184)_1$ in the $(B, \dot{B})$-plane. All integral trajectories are confined within the region bounded by the co-ordinate axes and the quarter-ellipse described by

$$\dot{B} = -2\sqrt{1-\frac{B^2}{k}}, \quad B \geqslant 0. \tag{6.191}$$

This curve, which we call the **outer ellipse** for short, intersects the coordinate axes at $(\sqrt{k}, 0)$ and $(0, -2)$. For each $\varphi_0 \in]0, \pi/2[$ there is an ellipse inside the outer ellipse, which is described by

$$\dot{B} = -2\sqrt{\cos^2\varphi_0 - \frac{B^2}{k}}, \quad B \geqslant 0; \tag{6.192}$$

we call each of these curves an **inner ellipse**. For a given φ_0, the corresponding inner ellipse intersects the B-axis at $(\sqrt{k_0}, 0)$, where

$$k_0 := k\cos^2\varphi_0. \tag{6.193}$$

We employ $k_0 \in [0, k[$ to label the inner ellipses. When $k_0 = 0$ the inner ellipse shrinks to the origin of the $(B, \dot{B})$-plane. When $k_0 \to k^-$ the inner ellipses approach the outer ellipse.

An integral trajectory hits the outer ellipse when $\varphi = 0$. Rephrasing Theorem 6.37 in this language amounts to saying that for $0 < k \leqslant 1$ an integral trajectory starting from the origin hits the B-axis again

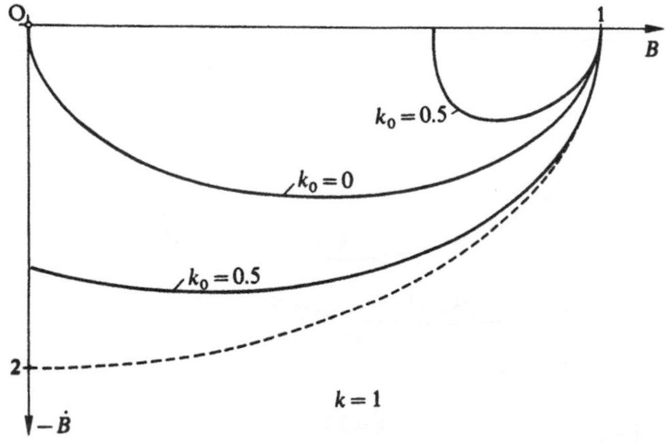

Figure 6.10

at $B = \sqrt{k}$, while for $k > 1$ it hits the outer ellipse first and does not reach the B-axis.

For $k = 1$, Fig. 6.10 shows the integral trajectory starting from the origin ($k_0 = 0$) along with two other integral trajectories starting from the inner ellipse with $k_0 = \frac{1}{2}$. The dashed line is the outer ellipse.

Figures 6.11 and 6.12 illustrate integral trajectories for both $k < 1$ and $k > 1$. In the latter figure one sees that the trajectory with $k_0 = 0$ hits the outer ellipse and stops there: this signifies that $B_k(0) < \sqrt{k}$.

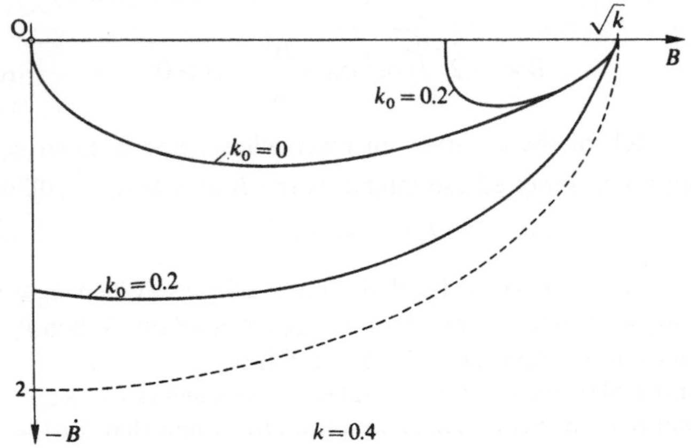

Figure 6.11

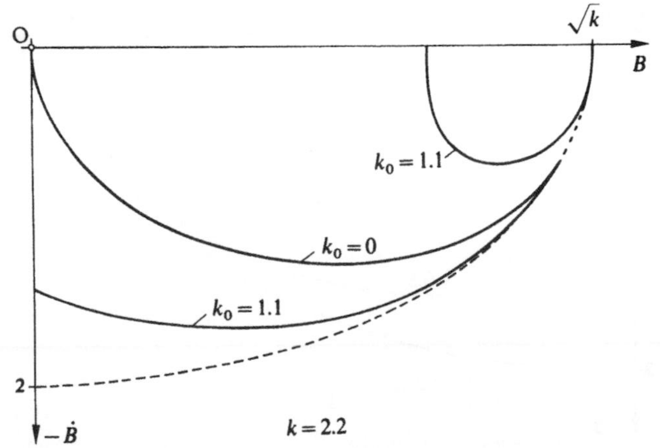

Figure 6.12

6.6 References

Almgren, F.J., Jr. and Lieb, E.H. (1988) Singularities of energy minimizing maps from the ball to the sphere: Examples, counterexamples, and bounds, *Ann. of Math.*, **128**, 483–530

Almgren, F.J., Jr. and Lieb, E.H. (1990) Counting singularities in liquid crystals, in: Variational methods (Proceedings of a Conference held in Paris, June 1988), *Progress in nonlinear differential equations and their applications*, Vol. 4, H. Berestycki, J.-M. Coron and I. Ekeland (eds), Birkhäuser, Boston, 17–35.

Ambrosio, L. (1990a) Existence theory for a new class of variational problems, *Arch. Rational Mech. Anal.*, **111**, 291–322.

Ambrosio, L. (1990b) Existence of minimal energy configurations of nematic liquid crystals with variable degree of orientation, *Manuscripta Math.*, **68**, 215–228.

Ambrosio, L. and Virga, E.G. (1991) A boundary-value problem for nematic liquid crystals with a variable degree of orientation, *Arch. Rational Mech. Anal.*, **114**, 335–347.

Brezis, H., Cornon, J.-M. and Lieb, E.H. (1986) Harmonic maps with defects, *Comm. Math. Phys.*, **107**, 647–705.

Calderer, M.C. (1990) Elasticity properties and kernel singularities of liquid crystal polymers, *Appl. Anal.*, **38**, 19–42.

Calderer, M.C. (1991a) On the mathematical modeling of textures in polymeric liquid crystals, in: *Nematics. Mathematical and physical aspects*, J.-M. Coron, J.-M. Ghidaglia and F. Hélein eds., Kluwer Academic Publishers, Dordrecht, 25–36.

Calderer, M.C. (1991b) On multiphase flows of liquid crystal polymers, *J. Rheol.*, **35**, 29–47.

Calderer, M.C. (1992) Stability of shear flows of polymeric liquid crystals, *J. Non-Newtonian Fluid Mech.*, **43**, 351–368.

Cladis, P. and Kléman, M. (1972) Non-singular disclinations of strength $S = +1$ in nematics, *J. Phys. (Paris)*, **33**, 591–598.

Cohen, R., Hardt, R., Kinderlehrer, D., Lin, S.-Y. and Luskin, M. (1987) Minimum energy configurations for liquid crystals: computational results, in: *Theory and applications of liquid crystals*, J.L. Ericksen and D. Kinderlehrer eds., IMA Volumes in Mathematics and its Applications, **5**, 99–121.

Cohen, R., Lin, S.-Y. and Luskin, M. (1989) Relaxation and gradient methods for molecular orientation in liquid crystals, *Computer Phys. Commun.*, **53**, 455–465.

Cohen, R. and Taylor, M. (1990) Weak stability of the map $x/|x|$ for liquid crystal functionals, *Commun. Partial Differ. Equations*, **15**, 675–692.

Dafermos, C.M. (1970) Disinclinations in liquid crystals, *Quart. J. Mech. Appl. Math.*, **23 S**, 49–64.

De Gennes, P.G. (1969) Phenomenology of short-range-order effects in the isotropic phase of nematic materials, *Phys. Lett.*, **30A**, 454–455.

De Gennes, P.G. (1971) Short range order effects in the isotropic phase of nematics and cholesterics, *Mol. Cryst. Liq. Cryst.*, **12**, 193–214.

Doi, M. (1981) Molecular dynamics and rheological properties of concentrated solutions of rodlike polymers in isotropic and liquid crystalline phases, *J. Polymer Sci.*, **19**, 229–243.

Ericksen, J.L. (1988) Static theory of point defects in nematic liquid crystals, Private communication.

Ericksen, J.L. (1991) Liquid crystals with variable degree of orientation, *Arch. Rational Mech. Anal.*, **113**, 97–120.

Fan, C.-P. (1971) Disclination lines in liquid crystals, *Phys. Lett.*, **34A**, 335–336.

Fan, C.-P. and Stephen, M.J. (1970) Isotropic–nematic phase transition in liquid crystals, *Phys. Rev. Lett.*, **25**, 500–503.

Giaquinta, M. (1992) Problemi variazionali per applicazioni vettoriali. Aspetti geometrici ed analitici, *Boll. Un. Mat. Ital.*, **6A**, 1–34.

Giaquinta, M., Modica, G. and Souček, J. (1989a) The Dirichlet energy of mappings with values into the sphere, *Manuscripta Math.*, **65**, 489–507.

Giaquinta, M. Modica, G. and Souček, J. (1989b) Cartesian currents and variational problems for mappings into spheres, *Ann. Scuola Norm. Sup. Pisa Cl. Sci.*, **16**, 393–485.

Giaquinta, M., Modica, G. and Souček, J. (1990) Liquid crystals: relaxed energies, dipoles, singular lines and singular points, *Ann. Scuola Norm. Sup. Pisa Cl. Sci.*, **17**, 415–437.

Giaquinta, M., Modica, G. and Souček, J. (1991) Cartesian currents and liquid crystal dipoles, singular lines and singular points, in: *Nematics. Mathematical and physical aspects*, J.-M. Coron, J.-M. Ghidaglia and F. Hélein eds., Kluwer Academic Publishers, Dordrecht, 113–127.

Goossens, W.J.A. (1985) Bulk, interfacial and anchoring energies of liquid crystals, *Mol. Cryst. Liq. Cryst.*, **124**, 305–331.

Hardt, R. and Kinderlehrer, D. (1987) Mathematical questions of liquid crystal theory, in: *Theory and applications of liquid crystals*, J.L. Ericksen and D. Kinderlehrer eds., *IMA Volumes in Mathematics and its Applications*, **5**, 151–184.

Hardt, R., Kinderlehrer, D. and Lin, F.-H. (1986) Existence and partial regularity of static liquid crystal configurations, *Comm. Math. Phys.*, **105**, 547–570.

Hardt, R., Kinderlehrer, D. and Lin, F.-H. (1988) Stable defects of minimizers of constrained variational principles, *Ann. Inst. Henri Poincaré Nonlin. Anal.*, **5**, 297–322.

Hardt, R., Kinderlehrer, D. and Lin, F.-H. (1990) The variety of configurations of static liquid crystals, in: Variational methods (Proceedings of a Conference held in Paris, June 1988), *Progress in nonlinear differential equations and their applications*, Vol. 4, H. Berestycki, J.-M. Coron and I. Ekeland eds., Birkhäuser, Boston, 115–131.

Hardt, R. and Lin, F.-H. (1986) A remark on H^1 mappings, *Manuscripta Math.*, **56**, 1–10.

Hardt, R. and Lin, F.-H. (1993) Harmonic maps into round cones and singularities of nematic liquid crystals, *Math. Z.*, **213**, 575–593.

Hardy, G.H., Littlewood, J.E. and Pólya, G. (1952) *Inequalities*, second edition, Cambridge University Press, Cambridge.

Hélein, F. (1987) Minima de la fonctionelle énergie libre des cristaux liquides, *C.R. Acad. Sci. Paris*, **305**, 565–568.

Hocking, J.G. and Young, G.S. (1961) *Topology*, Addison-Wesley, Reading, Mass.

Kinderlehrer, D. (1991) Recent developments in liquid crystal theory. in: *Frontiers in pure and applied mathematics*, R. Dautry ed., North-Holland, Amsterdam, 157–178.

Kinderlehrer, D. and Ou, B. (1992) Second variation of liquid crystal energy at $x/|x|$, *Proc. R. Soc. Lond.* **A437**, 475–487.

Kinderlehrer, D., Ou, B. and Walkington, N. (1993) The elementary defects of the Oseen–Frank energy for a liquid crystal, *C.R. Acad. Sci. Paris*, **316**, 465–470.

Lemaire, L. (1978) Applications harmoniques de surfaces Riemanniennes, *J. Diff. Geom.*, **13**, 51–78.

Lin, F.-H. (1989) Nonlinear theory of defects in nematic liquid crystals; phase transition and flow phenomena, *Comm. Pure Appl. Math.*, **42**, 789–814.

Lin, F.-H. (1991a) Nematic liquid crystals with variable degree of orientation, in: *Nematics. Mathematical and physical aspects*, J.-M. Coron, J.-M. Ghidaglia and F. Hélein eds., Kluwer Academic Publishers, Dordrecht, 247–259.

Lin, F.-H. (1991b) On nematic liquid crystals with variable degree of orientation, *Comm. Pure Appl. Math.*, **44**, 453–468.

Lin, F.-H. and Poon, C.-C. (1993) On Ericksen's model for liquid crystals, to appear in *J. Geom. Anal.*

Macmillan, E.H. (1992a) On the hydrodynamics of biaxial nematic liquid crystals. Part 1: General theory, *Arch. Rational Mech. Anal.*, **117**, 193–239.

Macmillan, E.H. (1992b) On the hydrodynamics of biaxial nematic liquid crystals. Part 2: Steady-state analysis, *Arch. Rational Mech. Anal.*, **117**, 241–294.

Meyer, R.B. (1973) On the existence of even indexed disclinations in nematic liquid crystals, *Phil. Mag.*, **77**, 405–424.

Mizel, V.J., Roccato, D. and Virga, E.G. (1991) A variational problem for nematic liquid crystals with variable degree of orientation, *Arch. Rational Mech. Anal.*, **116**, 115–138.

Roccato, D. and Virga, E.G. (1992) On plane defects in nematic liquid crystals with variable degree of orientation, *Continuum Mech. Thermodyn.*, **4**, 121–136.

Schoen, R. and Uhlenbek, K. (1982) A regularity theorem for harmonic maps, *J. Diff. Geom.*, **17**, 307–335.

Schoen, R. and Uhlenbek, K. (1983) Boundary regularity and the Dirichlet problem for harmonic maps, *J. Diff. Geom.*, **18**, 253–268.

Simon, L. (1983) Asymptotic for a class of non-linear evolution equations with applications to geometric problems, *Ann. of Math.*, **118**, 525–571.

Virga, E.G. (1991) Defects in nematic liquid crystals with variable degree of orientation, in: *Nematics. Mathematical and physical aspects*, J.-M. Coron, J.-M. Ghidaglia and F. Hélein eds., Kluwer Academic Publishers, Dordrecht, 371–390.

Volovik, G.E. and Lavrentovich, O.D. (1983) Topological dynamics of defects: boojums in nematic drops, *Sov. Phys. JETP*, **58**, 1159–1166.

Ziemer, W.P. (1989) *Weakly differentiable functions*, Springer, New York.

6.7 Further reading

Classical defects

Almgren, F.J., Browder, W. and Lieb, E.H. (1988) Co-area, liquid crystals, and minimal surfaces, in: *Partial differential equations*, S.S. Chern ed., *Lecture Notes in Mathematics* No. 1306, Springer, Berlin, 1–22.

Bethuel, F., Brezis, H. and Coron, J.-M. (1990) Relaxed energies for harmonic maps, in: Variational methods (Proceedings of a Conference held in Paris, June 1988), *Progress in nonlinear differential equations and their applications*, Vol. 4, H. Berestycki, J.-M. Coron and I. Ekeland eds., Birkhäuser, Boston, 37–52.

Bethuel, F., Brezis, H., Coron, J.-M. and Hélein, F. (1990) Problèmes mathématiques des cristaux liquides, *Courier du CNRS*, *Images des Math.*

Bouligand, Y. (1981) Geometry and topology of defects in liquid crystals, bibliographical notes, in: *Physics of defects*, R. Balian, M. Kléman and J.-P. Poirier eds., North-Holland, Amsterdam, 665–711.

Brezis, H. (1987) Liquid crystals and energy estimates for S^2-valued maps, in: Theory and applications of liquid crystals, J.L. Ericksen and D. Kinderlehrer eds., *IMA Volumes in Mathematics and its Applications*, **5**, 31–52.

Brezis, H. (1989) S^k-valued maps with singularities, in: *Topics in Calculus of Variations* (CIME Course, 1987), M. Giaquinta ed., *Lecture Notes in Mathematics No. 1365*, Springer, Berlin, 1–30.

Brinkman, W.F. and Cladis, P.E. (1982) Defects in liquid crystals, *Physics Today*, May, 48–54.

Choi, H.I. (1987) Degenerate harmonic maps and liquid crystals, in: *Theory and applications of liquid crystals*, J.L. Ericksen and D. Kinderlehrer eds., *IMA Volumes in Mathematics and its Applications*, **5**, 63–71.

Hardt, R. (1991) Axially symmetric harmonic maps, in: *Nematics. Mathematical and physical aspects*, J.-M. Coron, J.-M. Ghidaglia and F. Hélein eds., Kluwer Academic Publishers, Dordrecht, 179–187.

Hardt, R., Kinderlehrer, D. and Luskin, M. (1986) Remarks about the mathematical theory of liquid crystals, in: *Proceedings of the*

Conference 'Calcolo delle Variazioni ed Equazioni alle Derivate Parziali' in honour of H. Lewy, Trento, Italy, June 1986.

Hardt, R., Lin, F.-H. and Poon, C.-C. (1992) Axially symmetric harmonic maps minimizing a relaxed energy, *Comm. Pure Appl. Math.*, **45**, 417–459.

Lavrentovich, O.D. and Nastishin, Yu. A. (1989) Defects with non-trivial topological charges in hybrid-aligned nematic layers, *Sov. Phys. Crystallogr.*, **34**, 914–917.

Lavrentovich, O.D. and Terent'ev, E.M. (1986) Phase transition altering the symmetry of topological point defects (hedgehogs) in a nematic liquid crystal, *Sov. Phys. JEPT*, **64**, 1237–1244.

Mermin, N.D. (1979), The topological theory of defects in ordered media, *Rev. Mod. Phys.*, **51**, 591–648.

Sunil Kumar, P.B. and Ranganath, G.S. (1989) On certain liquid crystal defects in a magnetic field, *Mol. Cryst. Liq. Cryst.*, **177**, 131–144.

Variable degree of orientation

Ericksen, J.L. (1984) A thermodynamic view of order parameters for liquid crystals, in: *Orienting Polymers, Lecture Notes in Mathematics No. 1063*, Springer, Berlin, 27–36.

Gramsbergen, E.F., Longa, L. and De Jeu, W.H. (1986) Landau theory of the nematic–isotropic phase transition, *Physics Reports*, **135**, 195–257.

Longa, L., Monselesan, D. and Trebin, H.-R. (1987) An extension of the Landau–Ginsburg–de Gennes theory for liquid crystals, *Liq. Crystals*, **2**, 769–796.

Longa, L. and Trebin, H.-R. (1989) Integrity basis approach to the elastic free energy functional of liquid crystals. I. Classification of basic elastic modes, *Liq. Crystals*, **5**, 617–622.

New defects

Haas, G., Fritsch, M., Wöhler, H. and Mlynski, D.A. (1989) Polar anchoring energy and order parameter at a nematic liquid crystal–wall interface, *Liq. Crystals*, **5**, 673–681.

Hardt, R. (1990) Point and line singularities in liquid crystals, in: Variational methods (Proceedings of a Conference held in Paris, June 1988), *Progress in nonlinear differential equations and their applications, Vol. 4*, H. Berestycki, J.-M. Coron and I. Ekeland eds., Birkhäuser, Boston, 105–113.

Virga, E.G. (1990) Disclinations and hedgehogs in nematic liquid crystals with variable degree of orientation, *Rend. Mat. Acc. Lincei*, **1**, 275–280.

Virga, E.G. (1993) New variational problems in the statics of liquid crystals, in: *Microstructure and phase transition*, D. Kinderlehrer, R. James, J.L. Ericksen and M. Luskin eds., *IMA Volumes in Mathematics and its Applications*, **54**, 205–215.

Disclinations

Anisimov, S.I. and Dzyaloshinskiĭ, I.E. (1973) A new type of disclinations in liquid crystals and the stability of disclinations of various types, *Sov. Phys. JEPT*, **36**, 774–779.

Balinskiĭ, A.A., Volovik, G.E. and Kats, E.I. (1984) Disclinations symmetry in uniaxial and biaxial nematic liquid crystals, *Sov. Phys. JEPT*, **60**, 748–753.

Bethuel, F., Brezis, H., Coleman, B.D. and Hélein, F. (1992) Bifurcation analysis of minimizing harmonic maps describing the equilibrium of nematic phases between cylinders, *Arch. Rational Mech. Anal.*, **118**, 149–168.

Ericksen, J.L. (1970) Singular solutions in liquid crystal theory, in: *Liquid crystals and ordered fluids*, J.F. Johnson and R.S. Porter eds., Plenum Press, New York, 181–192.

Gartland, E.C. Jr, Palffy-Muhoray, P. and Varga, R.S. (1991) Numerical minimization of the Landau–de Gennes free energy: defects in cylindrical capillaries, *Mol. Cryst. Liq. Cryst.*, **199**, 429–452.

Maddocks, J.H. (1987) A model for disclinations in nematic liquid crystals, in: *Theory and applications of liquid crystals*, J.L. Ericksen and Kinderlehrer, D. eds., *IMA Volumes in Mathematics and its Applications*, **5**, 255–269.

Saupe, A. (1973) Disclinations and properties of the directorfield in nematic and cholesteric liquid crystals, *Mol. Cryst, Liq. Cryst.*, **21**, 211–238.

Schopohl, N. and Sluckin, T.J. (1987) Defect core structure in nematic liquid crystals, *Phys. Rev. Lett.*, **59**, 2582–2584.

Williams, C.E., Cladis, P.E. and Kléman, M. (1973) Screw disclinations in nematic samples with cylindrical symmetry, *Mol. Cryst. Liq. Cryst.*, **21**, 355–373.

Index

Printed and bound by CPI Group (UK) Ltd, Croydon, CR0 4YY
23/10/2024
01778227-0001